STELLAR STERILITY

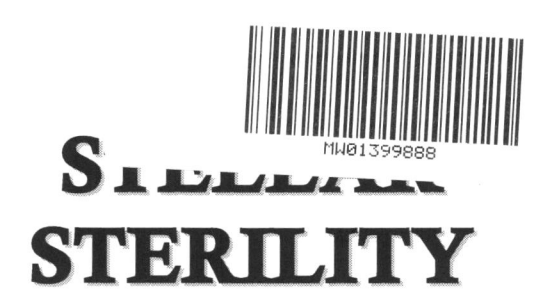

"The Search for Alien Presence"

By
Ross Marshall

Stellar Sterility
"The Search for Alien Presence"
Volume – 1

By
R. S. Marshall

2013 © Ross S. Marshall, Weirdvideos.com ISBN Number:
Cover Art credit by CreateSpace, Amazon Books
Interior book design by BEAUTeBOOK
Printed in Anacortes, WA USA

All rights reserved. No part of this publication may be reproduced, distributed, or transmitted in any form or by any means, including photocopying, recording, or other electronic or mechanical methods, without the prior written permission of the publisher, except in the case of brief quotations embodied in critical reviews and certain other noncommercial uses permitted by copyright law. For permission requests, write to the publisher, addressed "Attention: Permissions Coordinator," at the address below.

Weirdvideos.com c/o R. S. Marshall
P. O. Box 1191
Anacortes, WA 98221
www.weirdvideos.com

Ordering Information:

Quantity sales. Special discounts are available on quantity purchases by corporations, associations, and others. For details, contact the publisher at the addresses above.

Orders by U.S. trade bookstores and wholesalers. Please contact Weirdvideos Distribution: Tel: (360) 421-7195; or visit www.Weirdvideos.com

Printed in the United States

STELLAR STERILITY
"The Search for Alien Presence"

By
Ross S Marshall

A Study in Selenography, whereby an attempt is made to determine whether there is evidence for extraterrestrial activity of space aliens. Illustrated with hundreds of NASA planetary pictures!

"When a person is honestly mistaken and hears the truth, they will either quit being mistaken, or they will cease to be honest."

-- Anon

Weirdvideos.com

Dedication

This book is dedicated to George Leonard, Fred Steckling, Richard Hoagland, William Saunders, Mark Carlotto, Mike Bara, Rob Shelsky, George Kemplani, Erich von Daniken, Zecharia Sitchin, David Childress, Josephus Goodavage, Jack Swaney, Felix Bach, and NASA.

ALSO, Don Ecker, Vito Saccheri, Rene Barnett, M. J. Craig, Victor Bertolaccini, Doug Turnbull, Tom Lehmann, Stephen Baxter, Ray Villard, Giorgio Tsoukalos, Stanley V. McDaniel and Monica Rix Paxson, Jason Martell, Philip Coppens, Louis Proud, Timothy Good, Alan McGregor, Stephen Webb, Maximillian de Lafayette, Whitley Strieber, Stanley McDaniel, Monica Paxson, Jim Marrs, Alex Milway, Adam Moon, Robin Moore, Xaviant Haze, Roc Hatfield, Mac Tonnies, George Haas, Brube Rux, Nick Redfern, Arthur C. Clarke, Gene Roddenberry and Lewis Carroll, with special thanks to Douglas Woodward.

CONTENTS

PREFACE

SECTION I
INTRODUCTION
THE PLURALITY OF INHABITED WORLDS............15
Arguments against aliens presence..............................40
Interstellar space travel is Poppycock!..........................49
Arguments against life beyond Earth............................82
The Moon is Dead...82
Mars is Dead..91
Meteoric evidence...100
The morphological argument......................................102
Carbonates conflict...104
Mars has little Blue Balls..106
"PAH" ! Tommy Rot!...109
Mars is a lot of hot (Methane) air................................112
If there is water, there are little extremophiles?........118
Alien cosmology, lies, little green idols......................132
Finding alien artifacts no proof of evolutionary theories...134
Validity of E. T. archaeology science..........................135

SECTION 2
ARE THERE ALIENS ON THE MOON?
INTRODUCTION
   1   The Sea of Tranquility Habitability....................148
   2   Oval Blown Boulders..................................152
   3   Tools in Tranquillitatis................................157
   4   Fra Mauro Hideaway-Lands.......................160

Stellar Sterility

5	Tackling Tycho Technology	165
6	White Rays of Tycho	173
7	The Big Screw	179
8	Crosses of Kepler	183
9	Blair Cuspids, E.T. Towers	188
10	The Moscow Monuments – Luna 9 still missing!	201
11	The L-Bow shaped Obelisk	209
12	Out of Control Wheels	215
13	Crater Rings, Domes and Speakers	218
14	Blobs and Black Crosses	220
15	Procellarum Pyramids	224
16	Alpine Construction Co.	230
17	Causeways in Crisis	238
18	Bridges on the Far Side	246
19	Alien Artifacts are the Pits	249
20	Lubinicky Junkyard	252
21	King of the Farce Side	259
22	Regolithic Rovers	264
23	X-Drones of Crater King	268
24	Dinosaur Crater	271
25	Clouds over Lobachovshy	277
26	Super-Size Me Rigs of Crater Pasteur-D	282
27	Super Treads and Lunar Ladders	287
28	Things George Leonard Missed	290
29	Lunar Graphics Experts or experts liars	296
30	Crater landslides and rim-slides	302
31	Mr. Vito vetoed! Trying to bridge the "gap"!	305
32	Conclusions	310

SECTION 3
Did We Discover Alien Bases on the Moon? NO!
FRED STECKLING Stuck It to Us!

Stellar Sterility
A Refutation of Fred Steckling's "We Discovered Alien Bases on the Moon"

INTRODUCTION
Evidence shows the solar system is sterile!

1. Alien UFO Hangers, Ponds and Lunar Blobs......
2. Moon Microbes...
3. Irrigation Ponds.......................................
4. Clouds over Vitello...................................
5. Alien Housing Projects and Crater Rim Domes....
6. Fred's Damn'd Damoiseau Walls and Rille Rivers..
7. Crater Cigars, UFO's and Airplanes.................
8. Double Craters and Listening Devices...............
9. Planetary Pies and Pie-cut Mounds..................
10. Crater Cabell's Alphabet Soup Letters...............
11. Tanks and Towers.....................................
12. Rolling Boulders and Mining Machines............
13. Crater Lakes...
14. Tsiolkowsky Water Crisis............................
15. Molehills, Mushrooms and Microbes................
16. Conclusions..

SECTION 4
JACK SWANEY'S "OBJECTS ON THE MOON"
Intermediate Idiotic Incidentals of Intergalactic Garbage

1. Mare Crisium Contraptions................
2. More Alien Crisis on the North Shore.........
3. The Gigantic Block in Endymion............
4. Robotic Bunnies in Crater Clavius............
5. Monumental Mountains of Moon Metal......
6. Cranes and Booms of Julius Caesar.........

7 Conclusions..................................

SECTION 6
RICHARD HOAGLAND'S MONUMENTS MADNESS
A POLEMIC DISCUSSION OF RICHARD HOAGLAND'S
"THE MONUMENTS OF MARS: A City on the Edge of Forever" or "Nowhere."

    1   The Ukert Baseball Diamond................
    2   Return of the Ukertians.......................
    3   Dangling Crystals of Ukert City.............
    4   A Glass Sphere Covering the Whole Moon?
    5   Ukert Crater Triangle.........................
    6   Crystal Dome of Sinus Medii..................
    7   Sinus Medii Dome Debunked.................
    8   The Shard Location............................
    9   The SHARD......................................
   10   The "Tower" and "Cube" Location............
   11   Where is the Shard?...........................
   12   Lunar Reconnaissance Orbiter Camera.............
   13   Rebar Beams of Crater Manilius................
   14   The Castles of Ukert.............................
   15   The Mare Crisium Dome and Crystal Spire...
   16   More Ukert Crystals............................
   17   NASA Cover-up or Touch-up?...............
   18   Photo Reference History of Mare Crisium...
   19   Conclusions....................................
FOOTNOTES..................................
BIBLIOGRAPHY.............................
SOURCES, LINKS, ADDRESSES, ADS...........

PREFACE

"A truth's initial commotion is directly proportional to how deeply the lie was believed. When a well-packaged web of lies has been sold gradually to the masses over generations, the truth will seem utterly preposterous and its speaker, a lunatic." -- Dresden James

If you are reading this at the moment, then you must be one of those persons that is interested in space science. From the title of this book you must as well be intrigued to some extent with whether aliens exist. Furthermore, if you are considering that aliens do exist, then it is not that far removed from your mind to be considering that these aliens (if they exist) must have left evidence behind in their activities just as humans do.

The shape of one's eyes, whether they be human or olive shaped does not change the fact that all creatures leave effected trails. No matter where a living thing slithers, it changes the very cosmos around them. Even the simple observation of an object, in some way, changes the object being observed.

In quantum mechanics, which deals with very small objects, it is not possible to observe a system without changing the system, so the observer must be considered part of the system being observed. This is called the "uncertainty principle" and it has been frequently confused with the observer effect, evidently even by its originator, Werner Heisenberg.

The uncertainty principle in its standard form actually describes how precisely we may measure the position and momentum of a particle at the same time.

Stellar Sterility

If we increase the precision in measuring one quantity, we are forced to lose precision in measuring the other. An alternative version of the uncertainty principle, more in the spirit of an observer effect, fully accounts for the disturbance the observer has on a system and the error incurred, although this is not how the term "uncertainty principle" is most commonly used in practice (Wiktionary, under the term "observer effect").

Well, whether it be the "observer effect" or the "uncertainty principle", apparently, things get changed when living (observing) beings come in contact with them. Thus, it is without a doubt, that if aliens have traversed the surfaces of our neighboring planetary systems and have done more than just "looked," we should find evidence. And, if they have dug up the surface and built gigantic structures, colonial outposts, highways and tunnel systems, swimming pools, cafes and theaters, then we surly should find some evidence.

Needless to say, the "uncertainty principle" and especially the "observer effect" has been proved once and for all in the "observations" done by the following alien artifact hunters. The sterile lunar surface (as proved from lunar photographic sources and the regolith samples as obtained from our unmanned spacecraft and Apollo Missions) has been polluted with uncountable changes. The lunar surface is now riddled with (supposed) archaeological artifacts!

There is tons of lunar data to be studied and once we have collected the selected data we need, all we have to do is add it up to see whether the summation points to an alien presence or not. Furthermore, we must also ask just how much evidence is needed to prove one way or the other. If we find one conclusive piece of evidence, just one fragmentary artifact of non-human origin, then we need no further proof.

Stellar Sterility

On the other hand, if we dig up both all of Mars and the whole of the Lunar surface and we find nothing, then even though it is still possible that aliens might exist elsewhere, it is most probable that they do not, at least on the Moon and Mars. In other words, one small "alien" step (footprint) would prove to us that man was not the first on the Moon. Alternately, the silence of one million billion tons of Martian and Lunar dirt sifted, observed, and classified as to bearing no evidence of alien activity, begs the possibility question and reduces the probability factor down to almost nothing.

Since, the most likely planetary systems that humans might be able to dwell upon are the Moon - being the closet, and Mars, the most potential planet to inhabit, unlike all the others, then we can expect to wonder if aliens might have done so long before us. After receiving pictures from the Moon and Mars missions, the question comes up as to what the many anomalies are that we find.

For example, the famous Face on Mars. Ever wonder what it is? Many people say this surface feature is evidence of alien activity! Others say it is a pile of rocks with a play of light and shadow giving an illusion of artificiality. Critics argued that light and shadow can play tricks on the mind, especially minds that are overly imaginative and educated. In other words, take for example, they say, where clouds often take the shapes of animals and human faces. The same is true of rock formations, such as the Great Stone Face supposedly on Mars.

Even before the latest Mars observation missions transmitted the Face's true constitution back to us, NASA scientists were telling us that it was just a blurry shadowy and highlighted mountain - a huge pile of rocks. This of course was based on the lack of evidence that it could be anything otherwise. The tendency of chaotic shapes to form patterns vaguely resembling familiar things is responsible for people making identifiable things out of anomalous things, and also the cause of many absurd books: rods, chemtrails, UFO's, ghosts, and other such nonsense.

Moreover, speaking of books and alien artifacts, the most absurd book ever written about Lunar and Martian alien artifacts is THE MONUMENTS OF MARS by Richard Hoagland, which we shall deal with later in this book. It may not be the funniest book ever written, but it is the most absurd one ever published. It is the most absurd only because it is the most well written and, unfortunately, the most dangerously tempting to believe.

More recently, UFO enthusiasts and mad Moon and Mars monument hunters have been playing 'find the hidden artifact' game with our planetary system. They pore over thousands of photographs of cratered surfaces until -- aha! -- they find something (questionable) suggesting the presence of outer space creatures. This is believable, if you have a vivid imagination, are a sucker for lies and can swallow the contorted, blurry, and twisted chunks of crater ejecta as unnatural artificial alien manufactures.

Notice, that before computers, it was all based on twisting and misinterpreting selenography; superimposing presuppositions over shady and overly highlighted lunar surface features and mistaking emulsions blobs and development splatters for alien artifacts. Since computers have come in, and since most of the aliens artifacts based on previous research has been disproved as photographic processing and development mistakes, it is now jpeg compression creations from pixel perversions that are the alien artifacts.

As early as 1976, George Leonard, in *"Somebody Else Is on the Moon"* (Pub. by David McKay, 1976), carried this kind of speculation to such extremes that he managed to write one of the funniest books ever written on the subject. Leonard was an amateur astronomer and retired public-health official in Rockville, Maryland. In his book, George claimed that bridges on the moon were among the least controversial things about the moon. Really?

Well, needless to say, all bridges vanished when the Apollo photographs were obtained. The "bridges" were nothing more than illusions created by sun-light and shadows, yet this wonderful enlightenment did not stop the astro-myth spinners. The myth of alien artifacts and structures still persists in alien presence based literatures to this day.

The same thing happened to the mysterious "spires" on the moon. Photos in 1966 of the Lunar surface showed objects casting such long shadows that alien artifact hunters decided they had to be rocket ships or radio beacons -- at least "something" built by aliens. A Russian periodical called *Technology and Youth* featured a wild article about these spires in its May 1968 issue. The spires turned out to be ordinary boulders, because it was found that their long shadows were caused by the sunlight hitting them at extremely low angles. If it was not Clementine's new pictures exposing the alien artifact hoax, it is now the Lunar Reconnaissance Orbiter. No more can we say "Somebody Else is on the Moon" as we shall see.

Leonard quotes an unnamed NASA scientist in his interview in the tabloid *Midnight* (February 8, 1977): *"A lot of people at the top are scared."* He thinks the aliens live underground and that seismic quakes on the moon are caused by their undersurface activities. He said, "*NASA is simply lying to the American people about UFOs,*". He suspects the aliens are waiting patiently to take over the earth after we blow ourselves up. Photos of the Moon's surface, he insisted, show rims of craters sliced away by giant machines, jets of soil spraying out (caused by mining operations), and tracks of huge vehicles. "*No, I do not know who they are,*" Leonard said, or "*where they come from or precisely what their purpose is. But I do know the government is suppressing the discovery from the American people.*"

Really? Have we ever seen a detailed close-up of these areas? The Lunar Reconnaissance Orbiter is now showing us "in detail" in each magazine of pictures that there is nothing more than lunar dirt on the Moon with no hint of alien activity.

In 1981, Fred Steckling published what is to be known as the second best funny-book ever written on Moon artifacts – "We Discovered Alien Bases On The Moon." Fred pawns off the same nonsense as George Leonard: Alien ponds, lakes, digging machines, pipes, tubes, screws, and oil tank fields, crater rim space craft landing pads, space craft hangers, and on and on. It is still published by the Adamski Foundation.

Seeing familiar anomalies on Mars has been common ever since the invention of the telescope. Percival Lowell found the red planet's surface so honeycombed with canals that he wrote three books about how the Martians, desperately in need of water, built the canals to bring water from polar regions. Now, of course, we know the canals were only a figment in Lowell's mind, distinguished astronomer though he was.

Unfortunately, this has not deterred seemingly intelligent people from similar self-deception. Even though such canals were later proved to be illusions derived from the play of shadows and the sun light reflecting off the dust storms on the Martian surface, Martian Oopart hunters continue to make alien mountain motels out of crater ejecta hills.

Here and there on Mars are formations with grid-like structures. "Did NASA Photograph Ruins of an Ancient City on Mars?" is the headline of a *National Enquirer* article (October 25, 1977). A photo of a region near the Martian south pole shows a series of square-like formations called "Inca City" because they somewhat resemble a decayed Indian village.

In 1977, electrical engineer Vincent DiPietro came across a 1976 photograph taken by the Viking spacecraft that orbited Mars. At first he thought it was a hoax. The photograph showed a remarkably human-looking stone face about a mile wide. NASA had released the photo shortly after it was taken in 1976, and planetary scientists emphasized that it was a natural formation. DiPietro thinks it isn't. Computer scientist Gregory Molenaar used image-enhancement to explore details of the face, and in 1982 DiPietro and Molenaar published a 77-page funny book, "*Unusual Martian Surface Features*," about their results

The authors concede that the face may have been produced by erosion but they suspect otherwise. They claim that computer enhancement shows an eyeball in the Face's right eye cavity, with a pupil near the center, and what looks like a teardrop below the eye. "*If this object was a natural formation,*" they write, "*the amount of detail makes Nature herself a very intelligent being.*" The more contemporary Richard Hoagland, if I remember correctly, believes that there

Stellar Sterility

are teeth to be found in the Face's mouth peeking through the shadows.

West of the big stone face, in the shadow of a pyramid-like formation, is a grid-like pattern suggesting a lost city with an avenue leading toward the face. (See "Metropolis on Mars," an unsigned article in *Omni*, March 1985). Skeptics have pointed out that the so-called pyramid is much cruder than scores of pyramids found as natural rock formations in Arizona.

But no, the alien propagandists are so compelling in their arguments that even Hollywood was convinced enough to make a movie in support of this fiction - "Mission To Mars." In this fictional account trip to Mars, the Face turns out to be an ancient alien UFO hanger, built eons before mankind ever existed. The fiction of course supports and propagates evolution by making the aliens the forefathers of mankind and the creators of the apes we came from.

Top drum-beater for the view that the stone face proves that an alien race once flourished on Mars is writer Richard Hoagland. He has published a book called The MONUMENTS OF MARS which tries to prove this. There are mathematical calculations all through, with descriptions, myths, inter-dimensional geometry and plenty of Martian topographical language to authenticate it.

The Mars Face, Pyramid and City of Cydonia Complex are especially important in his book. Even though the latest Martian pictures reveal this Face to be just a piles of rocks - and Mr. Hoagland should be happy that NASA fulfilled his dream of obtaining these better images - Richard continues to insist that it is an alien sculpture and that NASA has tampered with the new images!

Stellar Sterility

Fred Golden, writing the "Skeptical Eye" page in *Discover* (April 1985), ridiculed Hoagland's claims and ran a photo of another spot on Mars, where the topography resembles Kermit the Frog. Many other Martianologists have found, and are still finding, what they believe are anomalous objects of alien origin.

The Internet now is teaming with wild and crazy web artifact web sites. It has apparently come to the point of pure competition as to who can dream up the best nonsense and with no care at all as to validity or factuality. Many of the blogs and web sites set rules to disallow negative and critical postings – all posts must support the competition.

If you search any kind of chaotic data, it is easy to find combinations that seem remarkable. Every page of a book of random numbers contains patterns with enormous odds against them if you were to specify the pattern before generating the random numbers. Every bridge hand you are dealt would be a stupendous miracle if you had written down its exact pattern before the deck was shuffled. But no one seems to have done either.

Furthermore, about every 20 years all previous pro-alien speculation upon the data collected prove to be just that - "speculation." Nevertheless, with new technology, new and more exciting "facts" are manufactured and pumped out of the spin doctor factories in support of the alien presence. Thank (the computer) god we now have pixel distortion!

Let someone close his eyes and talk for 15 minutes about a scene he imagines. You'll have no trouble finding amazing correlations between his description and any randomly selected scenic spot. Let a psychic crime-solver rattle on for an hour about clues to a missing corpse. It's inevitable he/she will have made

Stellar Sterility

some lucky hits if and when the body is found. Now, give a group of artifact hunters a planet with disturbed surface features and inevitably they too will shovel-up "evidence" of the existence of aliens. This is unbelievable but it is very explainable why they do what they do.

Close your eyes and thumb-press your retinas, and you will see all sorts of things, and every other imaginable thing, other than real tangible things. For, the creative human imagination is a wonderful device, but when used to create illusions to pawn off as reality, and make a buck ta' boot, that imagination is a con-artist.

This way of seeing "alien" things and selling the stooge a pro-alien artifact book for a fast buck is not too far removed from the ancient (extortion) arts of Tesseography, haruspicy, extispicy, hepatoscopy or hepatomancy and kypomancy (divination by tea leaves, animal guts and coffee grounds) and especially, when dealing with the geologic features and materials of mountains, sand, rock and dust - oromancy, abacomancy, aichmomancy, amathomancy. No matter what the lunamantic diviner sees in moon rocks, cracks, shadows and highlights, the poor under-developed phren believes it, pays up and toddles off with his newly purchased enchantment.

The great and acclaimed raving chresmomancist Richard Hoagland has mastered the shadowy arts of the ancients, taking the royal position as Vicar of alien artifact morpho-divination — the new selenomantic of lunar research. With exoenthusiasm and the calligraphic whisk of chartographomancy, the new Hoaglandomantic gut-busting geloscopeable topomological science was born, published and dumped on the market, and the green-backs roll in as the new pseudomology spreads like wildfire within society.

The list of alien artifact con artists, entertainers, pabulum producers, paradigm programmers, pseudoscientists and pseudomantics is practically endless. The alien artifact pulp rags continue to pile up even all the way to the moon! If the cow never made it, Hoagland's book "Dark Mission" surly did.

The publishers range from blind pseudo-scientific writers such with Leonard, Steckling, Hoagland, David Childress, Erick von Daniken, Sitchin all the way to pure science fiction writers and game designers like Doug Turnbull, Tom Lehmann and Stephen Baxter. The first believe their own fictions, while the latter wish their fictions were believable. Nevertheless, in both cases, the idea of space intruders is advanced at the expense of truth.

Some last words on these people must be noted. In almost every case, it will be noticed that these rags start with the presupposition (false premise, idea) that aliens exist. From this porcus-swill all data is bent to support such nonsense. This is the easiest way to pawn off a fast buck-maker. To approach it any other way is time consuming, tedious, technical and requires much source tracing and testing. This is the preferred way used in this book – start with the data and work towards a conclusion.

There are many reasons why pseudo-science books are written the way they are. First is to sell the book on a large scale quickly. Secondly, to reinforce the idea that aliens exist and lastly, to replace good-ole' scientific and religious truth. There are two witnesses against alien presence: Sacred Scriptures and government space agencies. Science has come to the

Stellar Sterility

conclusion that the probability of finding evidence for the existence of aliens is zilch!

They have not ruled out the possibility, but the probability has decreased tremendously. Religions have, alternately, pointed this out by defining the probability factor as being non-physical, spiritual beings called good and bad or unfallen and fallen angels. The religious view more closely supports the hyper-dimensional interpretation.

In his book we will not debate whether aliens really exist. This writer actually believes that "something" alien to us exists, but proof is lacking as to their physicality. What will be analyzed is whether what evidence we do have supports the theory that aliens have visited our solar system. It is most probable that my readers have waffled through other alien artifact books supporting the alien presence. It is now time for alienomantic enthusiasts to hear the other side of the argument.

This writer once believed "We Never Went to the Moon" (Ref: Bill Kaysing) only to find out later that we probably never left. Recently the facts have dictated that we did go and that there is no real need to go back but for a few more samplings or until technology advances further to support human occupation.

The next deception came with "The Face on Mars." Mr. Hoagland and Carlotto did a wonderful job in convincing the public of an alien presence on Mars. But no one was happy with the blurred pictures. It was not until later, after further Martian spacecraft photography that the Hoaglandite belief came tumbling down into a pile of rubble. The face was only a misinterpreted heap of rocks. But, this did not stop the con-game of pawning off the theory of E. T. intruders.

As soon as the more recent Martian pictures came in, and quickly after the Mars face was debunked, thousands of new artifacts popped up. With the new LROC and the exposure of the moon to higher resolution photography, the public eye quickly shifted from lunar to Martian observations. But not with this writer. I soon realized that the redirecting of the public from lunar searches — since there was nothing to be found there, one could continue selling the gullible the same belief system. As moon photography was to Leonard, Steckling, so the Martian photography is to us today.

Now, it is believed by this writer, as the lunar pictures are to us today — sterile of any evidence of alien presence - so one day will be the Martian photography. Mars will be another study as the photography gets better. For now, let us see what the higher resolution LROC photography has revealed about aliens on the moon and what science has determined about interstellar space travel. The likelihood that aliens have visited us from another star system appears to be highly improbable. In the process of collecting data keep track of all the "zilches!"

SECTION-I

INTRODUCTION
PLURALITY OF INHABITED WORLDS DECEPTION

"Because false doctrine is so abundant: "The Truth will be the Strangest Thing you will ever Hear" - **Paul Smith**

The belief in the plurality of inhabited worlds, which for centuries has had its advocates among famous authors and philosophers, took deep root in many cultured minds several generations ago. During the last 50 years, however, progress in astrophysics, biology, planetary studies, and [more recently], satellite planetary exploration eliminated, first of all, the stars as a whole, and then nearly all the planets and their satellites as possible abodes of life, as we know it on earth.

Nevertheless, in recent times, a rage of E.T. artifact madness has taken a strong grip upon a large segment of man. Hence, opinion has split into two opposing camps, both arguing from a more or less well founded metaphysical basis, and using the same accumulated space science data for supporting their positions. Evidence seems to be plentiful, but irrefutable conclusive proofs are hard to find.

Stellar Sterility

The first points out the exaggerations to which belief in life on other planetary bodies has given rise. The Second, on the contrary, stresses the amazing variety of life forms on Earth, and the powers of adaptation, and is ready to extend this existence as a kind of 'generalized' life, in which life on Earth would only be a 'special' case.

As to which one is correct, the first makes the mistake of assuming ANY and all life, [if it exists at all in space and on OTHER planets,] must fit within the narrow limits of life giving properties, as Earth defines them for humanity. This view derives its conclusions primarily from physics, astrophysical researches, and planetary probes. The second commits an equal mistake of presupposing the "universality of life", assuming no minimal conditions for its manifestation.

This view derives its conclusions mostly from visual observations. The first denies all but carbon based creatures, while the latter extends the basis of life to silicon and other types, even to extra-dimensional life forms.

Obviously, each are completely objectionable to the other, yet both share some commonalities. They both are mutually antagonistic, and both fall victim to an over emphasized non-revisable presupposition. In any case, the only field where these ideas could possibly be confronted with observational and experimental data, is that of Planetary Physical Astronomy.

In recent years, the only planetary bodies to be pin-pointed as probable locations for ET artifacts are the Moon and Mars. And this brings us to this research presentation. To discuss the evidences as to whether there is any ALIEN ACTIVITY on other planetary bodies, other than Earth.

Stellar Sterility

It is NOT the intent of this study to re-open the discussion in the above extreme, but, on the contrary to try to present materials to clarify the matter. This book is not meant to confuse it, as many fellows have done, but rather to simplify, and bring the reader back to a fresh and unbiased mind to the FACTS, as to whether the evidences found point to E.T. activity on our planetary neighbors.

Our approach is simple and scientific and is as follows: 1.) We must return to the original sources, where possible. 2.) We must use what is authentic and discard what is erroneous. 3.) We must properly interpret the area of study and gain the TRUTH as to its nature, and discard what is obviously erroneous; or to conclude a neutrality, until further research can be obtained to decide, [without a SHADOW of doubt], whether it is NATURAL or ARTIFICIALLY created. 4.)

We must present the FACTS and evidences to the viewer with what we have found to be true [or, at least possible], and then allow the viewer to decide for himself. 5.) We must supply the viewer with resource connections and outlets for the viewer's personal pursuit of the matter. Only this way can we come to terms with who is blowing Lunar dust and who is not.

The above numbers 2 and 3 are of extra importance, when dealing with data. The question arises as to what is authentic and conclusive and what is not and is anomalous. And how do we go about deciding the difference to come to a conclusive "conclusion," that there are truly alien artifacts on the Moon?

I would like to present a mathematical formula. Let us take a set of data, say 100 pieces of material (in this case, an analogy for NASA data, photographic materials, eye witness testimonies, etc.). Let us "tag" each item with a positive or a negative symbol. The *positive* symbol is to refer to those pieces of data as conclusively "of non-human origin" or "artificially created by non-humans."

The negative symbol is to refer to all those pieces of data that are non-conclusive of the above. Now, let's take that set of 100 pieces of data (try maybe 1 million pieces?), for example, 100 photographs of the Lunar surface . If each piece is decided to be non-conclusive as to supporting alien presence, we have 100 negatives. Let us ADD up all 100 negatives. Do we get a positive answer? Does -1, plus -1, plus -1 = equal +3? The answer is no.

Now, let us do this with all 100 negative pieces. You will get -100. Adding up 100 inconclusive pieces of data gets us a big fat negative pile of inconclusiveness. How then can people say that all these negatives add up to proof that aliens are on the Moon?

The opposite result would get us 100 positives, meaning we have 100 totally conclusive pieces of data telling us that there are aliens, and that they have been, and maybe still are on the Moon. Hell,... even ONE positive piece of data would be enough! Should we say any more? Please, can someone bring forth just ONE positive piece of evidence?

 For centuries, the planet Mars and the Moon, have been the "FIRE" in many planetary observers. Many have devoted their whole lives to such studies and have even built observatories to delve deeper into the many unanswered questions. Along this path of planet watchers we will find many wild theories and tales to read.

If we apply the above rules, we will determine quickly the difference between falsity and truth, and

Stellar Sterility

also whether there is any alien presence or "life" extant in our planetary system other than on planet earth.

When it comes to data, resources, artifacts and testimonies, one will find it is no easy task. There is so much emotional and bias thinking, as well as outright deception, that even the best scientist can be fooled. Astro-archaeology is as speculative as literary archaeology, and can sometimes be a quandary as to deciding what is real and what is poetic and/or fictional.

A good example of confusion in decision making over, for instance, ancient claims and testimonies of extraterrestrial activity, or at least ancient advanced human technological achievements can be sampled in the following.

The ancients themselves "claimed" many, surprisingly modern sounding stories of space travel and heavenly contacts. The ancient Hindu's, in their astronomy book the Surya Siddhanta ["The God Surya's Astronomy," 500-2000 BC] claim that their older ancestors went to "MAYA" the Moon and that the "Siddhas" and "Vidvaharas," or philosophers and scientists, had the ability to "examine the regions below the moon but that are above the clouds" They found [2000-4000. BC?] that the Moon had NO AIR, was DRY as powder, and looked as if it was covered with "*Burnt Ash and Dust*" (The Surya Siddhanta. See Andrew Tomas, "We Are Not the First", London: Sphere, 1971, p. 149).

In the 3rd century B.C., Chuang Tzu, in a work entitled "Travel to the Infinite," relates a trip the Chinese made into space some 32,500 miles from the earth. The following is a synopsis: In the year 2309 B.C. the engineer of Emperor Yao "Hou Yih" decided to go to the moon. The "*celestial bird*" provided Hou Yih with information on his trip. He explored space by "*mounting the current of luminous air*" (i.e. the exhaust of a fiery rocket). Hou Yih flew into space where "*he did not perceive the rotary movement of the sun.*"

On the moon he saw the '*frozen-looking horizon*" and erected a 7 STORY building, which they called "*the Palace of Cold*'. (This statement is of some importance in corroborating the story because it is only in space that man cannot see the sun rise or set.) His wife Chang Ngo likewise made a trip to the moon, which she ACCURATELY DESCRIBES as a "*luminous sphere, shining like glass, of enormous size and very cold; the light of the moon has its birth in the sun,*" she declared.

Chang Ngo's moon exploration report appears to be surprisingly correct.

The Apollo Mission astronauts found the moon desolate with a glasslike soil with parts of it even paved with pieces of glass - probably melted from the intense heat of the Sun. Most of the Moon, at any given time, is in the extreme cold. It plunges to minus 250 degrees Fahrenheit at midnight. But in the day time, it swings to just the opposite, 250 degrees above. That's hot! No need to argue whether the Moon can be hot and dry as powder. Not to digress to far here, but notice that none of these accounts mention any alien presence or extraterrestrial threats.

In Tibet and Mongolia, ancient Buddhist books speak of "*iron serpents which devour space with fire and smoke, reaching as far as the distant stars.*" The Epic of Etana (4,700 years old) supplies us with very accurate descriptions of the earth's surface from progressive altitudes – descriptions which were not verified in our own era until the high-altitude aerial

Stellar Sterility

flights of the 1950s and the first space shots of the 1960s.

The description of this ancient space flight depicts exactly what happens when man leaves the earth (the concept of the round earth which becomes small, due to perspective as distance increases, and changes into particular colors). The ancient Book of Enoch said that in space "*it was hot as fire and cold as ice*" (where objects get hot on the side illuminated by the sun and icy cold on the shaded side) in the "dark abyss."

Yes, these claims are very suspicious of a highly advanced technology lost in past times. Mythology claims even that the ancients built Moon bases and cities, then blew them up. Hence the lunar glass dome theory and the atomic bomb theory for lunar crater formation! For, after all - "*The Gods and demons behold the Sun, after it is once risen, for half a year; the Fathers (pitaras), who have their station in the Moon, for half a month;*" (Text Book of Hindu Astronomy. xii-74. F. Burgess & W. D. Whitney).

On the other hand, these stories could be no more accurately depicting the historicity of ancient space travel or alien activities than the works of: Cyrano de Bergerac, in his "L'autre Monde Ou Les Etats et Empires De La Lune" 1657. - "*Here is how I betook myself to heaven,*" Bergerac writes. "*I attached to myself a number of bottles of dew, and the heat of the sun, which attracted it, drew me so high that I finally emerged above the highest clouds. But the sun's attraction of the dew drew me upwards so rapidly that instead of approaching the Moon, as I intended, I seemed to be farther from it than when I started. I broke open some of the bottles and felt my weight overcome the attraction and bring me back towards the earth.*"

Stellar Sterility

Or, Baron Karl Friedrich Hieronymus Münchausen, in his "The Surprising Adventures of Baron Münchausen" 1785. - Like de Bergerac, the Baron was both an historic and literary figure. In life, the baron was a German nobleman. Yet, in the pubs, the notable Baron was a story teller of tall tales: For, the Baron *"rode a cannonball," "danced in the belly of a whale,"* and *"traveled to the moon"* – and he did it twice! The famous "story" tells us he did it with multiple strips of bacon tied to a long rope, swallowed by (so many) geese, and with the blast of a pistol, off he went.

Or, Jules Verne, in his famous "From the Earth to the Moon" 1865. - Jules Verne's humorous science fantasy begins when members of the Baltimore Gun Club devise a plan to manufacture a giant cannon that will shoot a "space-bullet" from Florida to the Moon. Verne deserves applause for his visionary calculations.

When the Moon landings were finally achieved nearly 100 years later, the calculations from his story were startlingly close to modern scientific findings and, last but not the least; H.G. Wells, in his fabulous "The First Men in the Moon" 1901. In this tall tale of scientific sounding (at that time) fiction, we see impoverished Englishman Mr. Bedford pair up with Dr. Cavor, a scientist working on a gravity-shielding material he calls 'cavorite'. After discovering a peculiarity in the material's behavior that makes the air above it weightless, the men build a spherical spaceship and travel to the moon where they are captured by insectoid moon-creatures. Well, as we reach the 1930's and 40's, this tradition of fictional story-telling increases until every science fiction book, comic strip, and (especially) the movie industry contained many parallels to modern space travel. Apparently, it is not all that impossible to foretell

Stellar Sterility

technological advances ahead of time as the imagination is a wonderfully creative machine.

Because of the close proximity of the Moon to the Earth, Governments today - just like the ancients of old - are infatuated and addicted to the mysterious world between the stars. Look at the world wide efforts in building scientific space craft.

Attempts to reach the stars dates back centuries, with one hilarious example being a Chinese Emperor who blew himself to smithereens from attaching solid fuel rickets to his royal bed figuring the engines would carry him to the stars. He did make it somewhere and we hope he made it into heaven but there is no proof he ever made it to the stars.

It is no wonder that many persons and Governments have a permanently fixed vested interest in space travel, colonization and now, radical support for ET occupation of the solar system. Without the "mythos" as a fire, there is no strong reason for the masses of humanity to pursue science, other than to escape the planet earth. Raw science is somewhat dead to the common man. Only a true scientist can extract for other scientists, sweetness from the sour grapes of EMPTY LIFELESS SPACE - and that is exactly what EMPTY LIFELESS space is it all about for billions and billions of light years distance: Nothing, just lifeless space!

Nevertheless, these are no sweet grapes to the mythologists: They reason *"There MUST be ALIENS out there. Or, we don't care about Space Science! What's it to us?.. We'll never go there! So, why should we waste our precious time (and money) on unimportant (unprofitable) things?"* So, if we are to invest into it, we will have to stick a few aliens in space, plant a few monuments on the Moon, and people will

Stellar Sterility

get interested. Why? Because NOW the masses have HOPE, whether they ever go for a space ride or not.

Could the ALIEN FACTOR be just an alternative HOPE mechanism to traditional religion? Could the argument for ALIEN technology be just a convenient alternative to Human technology, which seems to be 200 years behind getting us all off this DYING PLANET? Could the subconscious desire be that, since humanity cannot get off the planet, then ALIENS can do if for us? "*Let's make a deal with'em!*" Maybe this is another topic for another book investigation? Or, maybe, because without some E.T. mythos, the Government (that coincidentally, neither cares to prove or debunk the urban legend) would not have near the publicity and funding it does.

Before the advent of artificial probes to the moons and planets of our Solar System, in the 1960's, people had only the benefit of Earth bound observations. The opinions were the same, though. Some were adamant about E.T existence on other planetary worlds, while the others were of the conservative belief that, if there is, it is of a more primitive type, like say, ORGANISIMS, or LEACHENS. Some went so far as to allow, mosses and plants and grasses to exist on certain locations of the Moon. Even before this time, others were digging canals on Mars, until they were proved to be changes in atmosphere caused by huge dust storms. (We may forgive these people for the lack of technology to prove otherwise).

For example, in the 1700's, Emanuel Swedenborg claimed to have ascended to the planets and found that ALL the planets had life on them. The Moon even had huge TALL Quaker-Like people on it. Now, this guy was the first to theorize the nebula theory, was an inventor, and a very smart man in many fields. Yet, I

Stellar Sterility

guess for pleasure (and profit?), he pawned off the alien theory to promote the sale of his products as well as his religion.

In this century, supporters have varied from the wild claims of ET fanatics, to the scientific Government opposition of the E.T. presence. In the last century and in the early part of this century, many opposing debates sounded throughout the scientific world. Astronomers such as Dr. Percival Lowell propounded that life does exist on other planets, like MARS. He *believed* and tried to prove that Mars had irrigation canals, green "seasonal" plant life growths, canal "pumping stations" and finally, of course, a highly advanced race of intelligent Martians living on the planet. Naturally, this led to such Sci-Fi entertainment's as "The War Of The Worlds." The brilliant Orsen Wells had a blast deceiving the public - all the while he stuffed his gut full with the proceeds! As the decade went along, Hollywood even went so far as to put a Robinson Family into Space.

Nevertheless, Lowell had his opponents, such as Alfred Russell Wallace in, "Is MARS Inhabitable?" (Pub., 1907), who proves him to be more imaginative than scientific. Referring to Lowell's fanatic Martian myth making, he said *"...that animal life, especially in its higher forms, cannot exist on the planet. Mars, therefore,* [After 110 pages of arguments] *is not only uninhabited by intelligent beings such as Mr. Lowell postulates, but is absolutely UNINHABITABLE."*

Now, whether it is uninhabitable or not, we'll leave it up to future colonists to decide. As to the evidences of Alien occupation, well, this is what we are centering upon. CAN WE FIND ANY evidence of Alien Activity? Patrick More in his, "The Planet Mars" (1950), rather suggests, after 90 pages of arguments, that *"Without speaking of the 'Martians', we cannot therefore say that there is or is not life on Mars; but we can say,...that the immense majority,... would find it impossible to accommodate themselves to the conditions, which prevail on Mars. In particular, we ourselves could not live there in the open air."*

Mr. More obviously means, Mars is INHABITABLE, if we artificially accommodate ourselves to the planet's environment. A hidden desire to believe, that if there are ET's, that they might inhabit Mars under artificial conditions? Mr. More hasn't said this, but the door is left open. And wide open it is, for in the classic 1955 book, "There IS Life On Mars", Mr. Nelson begins, what has turned into a full-fledged alien infestation of our Solar System. He wedges the door open by saying, *"It is most unlikely that intelligent life exists there [on Mars], although even this is not an impossibility...Even IF intelligent creatures did exist,...they would probably bear no remote resemblance to us..."*

As a conservative scientific alternative to this open door, NASA, in their publication "The Book of Mars" (NASA SP-179. 1968 Edition, Ref.1), had concluded from the Mariner probes, that, *"In view of the almost complete lack of oxygen and the absence of any significant quantity of liquid water, it is highly improbable that there are now any advanced life forms on Mars."* Such statements about Mars coincided with what we were learning about the Moon too. Up to 1969, nothing had landed on the Moon but artificial satellites. The Government had only landed probes and orbited the Moon taking photographs of the surface. Just as Mars was to be photographed later, the Moon had been under the photographic eyes of the Ranger, Surveyor and Lunar Orbiter Missions, and the preliminary

Lunar orbits of the Apollo Mission. Many photographs were taken. The numbers range into the 10's of 1000's of frames. Unfortunately for alien artifact hunters, there is not one shred of conclusive evidence to alien presence to be found.

As to the silence of Government on the issue, the myth spinners say, the Government is erasing all evidence, only because proof of alien presence would destroy human civilization. This is bunk! Over half of humanity "believes" in aliens! The truth spinners say, actually, the reason why Government does not come right out and fully debunk alien presence, but let the Hoaglandites rattle off the *bull*, is because the circus barkers bring them lots of votes and money to boot! Thanks to the World Wide Web, NASA has grown into a vast scientific social network unlike the dry laboratories of the 60's and 70's. It was during THIS period, from the 1960's up to the present, that the specialized field of ALIEN presence on planetary bodies really exploded into what it is. Notice also the financial support for NASA and other government space programs increased too. A few writers grabbed the NASA materials and began their studies to prove ALIENS were occupying the Moon. A scientific follow-up to the Hollywood manufacturing going on with Lost In Space, Star Trek, and the likes. ALL Fabian Socialist; all New World Order converts, and all Alien Agenda propagandists!

It was only a few years later that Mars would be dragged back into the Lowellian sphere of the "occupied Mars" syndrome. All the conservative kicking and screaming in the world, would do no good. Neither would the conspiratorial arguments, such as that of Bill Kaysing, who claims "We Never Went To The Moon," do one thing to stem the tide of the Moon

Stellar Sterility

Mars monuments madness. When ignorance rules the minds of audiences and laziness rules their ability to reach out and open a NASA book, people like Leonard, Steckling and Hoagland will always get away with pawning off mythos as science - there is a grand pay check involved here!

The Moon had already been established as a base for aliens, according to George Leonard and Fred Steckling. Now Mars was to undergo another round of treatment. If it wasn't Fred Steckling "sticking" us with George Leonard's older nonsense of "Someone Else Is On The Moon," it is now Richard Hoagland hosing off the old Lowellian dream of Martian cities and canals, and with Egyptoid FACES and Sphinx to boot! And, no doubt eventually, some hieroglyphics popped up as proof, as were supposedly found on Roswell UFO Crash metals.

Roswell Crash Metals? Now, is there alien evidences on our planet and the Moon? Has the Government sinisterly covered up all the "facts" that would prove such existence? With all the studies and claims floating around, WHO IS HOAXING WHO? Who is denying the other the truth? Who is falsifying and who is really doing the fact-finding?

Yes, we should all agree that a man has to eat and make a living, and if he writes a book to do it, that's O.K. If he produces a video, and tells a story, that's O.K...and, if he cooks for a living, that's O.K. too. But, ...IF I go and order a "real" hamburger from a vendor, I expect a real one out of the oven, not the plastic one in the glass show case. There are real hamburgers and there are rubber "sodium-nitrate" ones.

However, science and technology has increased so far, that sometimes the fake is confused for the real, so

Stellar Sterility

much so, that sometimes even the cook does not know the difference. Scholastic bowel troubles and stomach problems? Yes. It has gotten so bad that it hurts to dig into the situation and really find out. Nevertheless, if you are going to really get your stomach full, you have to bite into it and eat it to find out. So, let us MUNCH OUT on MOON MARS MONUMENT MADDNESS and see how eatable it is, or rather how reliable the information is.

Let us find out whether the supporting evidence actually supports the theory of alien occupation and habitation of our neighboring planetary orbs.

There are a number of resources to consider when we pursue this subject, and NASA has tons of it to buy at a cheap price. There are articles on crater development, lunar morphology, selenology, volcanism, dirt, rock, soil and meteorite thin slice studies, and Lunar Transit Phenomenon, [LTP's: Ref.3], which are plentiful. This last source of anomalies dates as far back in historical times [not to exceed beyond the 1700's] as the late 1700's; anomalies that resemble very new Lunar Transient evens, such as an object passing before the Moon.

This object, by the way, was recorded by the Apollo-15 Mission. [1] What the Hell is it? And, who does it belong to? No earth government has claimed it yet. this makes you wonder if it might be alien in origin. This is a real fact to interpret and play around with. Could it be an orbiting alien spaceship? What about it being an orbiting fragment of rock? Look it up on "YouTube" and decide for yourself. [1a]

The Moon has been, for a long time, an object of Mystery. Early Moon watchers claim to have seen many wondrous things,... things that only today seem to easily fit into the ALIEN PRESENCE THEME. To them it was a little different. They had no answers, even though Swedenborg would have, and probably did, tell them that, what they were seeing were the lunatic religious Quakers. One early eye witness in 1790, a Mr. Frederick Herschel, was looking through his telescope and saw, in time of a total eclipse, "*many bright and luminous points, small and round.*"

In 1794, Dr. William Wilkins of Norwich, was amazed to see a light like a star appear on the dark disk of the Moon. He said, "*This light spot was far distant from the lighted part of the Moon...it lasted for 15 minutes...was fixed...(and) brightened. It was brighter than any light part of the Moon, and the moment before it disappeared, the brightness increased.*"[Ref.4]

Others from different succeeding periods saw many other things to add to the list of Lunar Transient Phenomenon. In 1824, Oct. Mr. Gruithuisen of Holland, a Selenographer, saw a light on the dark part of the Moon. In 1825, on Jan.22, two British Officers of the H.M.S. Coronation saw in the Crater Aristarchus, a light project from the Moon's upper limb and vanish.

In 1867, Thomas Elger claims he saw a flaming star-like light flash out from the dark part of the Moon. He said, he had seen lights on the Moon before, claiming this one was the brightest he had ever seen. He further stated, in his report to the Astronomical Register, that there were funny objects near Craters Birt, Sword shapes, crosses, and geometric shapes around other craters. In 1870 Lights were seen in the Crater Plato by English Observers.

In 1877, Monsieur Trouvelot, of the Observatory, near Paris, saw in Crater Eudoxus, a fine luminous LINE or cable drawn across the crater. Others in 1877 saw other strange things: Mr. Barrett saw a bright

Stellar Sterility

light inside Crater Proclus; Mr. Harrison, in New York, saw a light on the dark part of the Moon. A contemporary of his, a Mr. Dennett, affirmed this too.; Mr. Klein reported to his French government, that he saw a luminous triangle of lights appear on the floor of Crater Plato, they were followed by others which flowed across to the Crater Plato.

Between 1885 and 1919, many other sites were seen. LTP's such as lights, reddish smoke, curved objects, illuminated orbs, black areas whitened, luminous cables, black spots with white borders, floating spots, red shadows, dark round objects floating around the Moon's Face, dazzling white flares and light explosions. The reports are recorded into the 100's by prominent people, not to mention all the commoners who were never recorded. IS this evidence for ALIEN activity of the Moon?

Some possible explanations for LTP's can be found in a little publication called "LUNAR TRANSIENT PHENOMENA, by Sky & Telescope, March 1991. They list the following as possible causes for such LTP's. But, this list does not answer all the LTP's observed. LTP's can be caused by one or a combination of: Tidal Forces, Albedo Changes, Thermal Shock, Magnetic Solar Plasma, Ultraviolet Radiation, Solar Winds, Spectral Diffraction's, Meteoric Strikes, Moon Quakes, False Color and Piezoelectric Effects.

In more recent times, writers and researchers have advanced this conjecture even to the point of providing photographic proof for the existence of ALIEN ACTIVITIES on the Moon and Mars. Josephus Goodavage, George Leonard, and Fred Steckling are some. Others who have followed are Mr. Swanev, Felix Bach, Don Ecker, Vito Saccheri, Rene Barnett, David

Childress, Mike Bara and last but not least, Mr. Richard Hoagland.

All these and many more upcoming Moon Monument theorists propound the ALIEN OCCUPATION AND INHABITANCE theory of our Moon and Mars. We must wait still, before we see Venus and the other planets become added to the list. Emanuel Swedenborg would be proud to see this day as he propounded (ignorantly?) that all the planets housed intelligent life.

To get back to the 1000's of NASA photos that supposedly show alien occupation of the Moon and Mars and the tons of comic books written on the subject, we come to the first excellently done comical study on MOON MONUMENTS. Mr. George Leonard's work "SOMEBODY ELSE IS ON THE MOON."

Taking Mr. Leonard's and Mr. T. H. Huxley's advice, which is quoted by Leonard, in the front of his book and with added interpretations, let us, *"Sit down before fact as a little child, be prepared to give up every preconceived notion, follow humbly wherever and to whatever abysses nature leads, or you shall learn nothing."*

Let us also take this advice and apply ourselves to the re-evaluation of Mr. Leonard's lunar lunacy; to Mr. Fred Steckling's photographic phantasms, and move on to a few more - (chuckle) - "lunatics", such as the monumental revelator, Richard Hoagland and his mountain of Moon Mars misadventures.

Our search to find whether there are aliens in our solar system and to verify if there are E.T. artifacts on the Moon as well as Mars will begin with our first regolith radical, George Leonard and "Somebody Else In On The Moon." But first, let us take a walk through the real facts of space travel and the possibility of life existing beyond earth's atmosphere.

IS THERE ANY LIFE BEYOND EARTH?

Everyone seems to be in the dark about other life in the universe. People ask many questions about alien life. Are humans the only intelligence in the universe? Are there other beings on other planets like us? Have advanced beings visited us in the past. Are super races from distant worlds watching planet earth? Are there Selenite's hiding under the lunar surface? Are there any simple answers?

Yes, there are simple answers. Some answers are in religious texts, such as the Bible. Other answers are obtainable from scientific research. The first records there are only two forms of life. One is human and material, while the other is nonhuman and supernatural. There is no indication of physical beings from other planets. Many scientists on the other hand, wish to find evidence for extra-terrestrial beings to avoid the supernatural.

The subject of this book is the universe is sterile of life forms other than what is on planet earth and it is the unique center and origin of life, as we know it. That earth is the center of life makes perfect sense, just as it makes sense that the big bang is the center and single origin of matter.

So far, life is exists only on earth, while no evidence of life exists beyond it. Consequently, it is most probable that there is none elsewhere. People are tossing all sorts of inconclusive alien fictions into the arena, but until someone pulls some cosmic alien rabbit out of the outer space hat, the burden of proof will continue to rest upon the proponents of alien presence. So far, every claim to alien life forms has failed the test of science. There has not been one example of absolute conclusive proof presented since ET research began. The air is silent and the table is empty.

Fortunately, for the skeptic, science must argue from silence, and this silence speaks more against than for ET life forms. Moreover, all claims of moon-men, Venusians and Martian invasions have been fictional, and therefore alien artifacts must follow suit.

Let us review only a selection of material as a sample test of the validity of the alien presence theory and the existence of ET artifacts. What will be obtained from this study will be a negative result and this is unacceptable to most people. At least half of humanity hopes that there are alien beings and a smaller percentage actually believe they exist. Some few claim they exist from personal experience! The materials dealt with will be both physical and photographic. When it comes to the photographic evidence for alien artifacts, which is the bulk of this book, the reader will realize that people see what they want to see and not what they should.

Scientific exploration consists of two primary fields of study. The first is a purely materialistic and geological, and consists of the hopes for human colonization and the acquisition of new resources. This is justifiable, since human colonization of other planetary bodies will require other world resources to be successful.

The second is the continued survival of earth itself. The second field is a two-part determination of astrobiologists. It is first to prove the chemical evolution of man and secondly to prove that man can survive in space, therefore consequently proving the higher probability of alien intelligence.

To find any alien life form, other than earth based complex organisms, is to prove the evolutionary theory of life and thus the higher potentiality of ET existence. The hidden agenda behind this ideology is to undermine the supernatural alternative. The agenda of metaphysics, on the other hand, is to disprove materialism in favor of the supernatural origin of biological life. The former is predominately polygenetic by believing in life spontaneously popping up (out of lifeless matter) all over the universe, whereas the latter is usually monogenetic and holds a minority view that life "popped" up only on earth. Both do share and favor some original explosive and creative causal mechanism.

Nevertheless, the primary argument is what comes first as the cause of the other - lifeless matter or intelligence. Therefore, this is ultimately the question: Did life come from lifeless matter, or did all that is physical come from some universal mind? When a person really looks at the data available, it is more "religious" to believe that intelligence came from dead matter, than to believe that matter originated from intelligence.

The continuation of the search for the origin of life evidences a high probability that humankind is alone. The "mind" that is looking toward lifeless matter for the answer is at a loss. The evolutionist cannot have the "mind" as the originator of matter substance (the dirt, rocks, minerals and liquids), since that is where they believe intelligent life came from.

Materialists continue to look for the origins of life in their own way, despite the fact that all efforts so far reveal sterility. The evolutionary biologist finds only lifeless rocks and liquids no matter how deep they dig. Furthermore, for all the species to exist today there must necessarily be a plenitude of fossil records of "smart-rocks."

However, there are no fossil records of educated rocks. An intelligent rock, as the first mutation from dead matter, does not exist. The evolutionary astrobiologists will find only the same lifeless rocks and liquids, no matter how far out in space they probe. Eventually, the astrobiologist must face this fact about space just as the evolutionary biologist must face it with earth. There is no evidence that life (intelligent or otherwise) stems from dead matter. The gap (the missing evolutionary linkage) between lifeless rock and the first living organism is as vast and empty as space itself.

Admittedly, a sterile universe seems more unbelievable than the fancy of one teaming with life, but for the moment, the unbelievable is very true, while the fancy is not. Apparently, what is most probably true is the least desired, whereas the undesirable ends up the very bed we sleep in – all alone in the cosmic sack!

Unfortunately, it seems that silence will continue to beg the question until this circus of pseudoscientists parade a sideshow of alien corpses in the news. How long will that take? Will another thousand exploratory rovers find the illusive microbial PAUL? How many more billions of tax dollars are to be wasted looking for nothing?

People do not seem to tire easily chasing imaginations, fantasies, spooks and space creatures. History shows us many who died looking for the Fountain of Youth, the Pot of Gold over the rainbow, the Serpent Crown, the planet Vulcan and the Holy

Grail. It is a fair and just assumption that this will be the case with aliens and consequently, with alien artifacts.

The great silence points more toward a sterile universe, a monogenetic and geocentric origin for man, and a creative intelligence (big banger) for the cause of it all. The bustling alien occupied universe of the UFO nuts has no support. A sterile universe seems to be staring them in the face with no intelligent "cosmic rabbit" in the empty hat of space to demand otherwise. The earth is unique in the universe as the only center and origin of life, and it took divine intelligence to make it happen. So, hats off to you, with no hard feelings, when you hear the bell toll *"And all things came into being by Him, through Him, and for Him"* [1].

Do angels, demons, the Jinn and other supernatural beings exist? Are they extra-dimensional beings or real flesh and blood creatures? There is no physical proof to their existence, though our histories are replete with records of them. Yet, this does not mean they do not exist in some "spiritual" dimension or plain of existence.

Nevertheless, to the metaphysician there is more evidence for ghosts than for all the above. So, what we are talking about in astrobiology? Astrobiology is the study of alien life forms and not angels, demons or Jinn.

Therefore, if aliens exist, they surly claimed and occupied the moon as well as other parts of the solar system or they have not. Moreover, because aliens are unproven to exist on the moon or Mars, some scientists say they reside in the asteroid belt. If they are not in the asteroid belt, they must be elsewhere – because they are out there.

Stellar Sterility

Logic dictates that in traveling, colonization, hunting for resources and establishing outposts, intelligent creatures surly leave conclusive evidence behind as a signature to their existence. Forensic science dictates that every criminal leaves a trace of evidence behind such as a fingerprint, DNA, a fingernail, maybe even bags of litter, trash, garbage, waste, foot tracks or varied other clues. Native Indian "trackers" can follow an animal for miles according to the disturbances made in animal movements.

Now, there is tons of testimony, claims, visions, apparent trace-fossils, supposed implants and even accounts of abductions, but nothing forensically conclusive to prove the existence of alien life forms of any kind.

This great silence is a lesson in adding up negatives, where the result is never positive. Statistical mathematics in exobiology favors more the argument against an alien presence by its stacking negatives and never summing up one positive. The probability of ET life existing slowly decreases to zero the more the negatives pile up. For example, the probability of alien life decreases each time a space mission returns a negative result.

The following evaluation of astrobiology will attempt to disprove the chemical evolution of life by showing the lack of evidence of alien life forms in our solar system. A list of scientific complications for interstellar travel will be enough to show no aliens have visited our planets.

A further examination and comparison of the geology, rocks, minerals and fossils of earth with those of other planets will show the reader that there are no ETs in our solar system. The supposed space bacteria, meteorite fossils, moon lichens, lunar bases, cities, and

Stellar Sterility

the many lunar and Martian photographs taken by space probes will demonstrate the same.

This book will conclude, from the complete lack of any biogenetic (biological fossil) evidence from ET materials, that there is practically zero probability in finding life beyond earth. This will establish the falsity of the chemical evolution of man as well as disprove the evolution of life forms in the universe other than that on earth.

This book will conclude with affirming the sterility of the moon and Mars, and that the likelihood of life elsewhere is practically zero, and consequently, that the belief in alien artifacts, is just that, a belief.

There are good arguments for the sterility of the universe. Dr. Dennis Bonnette, author of 'Origin of the Human Species' said it the best. *"Inter-specific evolution [also "naturalistic evolution"] would appear impossible for two reasons: 1.) because the effect (a new and higher form) cannot come from an insufficient cause (the prior and lower form), and 2.) an agent in a given species tends only to the production of effects that remain within that same natural species. For both reasons, non-living agents cannot produce living effects, non-sentient organisms cannot give rise to sentient ones, and non-intellective primates cannot give rise to intellective ones."* [2]

The following negatives indicate that a sterile universe is most probable. Until the advocates of the "alien presence" fiction prove otherwise, the silence of positive evidence will continue to stand against them.

ARGUMENTS AGAINST ALIENS IN OUR SOLAR SYSTEM AND THE IMPRACTICALITY AND IMPOBABILITY OF INTERSTELLAR SPACE TRAVEL

Society is bombarded with what seems to give credibility to alien intelligence. Front covers of popular magazines proclaim all sorts of claims of alien contacts and proofs of ET visitors. Movies like Close Encounters, E.T., PAUL, and others with their dazzling special effects, also fuel the imagination.

The list is endless with stories that further credence to the idea that there are other beings out there. With all the hype, it is tempting to jump on the ET bandwagon no matter who you are. The ET syndrome is a cultural paradigm that seems to span the entire spectrum of society with impunity.

The stars above us are such a beauty that men have fashioned whole mythologies around them. The stars are truly a brilliant sight and now that we have extended our reach to the moon, even landing on the moon, the natural progression is that we might want to travel to them. Such travel is a basic part of countless science fiction stories and films, and many come away with the impression that interstellar travel is an easy task, perhaps just around the corner for man.

Sadly, there are serious problems to consider. Problems that we will see are insurmountable and practically impossible to overcome. Since the laws of physics are the same throughout the universe, the following physical difficulties will show that it is highly unlikely that any other advanced race of beings have mastered space travel. It will also show the high

Stellar Sterility

probability that man will not travel deep space or colonize another planet.

A small list of some of the mired problems will suffice in showing the extreme complexity and near impossibility of interstellar space travel for both humans and aliens. The technology necessary to reach another livable planet within an aliens' concept of practicality cannot be too different from ours, according to the laws of physics. Just to consider accepting such a challenge would require the motivation of a doomsday scenario.

The documentary "Evacuate Earth" presents just such a scenario and just how almost impossible it would be for man to escape earth within a 75 years period from an approaching global killer neutron star. Just the idea of building a prefect (100 year plus) generation space ship (they call it "the Horizon Project") to reach another planet is, needless to say, ridiculous and totally impractical at this time.

When the details are really considered, beyond the lame simplicity of the docudrama, the whole idea of an interstellar generation spaceship should rather be considered a suicide ship. Just getting to another planet is "Treky" to begin with, not to mention determining the perfect planet to travel to. Moreover, creating the perfect ecosystem on the new planet (terraforming) to accommodate our species is stretching time twice that spent getting there.

Even though the above docudrama leaves us with some small bit of hope, and a false hope at the moment, what it forgets to mention, after solving all the trivial problems of building the ship, supplying it with all our provisions, is deep space protection from radiation, space cancer, artificial gravity deformation of human physics and interstellar disasters as we shall soon demonstrate.

The show, by the way, ends with the evolutionary rhetoric of the survival of the fittest "necessarily" succeeding in reaching the new livable planet in light of and in expectation of expanding out into the Universe forever – for they cannot see that there is an end to humanity as we know it.

If beings have come to our solar system from some distant star system, which does n't seem to "sound" ridiculous, it would define them as so advanced, that if they were observing us, we would not amount to much more than a curiosity, if not a hindrance to their colonization. Considering us even closely "intelligent" or at least valuable would be straining their better judgment.

Furthermore, if life is universal and abundant, and livable planets are at a premium, we would be expendable with no one to protect us. To such advanced races, we would be no more worthy to them than insects are to us. If they could even get here from such vast distances, believe me, we would be very expendable if their lives depended on it.

DISCOURAGING DISTANCES

Is it possible to build an engine practical and safe enough to get humans to other star systems? The obstacles in traveling the enormous distances in space are huge, so the prospect that space aliens are visiting earth is highly unlikely. Even if one assumed the existence of life somewhere else in the universe, a visit by ETs to earth, such as is claimed in UFO reports, seems impracticable, if not impossible. The distances

Stellar Sterility

and consequently, the likely travel times are unimaginably vast.

For instance: The first on the list of rest stops is the closest star system Proxima Centauri, which is 40.7 million, million kilometers (c. 25 million, million miles) away. This is 4.2 light years away. This measurement designates the time it takes light to travel one year, which is 186,000 miles per second. At the incredible speed of one-tenth of the speed of light, the trip would still take 43 years. Proxima Centauri is the third star in the Alpha Centauri star system, also known as Alpha Centauri C. Lucky for us, Alpha Centauri B has one confirmed planet. This was the hope behind the science fiction TV series, "Lost in Space." Good luck to whatever Robinson family-II foolishly sets off to that destination! [3] There is no assurance that the planet is habitable.

The distances to the other nearest star systems, especially those with livable planets are vast. In fact, the distances are out of this world and beyond practicality.

Other planetary star systems also offer very little hope, if any at all, of the likelihood of man colonizing outside this star system. There are no livable planets close by and what potentially livable ones there might be are too far away – and this is an understatement.

(2.) If Alpha Centauri is not good enough, the next closest star is Barnard's Star. It is 5.9 light years away. It is a faint red dwarf star, discovered in 1916 by E. E. Barnard. Recent efforts to discover planets around Barnard's Star have failed. Obviously, this is no good either. What about colonizing the next star system?

(3.) Wolf 359 is 7.7 Light years away. Wolf 359 is also a red dwarf. It is so small that if it were to replace our sun, an observer on Earth would need a telescope

Stellar Sterility

to see it clearly. The star is too small for any habitable planets to exist. Let us keep on trucking to the next hopeful star base.

(4.) Lalande 21185. Distance: 8.26 LY. While it is the fifth closest star to our own sun, it too has no planets. We could plant a space base and refueling station here as a stepping stone to the next planetary system. There are four more star systems with no planets:

(5.) Luyten 726-8A and B., 8.73 LY.

(6.) Sirius A and B. No known planets. 8.6LY.

(7.) Ross 154. 9.693LY, and

(8.) Ross 248, is a whopping 10.32 light years away.

(9.) Finally, there is the first star system with a known planet, Epsilon Eridani. Hot dog! Maybe we found a new home? Let's check it out. Epsilon Eridani. Distance: 10.5 LY. Eridani (tenth closest star to Earth) is the next closest star known to have a planet, Epsilon Eridani B. It is the third closest star that is viewable without a telescope. It would take about 100+ years to get there at 1/10th speed of light, considering no stops along the way or accidents. Nevertheless, if this planet offers no hope, we must travel further and further, and further, and keep on going.

(10.) Tau Ceti is next on the list, which is 11.8 light years is a single G8 star similar to the Sun. High probability of possessing a solar system type planetary system: Current evidence shows 5 planets with potentially two in the habitable zone. Let us round this off to 12 Light years away. This is 4 light years (25 Trillion miles) times 3 = 75 Trillion miles away.

(11.) Gliese 581, 20.3 LY away. Multiple planet system. The unconfirmed exoplanet Gliese 581g and the confirmed exoplanet Gliese 581d are in the star's

Stellar Sterility

habitable zone. This one is 5 times 25 trillion miles away = 125 trillion miles.

(12.) Vega is 25.0 LY away. It has at least one planet and is of a suitable age to have potentially evolved primitive life – if you happen to believe in the fiction of evolution. This one is about 150 trillion miles away. If it takes 44 years to reach the nearest star, it will take 6 x 44 years or about 265 years to reach this one. We do not want to calculate the fuel, food and multi-tons of other supplies needed to get there. Now, what if this is a dead end planet? We must keep traveling, right?

FICTIONAL FEASABILITIES

Space travel is amazingly difficult if not a fantasy in the minds of fiction writers. Since humans have traveled to the moon, and have sent probes on short trips within our Solar System, we tend to underestimate how big and how hostile interstellar and intergalactic space is. Though there may be physics we have not yet discovered that allows us to travel safely among stars, as of now we do not see any evidence that such physics exists.

The vast distances between planetary systems screams against the survival of any physical creature traveling outside their planetary system. A visit or colonization to our solar system by ETs is impractical and most improbable, no matter if possible. Just eliminate what is impossible and whatever is left is most probably the answer. It's too dammed far and costs too much and is purposeless.

As we mentioned, the distances and travel times are unimaginably vast. Even the closest star system Proxima Centauri is too far away. Mathematical

Stellar Sterility

calculations show Proxima Centauri to be 25.12 TRILLION miles from our sun. To understand this kind of distance let us use some comparisons.

If you could shrink the galaxy down to the point that our Sun is the size of a golf-ball, then Proxima Centauri would be a golf-ball about 30 miles away. How far is that really? This gives us a "visual" idea.

For example, the sun is 93 million miles away from earth, so let us compare that to Proxima Centauri. To get an idea how far this is, first picture the measurement of 93 million miles as a one-foot ruler. How many one-foot rulers would you need to lay in series to equal that distance? First, figure how many "sun to earth" distances there are by dividing 25.12 trillion by 93 million. How many rulers are you going to need? The answer is 270,108 rulers! That is actually 93 million miles TIMES 270,108 thousand! Just take the 108 times 93 million and you have about 10 billion miles. A trillion is 1000 billion. We are talking about 27 times this distance.

Well, we can all agree that it really was a great accomplishment for Neil Armstrong to set foot on the moon, and maybe a bigger step for mankind, but there will be Mules populating the moon first before humans ever step foot in another solar system!

Despite this impossible hump, and the sterility of Mules, many Mule-headed dreamers believe humans can do it if we have the right propulsion system. Yes, this sounds true, but it is more believable that the ass spoke to Balaam than to swallow that whale of a promise.

When busying ourselves in our man-cave watching this "Lost in Space" scenario to Alpha Centauri, let us not forget to bring plenty of popcorn. And plenty we will need. Just how much popcorn do we need to see

Stellar Sterility

this trip from start to finish? First let us calculate the time it will take us from the distance there and with the technology we have.

Calculating the time, size of family and how much each person must eat to stay alive during the trip, we should come up with the proper amount. But, would this be reasonable? The imagination is a wonderful gift from God for creative thinking and entertainment, but that is all it is good for, when planning a trip like this.

You are thinking this is silly reasoning, and that it is possible to join the Robinson family. Let us see which is silly. Imagine this: The Apollo flights took three days to get to the moon. At the same speed, one would need 870,000 years to get to our nearest star. A Mule would just as well get you there! Munchhausen used a string of geese, and he was not that far removed from telling us the truth.

Nevertheless, humans are inventive and he could increase the Apollo speed to one-tenth the speed of light. How long would it take humans to get there? At the incredible speed of one-tenth of the speed of light, the trip would take about 43+ years!

OK, so my son would live to enjoy landing on the new planet. I would have to leave here at the age of at least 25, because when we add 43 years to that, we are talking about being 68 years old when we get there. Of course, I would have to refrain from procreating until 20 years has lapsed, least I over populate the spacecraft, etc.

To continue our calculations, what is one-tenth of the speed of light in miles per hour? In imperial units, the speed of light is approximately 186,282 miles per second. This is 11,176,920 miles per minute, times 60 minutes in an hour = 670,615,200 miles per hour. My

Stellar Sterility

God! That is FAST! I just hope we do not hit a chuck hole along the way.

OK, so our flimsy Apollo spacecraft has to travel at 670 million miles per hour to get there within a reasonable length of time - (cough!). Let us just say that our spacecraft can travel at 93 million miles per hour or one trip to the sun in an hour. How many "earth to sun" distance "hours" does it take to get to this star? The answer is 270,108 hours. Divided by 24 hours this comes to 11254.5 days, which is 30.83 years.

So, we have calculated the distances, the time and speeds, and now we can determine the amount of popcorn to take along.

SPACE TRAVEL IS POPPYCOCK!

How much popcorn must we take to make the above trip, if only six people go, and each two people eat one pound of popcorn every two days, until they get there? 11.77 tons! Now remember, this is just popcorn. We have not even started adding up all the other foods and supplies we will need. We have also not considered that it will take more than six people to get there. Try calculating the amount of supplies for 100-200 and maybe more than 300 people.

Is it possible for man to set foot in another star system taking along that much popcorn, and of course arriving with his full faculties? It is pure poppycock to think so! Nevertheless, anything is possible on paper, in a movie and within the imagination of scientific fiction writers. Who knows what missing element can "make it so," as Captain Jean-Luc Picard said. I believe he is actually replying to the question of butter.

This brings up the vital question of how much butter we must take along with the popcorn, not to

Stellar Sterility

mention the cola to go with the sideshow. Of course, the purpose of the popcorn is to feed the travelers while they are entertained with movies. So, how many (new) movies would be needed for entertainment, if one movie is shown every two days? 7,847 movies and one short 30 minute commercial, of course showing us popcorn.

NASA and other scientists have all sorts of wild new fantastic "theoretical" propulsion systems on the drawing board. That is where they will stay too, on a drawing board. Nevertheless, just to be sure, why not rather ask the real first man on the Moon, the Baron, Hieronymus Carl Friedrich von Münchausen.

Seriously though, let us consult the best authorities at Acme Scientific Co., and see what engines are available. Once we exit science fiction theatricals, physics tells us what we can and cannot do, as well as what aliens may do and not do.

SPACECRAFT AND PROPULSION SYSTEMS

The Jet Propulsion Laboratory in California studied the feasibility of an ion-thruster vehicle as a forerunner of a star-ship. Called the TAU (Thousand Astronomical Unit) mission, its aim was to carry instruments to the edges of interstellar space. In calculating its capacity, they concluded that if traveling at 105 kilometers per second, it would take more than 10,000 years to reach our nearest star. This is one down! But, let's see what other options we have.

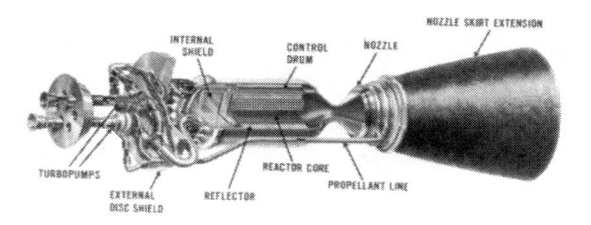

There are other kinds of space propulsion systems on the drawing board for traveling to another star system. Spacecraft for interstellar travel needs an efficient propulsion method. No such spacecraft exists yet, but there are many designs. Since interstellar distances are astronomical, a tremendous velocity is required to get a spacecraft to its destination in a reasonable amount of time. Acquiring such a velocity on launch and getting rid of it on arrival will be a formidable challenge for spacecraft designers.

There are many proposed designs for getting humans to another star system. Let's analyze each to see which one is most efficient and affordable.

Nuclear Rocket Engine:

One particular design is the Nuclear Rocket Engine. However, is this engine sufficient to catch up with the Robinson family? The high exhaust velocities necessary to reach high speeds, especially for interstellar distances, makes important engineering considerations with rocket concepts of all types very difficult.

Regarding the acceleration time issue, accelerating at 1 gravity (1g) for one year, a ship would reach just somewhat less than the speed of light. Thus, even a rocket with a theoretical top speed of 0.5c, but limited to a peak acceleration of 0.001g, would take ~500 years to reach maximum velocity. [4] The price listed in the Acme Scientific Co., Catalogue #1 is $600,000,000. Add $600,000,000 more for an extended warranty. This does not include the shipping cost of sending it back 27 trillion miles. Obviously, this one is too slow and too costly.

Thermal Fission Rocket Engines:

Fission-based thermal rocket concepts incorporate heating an exhaust gas such as hydrogen like the NERVA nuclear rockets studied during the 1960s, while potentially able to achieve high accelerations, have fairly low exhaust velocities (ultimately perhaps up to a few tens of times greater than the best chemical rockets) and are thus unpromising for missions over interstellar distances. [5] Acme Scientific Co., in Catalogue #2, discontinued this device. Obviously, not very promising.

Fission-electric Engine:

Nuclear-electric or plasma engines, powered by fission reactors, have the potential to reach speeds much greater than chemically powered vehicles or nuclear-thermal rockets. With fission, the energy output is approximately 0.1% of the total mass-energy of the reactor fuel and limits the effective exhaust velocity to about 5% of the velocity of light. This means that achieving start-stop interstellar trip times of less than a human lifetime require mass-ratios of between 1,000 and 1,000,000, even for the near stars. This is achievable by multi-staged vehicles on a vast scale.

The disadvantage is that it is too massive and too expensive. [6] Again, the Acme Scientific Co., Catalogue #3, priced this at $600,000,000, but does not expect any buyers. This one costs too much and is now discontinued.

Nuclear Pulse Propulsion

NPP systems contain the prospect of very high specific impulse and high speed, and therefore of reaching the nearest star in decades rather than centuries. **Project Orion** was a study of a spacecraft intended to be directly propelled by a series of explosions of atomic bombs behind the craft (nuclear pulse propulsion).

The Orion concept offered high thrust and high specific impulse, or propellant efficiency, at the same time. The unprecedented extreme power requirements for doing so would be met by nuclear explosions, of such power relative to the vehicle's mass as to be survived only by using external detonations without attempting to contain them in internal structures.

From 1957 until 1964 this information was used to design a spacecraft propulsion system called "Orion", in which nuclear explosives would be thrown behind a pusher-plate mounted on the bottom of a spacecraft and exploded. The shock wave and radiation from the detonation would impact against the underside of the pusher plate, giving it a powerful "kick". The pusher plate would be mounted on large two-stage shock absorbers that would smoothly transmit acceleration to the rest of the spacecraft.

However, interstellar travel would only be possible using advanced derivatives of present designs with cruising speeds of 8%–10% of the speed of light. [7] Present designs have exhaust velocities of 20–30 km/s, far too low to achieve reasonable interstellar cruising speeds. [8] A 1968 study estimated that it would take on the order of 1330 years for the energy-limited heat sink Orion design to reach Alpha Centauri.

The "Momentum Limited" Orion design was estimated at 133 years to reach the nearest star. (8a) Modern proposals utilizing Z-pinch fusion schemes are also under development, though again, the technology may be more appropriate for outer Solar System exploration than true interstellar flight.

As with rocket concepts generally, the high exhaust velocity vs. high acceleration issue noted above relative to specific exhaust power applies, and is to likely limit realistic interstellar applications. This one is too slow with present technology.

Fusion Rockets:

Fusion rocket star-ships, powered by nuclear fusion reactions, should conceivably be able to reach speeds of the order of 10% of that of light, based on energy considerations alone. In theory, a large number of stages could push a vehicle arbitrarily close to the speed of light. The maximum exhaust velocities potentially energetically available are correspondingly higher than for fission, typically 4-10% of c.

However, the most easily achievable fusion reactions have a significantly high energy loss. Thus, while these concepts seem to offer the best prospects, they still involve massive technological and engineering difficulties, which may turn out to be intractable for decades or centuries. This one is too difficult and has too many problems to resolve.

Ion Drives:

An ion thruster is a form of electric propulsion that creates thrust by accelerating beams of ions to create thrust. For Earth launch, the engine would have to be supplied with one to several gigawatts of power, equivalent to a major metropolitan generating station. Nevertheless, outside the earths' atmosphere, ion engines are practical, at least within proximity to our solar system.

A major drawback with ion propulsion systems is the lifetime of electrostatic ion thrusters. The engine becomes dysfunctional when holein the accelerator grid become so large that the ion extraction is largely affected. Grid erosion is unavoidable and is the major lifetime-limiting factor. Hall thrusters suffer from very strong erosion of the ceramic discharge chamber.

A good example of ion propulsion for inner solar system use is the Dawn spacecraft, launched on 27 September 2007, to explore the asteroid Vesta and the dwarf planet Ceres. Dawn's ion drive is capable of accelerating from 0 to 60 mph (97 km/h) in 4 days, firing continuously. [9] Apparently, this model is unsustainable and cannot be used for long distance.

Antimatter Rockets:

An antimatter rocket would have a far higher energy density and specific impulse than any other proposed class of rocket.

Energy resources and efficient production methods of antimatter make these rockets impractical and thus nonexistent. With the proper amount of antimatter, it would be theoretically possible to reach speeds approaching that of light.

Stellar Sterility

Two further problems need a solution. First, in the annihilation of antimatter, much of the energy is lost. Even so, the energy available for propulsion would probably be substantially higher than nuclear fusion, the next-best rival candidate. Second, heat transfer seems likely to deposit enormous wasted energy into the ship - gamma rays. Third, biological shielding is necessary to protect the passengers. Energy would inevitably heat the vehicle, and may thereby prove limiting. This requires consideration for serious proposals if useful acceleration is to be achieved.

The energy involved is also very large. The other problem is, the antimatter must be electromagnetically kept suspended and from touching matter – the inside walls of chambers, or it instantly explodes. If the magnetic suspension chamber fails and the two combine, the ship would be destroyed. So, for 100 years plus, not one failure must be allowed to happen. Anti-matter engines are at present are practically impossible and highly dangerous.

Magnetic Monopole Rockets:

If some of the Grand unification models are correct, we can construct a photonic engine that uses no antimatter, but uses a magnetic monopole, which hypothetically can catalyze decay of a proton to a positron. As a result, a hydrogen atom turns into 4 photons and only the problem of a mirror remains unresolved.

Nevertheless, most of the modern Grand unification theories predict no magnetic monopoles, which casts doubt on this attractive idea. [10] As far as anyone knows this is not possible.

Interstellar Ramjets are Inefficient.

In 1960, Robert W. Bussard proposed the Bussard ramjet, a fusion rocket in which a huge scoop would collect the diffuse hydrogen in interstellar space, "burn" it on the fly using a proton–proton fusion reaction, and expel it out of the back. Later calculations with more accurate estimates suggested that the thrust generated would be less than the drag caused by any conceivable scoop design. [11] Totally inefficient and useless for star travel. But, Acme Scientific Co., has the theoretical plans. PDF download if you can locate them.

Beamed Propulsion:

This diagram illustrates Robert L. Forward's scheme for slowing down an interstellar light-sail at the destination star system. A light sail or magnetic sail powered by a massive laser or particle accelerator in the home star system could potentially reach even greater speeds than rocket- or pulse propulsion methods, because it would not need to carry its own reaction mass and therefore would only need to accelerate the craft's payload.

Robert L. Forward proposed a means for decelerating an interstellar light sail in the destination star system without requiring a laser array to be present in that system. In this scheme, a smaller secondary sail is deployed to the rear of the spacecraft, while the large primary sail is detached from the craft to keep moving forward on its own. Light is reflected from the large primary sail to the secondary sail, which is used to decelerate the secondary sail and the spacecraft payload. [12]

A magnetic sail could also decelerate at its destination without depending on carried fuel or a driving beam in the destination system, by interacting with the plasma found in the solar wind of the destination star and the interstellar medium. Unlike Forward's light sail, this would not require the action of the particle beam used for launching the craft. Alternatively, a magnetic sail is pushed by a particle beam or a plasma beam to reach high velocity, as proposed by Landis and Winglee.

Beamed propulsion seems to be the best interstellar travel technique presently available, since it uses known physics and known technology developed for other purposes and would be considerably cheaper than nuclear pulse propulsion. Possible, but highly improbable.

Anti-gravity Propulsion:

So far no devices have surfaced and most probably never will. At present they are non-existent.

ENERGY / FUEL REQUIREMENTS

A traditional system involves burning fuel or reaction mass, but to reach another star, impractically vast quantities are required. One solution is to pick up fuel along the way. In the space between stars, there are not convenient asteroids and planets to land on and mine for fuel.

Luckily, space is not quite a vacuum, and there exist tiny atoms scattered far apart, mostly hydrogen. Going at a fast speed, these atoms could be gathered and used as fuel in an efficient reaction such as fusion (presuming we achieve fusion someday). To collect them, a huge scoop is necessary, and conservative

Stellar Sterility

calculations put it at least 2000 square km in area, which would cripple the ship with its drag and limit the speed to being slower than the space shuttle.

This system is horrendously inefficient and not viable considering that our sun is placed in a sparse region of space, providing a poor fuel source. The enormous energy necessary to reach Proxima Centauri would be roughly equivalent to the electricity output of the world's largest hydroelectric power station for four days.

One honest scientist warns, *"For a manned spacecraft weighing many tonnes, the energy requirements would exceed the world's annual electricity consumption. For the city-sized spacecraft in [the movie] Independence Day, the energy requirements would be even more staggering. In addition, when the spacecraft slowed again, it would need to use up almost this amount of energy in braking. If the spacecraft had to accelerate to c/10, slow down and start up many times, it is hard to imagine how enough fuel could be carried."* [13]

It seems that if it must take the total energy requirements of planet earth to get there, it would be more feasible to jettison the entire planet earth in that direction somehow.

The biggest problem contributing to interstellar travel is not so much the propulsion system, but the energy needed to obtain a reasonable travel time. The energy needed to accelerate the craft to reasonable travel speed is equal to the energy needed to decelerate. This means the craft has to have at least twice the fuel to reach its (one-way) destination.

The velocity for a manned round trip of a few decades to even the nearest star is thousands of times greater than present space vehicles. Millions of times

Stellar Sterility

as much energy is required for star travel. Accelerating one ton to one-tenth of the speed of light requires at least 125 billion kWh. The spacecraft must carry the necessary energy, as solar panels do not work efficiently in deep space. [14]

There is some belief that the magnitude of this energy may make interstellar travel impossible. In 2008, the Joint Propulsion Conference concluded that it was improbable that humans would ever explore beyond the Solar System. Brice N. Cassenti, of Rensselaer Polytechnic Institute, stated, *"At least 100 times the total energy output of the entire world [in a given year] would be required for the voyage* (to Alpha Centauri)." [15]

DANGERS OF LESS THAN LIGHT SPEED

Big bangs come in small packages! Micrometeorites can be extremely dangerous if not deadly. The kinetic energy of a particle with a mass of a tenth of a gram traveling at a tenth of the speed of light, calculated from the spacecraft's reference frame, is: $1/2mv^2$, which = $1/2 \times 10^{-4}$ kg $\times (3 \times 10^7$ m/s$)^2$, which is = 4.5×10^{10} J. In a chemistry lexicon (Roempp Lexikon), the combustion energy of TNT is given as: 4,520 kJ/kg = 4.52×10^9 J/tonne. Thus, 4.5×10^{10} J is equivalent to: $(4.5 \times 10^{10})/ (4.52 \times 10^9)$, which is = 9.95 tons of TNT. Therefore, the impact energy of one of those 0.1 g objects would be the equivalent of an explosion of about 10 tons of TNT. A one gram particle would be 100 tons of TNT.

What is a 10 ton of explosion like? If, you can imagine what it's like when dropping a big house brick on a tiny ant. The effect is the same. Another example is it is enough to obliterate a battleship. Well, a 10 ton

Stellar Sterility

TNT explosion is bad enough, but what about the following problems?

We know now what a one-tenth of a gram size particle will do to our spaceship. Let us add more "bang" for the buck to our argument to illustrate the absurdity of interstellar space travel. The question is, what possibility is there of survival with a larger size particle slamming into our craft.

Let us calculate: A one gram particle hit: .1g x 10 tons TNT x 10 = 100 tones TNT. A one ounce meteor hit: .1g x 10 tones TNT x 10 = 100 tons TNT x 28g = 2,800 tons TNT. This is equal to a small nuclear bomb.

Now, average ONE rock (about one ounce) hitting the craft every one million mile. We should figure one in a million miles could be a fair appraisal of the frequency of particles between 1/10th of a gram to a one ounce size hitting the "front" of the craft, but let us say, that the worst case is, 1 one-ounce rock hitting every one million miles.

Remember, the craft is traveling at 1/10th the speed of light, and that is 67 million miles per hour. This equals 67 x 2800 tons of TNT hitting our craft per hour = death! Well, at least it is 187,600 tons of TNT. This is a lot of restriction as well in traveling forwards. This would have to be counteracted by equaling if not superseding this amount in propelling the craft. So, one must figure not only the base fuel, but the extra for compensation.

We are being hit 67 times per hour with a 2.8 kiloton nuclear bomb! Check out the internet for video examples of 10, 100, 300 and 1000 kt bombs. Now, calculate 67 times per hour, times 270 thousand hours or 43 years. Notice the energy and danger facing us: 187,600 tons TNT x 270 thousand hours =

Stellar Sterility

50,652,000,000 tons of TNT! 50 billion tons of TNT? Come on people, "alien" or no, we cannot go.

Thus, we have only two choices. We take our time and die or we hurry up, and suffer the above, and die. Even the least dangerous brings us into contact with micrometeorites, which are very deadly traveling at extreme speeds.

We may argue that "most" particles will be made of magnetic materials, and deflectable with a high-energy electromagnetic field. Nevertheless, this does not protect us from those made of non-magnetic materials. The craft cannot dodge these at the speeds talked about. If we cannot deflect them, then what can we do to protect ourselves? We might travel in a plastic water-lined balloon and let them fly right through our spacecraft. The best plan is to let it pass through and hope no one is standing in the way!

The big question is, how long before the above frequency of hits tares our cosmic swimming pool apart. Speed therefore is not the answer. Going slower must be safer. Remember, every few million miles the craft slams into a small nuke.

However, the die-hard space traveler will say, the above is not fair because it is not true that every million miles the craft is hit. This is a good argument. Nevertheless, one hit is potentially deadly. There is no assurance that the one rock will not ever be the fatal rock. It is false hope that said humans can make it there without facing a fatality.

To travel to another star system humans need 100% safety. On earth, all we need is a large percentage, because we have other humans to call to the rescue. In the depths of space, there is no one to rescue a fatality. Besides, the only 100% guarantees that exists are death and taxes, and the latter is questionable with the right attorney.

What does Murphy's Law tell us? The probability of a negative happening depends upon the percentage of the possibility. Is it possible for one huge doomsday meteor to destroy an interstellar spacecraft? Is there a chance that you might win the lotto despite the odds against you? Playing the odds with the lotto is one thing; playing with the odds of making it alive to another star system is another. You have a better chance of winning the lotto than making to Alpha Centauri.

DANGERS OF LIGHT SPEED

Many stories include zany explanations of how faster-than-light travel is possible. The reality is that physics prevents this. There are no cheats. Even close-to-light travel runs into all sorts of interesting relativistic problems involving mass and energy, as well as increasing radiation problems. Staying within the realm of matter, traveling at almost light speed intensifies whatever problem one would have at lower speeds. It turns atomic particles in to mini atomic bombs. Light speeds increase the density of particle bombardment and does subject spacecraft occupants and equipment to extreme levels of radiation.

Dr. Oleg Semyonov said, *"When a ship accelerates to a relativistic velocity above 0.3c, interstellar gas becomes a flow of relativistic nucleons, hard radiation bombarding the starship, In addition, interstellar space contains high-energy cosmic rays and dust, all of which can present a huge problem if no proper protection is implemented... a radiation dose obtained in a non-relativistic space module...would be, approximately, 70 rems/year...(the safety level for a person is between 5 - 10 rems/year). The dose will, most likely, increase when accelerating to relativistic velocities...utilize extreme relativistic velocities...would present significant radiation hazards onboard."* [16]

DANGERS OF WARP SPEED

Warp speed is a sci-fi term for faster than light speed travel. Star Trek fans, prepare to be disappointed. Kirk, Spock and the rest of the crew would die within a second of the USS Enterprise approaching the speed of light. The problem lies with Einstein's special theory of relativity. It transforms the thin wisp of hydrogen gas that permeates interstellar space into an intense radiation beam that would kill humans within seconds and destroy the spacecraft's electronic instruments.

Interstellar space is not empty. For every cubic centimeter, there are fewer than two hydrogen atoms, compared with 30 billion x billion atoms of air here on Earth. According to William Edelstein of the Johns Hopkins University School of Medicine in Baltimore, Maryland, that sparse interstellar gas should worry the crew of a spaceship traveling close to the speed of light even more than Romulans de-cloaking off the starboard bow.

Special relativity describes the distortion of space and time for observers traveling at different speeds. For the crew of a spacecraft ramping up to light speed, interstellar space would appear highly compressed, thereby increasing the number of hydrogen atoms hitting the craft.

LIGHT SPEED / WARP SPEED DEATH RAYS

Another insurmountable problem is that the atoms' kinetic energy also increases. For a crew to make the 50,000-light-year journey to the center of the Milky Way within 10 years, they would have to travel at 99.999998 per cent the speed of light. At these speeds, hydrogen atoms would seem to reach a staggering 7 teraelectron volts – the same energy that protons will eventually reach in the Large Hadron Collider when it runs at full throttle. *"For the crew, it would be like standing in front of the LHC beam,"* said Edelstein.

The spacecraft's hull would provide little protection. Edelstein calculates that a 10-centimeter-thick layer of aluminum would absorb less than 1 per cent of the energy. Because hydrogen atoms have a proton for a nucleus, this leaves the crew exposed to dangerous ionizing radiation that breaks chemical bonds and damages DNA. *"Hydrogen atoms are unavoidable space mines,"* said Edelstein.

The fatal dose of radiation for a human is 6 sieverts. Edelstein's calculations show that the crew would receive a radiation dose of more than 10,000 sieverts within a second. Intense radiation would also weaken the structure of the spacecraft and damage its electronic instruments. Seriously, the potential damage is astronomical. It is equal to stepping into a nuclear furnace.

Edelstein speculates this might be the one reason why aliens have not paid us a visit. Even if Mr. E.T. has mastered building a rocket that can travel at the speed of light, he may be lying dead inside a weakened craft whose navigation systems have short-circuited.

Nevertheless, science fiction said there is nothing impossible for aliens. [17] Maybe their spacecraft runs off antigravity and is controlled by photon based quantum computers? (17a) maybe the aliens have mastered radiation shielding and solved food storage problems? Maybe if my grandmother had wheels, she would be a wagon too? – Engineer, Scotty.

ALTERNATIVES TO CONVENTIONAL PROPULSION

Alternatives that increase the theoretical possibility of such travel are also not practical. They are too dangerous or are technologically impossible. For instance, science fiction is offering ideas.

WORM HOLE SHORT CUTS

Our only possibility is to use wormhole portals. Such a wormhole would have to be carefully controlled, which is beyond our present capabilities, and we would have to somehow manage to create a twin wormhole far off at our desired destination, which might require someone else at the other end. Needing someone else to be there beforehand is not feasible for the first interstellar flight. Worse, the physical effects of traveling through a permanent or semi-permanent wormhole would warp and destroy any matter. You would arrive at your destination in the form of plasma.

TELEPORTATION – BEAM ME THERE, SCOTTY!

Classic teleportation involves a person activating a device and vanishing only to reappear simultaneously at some destination. There are a few movies that depict this fantasy: "They Live" and "Star Trek" are just two examples. This is not as straight forward as it first

Stellar Sterility

appears. First, the atomic structure of the person in the teleport machine is unassembled. Secondly, these atoms are teleported to their destination and reassembled.

The reassembly, however, requires a machine to be at the destination location and there are physical laws that do not permit the manipulation of matter at such a fine level, and over such vast distances. Teleportation can only be to places already visited. The reassembly is currently beyond our ability, but might be possible.

The atoms would still have to travel to another star, which might be faster than traveling as a body, but would still take years at least. The closest star to the sun is four light years away, so anything sent would take longer than four years to get there.

Alternatively, the reassembling machine could have a store of atoms from which to assemble the person, but this is in essence creating a copy and destroying the original – there is a movie that depicts this load of baloney. Many people would not be comfortable with this.

There is no assurance that the teleport machine would carry the soul with it. No human is assured the soul is purely a derivative of biological processes. It could be that biology is the product of "soul" or "spirit." Thus, there is no guarantee that the teleport machine is not an incinerator.

Another possibility is the "Fantastic Voyage" concept applied to outer space travel where the space ship and occupants are reduced in size to the molecular level and inserted into a laser beam and transported by the laser beam to the destination – this would be only possible after real humans landed there first and set up a receiver unit to re-enlarge the travelers. Pfff! We need to grab another drink and plug in the next episode of the Outer Limits.

GENERATION SHIPS

If faster-than-light travel is impossible or at least impractical, we might look towards generation ships. Driving like Hell usually leads to Hell. So, wisdom said, "Take your time." Even though our nearest star takes light only four years to reach, heavy objects would take much longer. Most stars would take hundreds of years to reach. Generation ships are an alternative designed for a population to live in for generations, until the descendants reach the destination.

There are several problems with a generation ship. The descendants might forget the original purpose of the mission as it fades into legend over the centuries. A cleverly designed computer system might be able to educate people born on the ship to avoid this, but it still becomes increasingly difficult to predict what might occur as the generations pass.

If there is a problem with the ship, the population, which has descended into savagery, will be helpless. This is very possible since computers can break down and people might forget how to fix them. Take vacuum tube technology for instance. With the passing away of the older generation it is becoming almost a dead science. As to nano-tube size circuitry and molecule size microchips?

TEST-TUBE "EGG" SHIPS

To remove as much uncertainty as possible in generation ships, designers have created the concept of "egg-ships." These egg carrying cosmic Noah's Arks would carry frozen human eggs (let us not forget the "sperm-bank"), which would be nurtured by carefully designed machines, acting as wombs, parents and educators. The human eggs can be grown for years, before the distant star or planet is reached, and computers could teach them about their mission, how to survive, and what to do, as well as their history.

Designing care-giving machines that would not emotionally stunt the new humans is well beyond our technology, but perhaps not impossible in the future. Again, there is another science fiction movie that depicts the computer defaulting, turning the mission into a horror movie.

However, like the generation ship, an egg ship does not help the individual, who wants to travel to the stars himself or herself. Waiting for artificially raised humans to live the dream of reaching the stars long after the parents have died is unacceptable to many people.

DEEP SPACE POPSICLES

The above pseudo-scientific solutions are for now impossible. However, there is space stasis or suspended animation, better known as cryogenics. In Laymen's terms, and fitting well with the popcorn and hot butter mentioned above, there is the freezing of humans into Popsicles. What we would have as a spacecraft is a flying space freezer. When longevity and using another generation are not possible, many films and stories depict humans kept in suspended animation to explain long trips.

There are many good alien movies where people are shown waking up to horrors beyond imagination. People would not be able to age in such a state, or would age very slowly, and it would be much like

Stellar Sterility

hibernation. Unfortunately, "telomeres" again present a problem. Our bodies always contain a small number of radioactive elements. These emit tiny amounts of radiation, which are harmless, because our cells continually replace damaged ones. If a person does not age in stasis, then their telomeres cannot be shortening and so their cells cannot be dividing.

It follows that any radioactive elements would cause permanent damage to the body, and if given enough time, could result in death. Even slow aging would not keep up with radioactive damage over a long time. We need our cells to divide at a normal rate.

LONGEVITY – SUPER LIFE EXTENSION

An alternative solution is genetically enhance people to live for hundreds or thousands of years so that they could make the journey in their lifetime, assuming the current problems of living in space were solved. Longevity and immortality are both subjects of much scientific research, but their biggest obstacle is *telomeres*.

Telomeres are sections on the ends of your DNA, which are cut slightly shorter each time your cells divide. Eventually the telomeres' lengths shorten and your cells begin damaging their own vital DNA as they divide. This means that our own DNA limits the number of cell divisions we can make. Cells divide to replace old or damaged cells, such as when you brush your skin on something or the constant replacing of your stomach lining cells due to the high acidity in the stomach [18]. The answer seems to be in keeping telomeres long, but generally, the only adult cells which can do this are cancerous.

DON'T BE COOL BE "Q"
IMMORTALITY AND INCORRUPTABLILITY

Traveling between stars will take the gift of immortality and incorruptibility in the biblical resurrection to accomplish. Yet, even with both of these gifts, there might still be problems getting to another star. Even though we cannot die or go insane, we are still corporeal beings, not magical beings or omnipotent masters over matter like the Krell, in Forbidden Planet. But, then again, look what they did to themselves! We could not snap our fingers and transmute matter nor transport ourselves over long distances without propulsion systems. We might be immortal and incorruptible, but we not the "Q" as depicted in the Star Trek series.

Both of the above must be so, to attempt interstellar space flight, for each is different in function. It is true that being immortal is not ever dying, but that does not mean you are immune to suffering corruption, misery and pain.

Alternately, to be incorruptible does not mean you are immune to death and dying. One refers to the physical, while the other refers to the mind or spirit. Without incorruptibility, sin can run just as rampant in immortals with mental illness chasing behind it as it does with a mortal. Check out the insanity of "Q."

On the other hand, without immortality, one could live a perfectly happy, joyful, and painless, but temporary life. Immortality gives you all the time in creation to get there, but does not insure you will appreciate it when you do - it does not filter out evil. It gives you the quantitative value, but not the qualitative aspect+ to get there.

Stellar Sterility

Qualitatively speaking for the incorruptible mortal, he would know better than to try, unless he preferred adventure and an imminent death, as opposed to comfort and a longer life. The former immortals would "become" unsuccessful suicidal maniacs, if they attempted such travel, whereas the incorruptible would have to be insane to even start, and thus would not choose so to begin with.

Biblical references give us the perfect example of this kind of scenario with the angels or celestials. Angels were and are immortal, but very corruptible as is demonstrated in some percentage falling, whereas the Christ, who was Jesus, was incorruptible, but very mortal and vulnerable to physical death. Finally, we have man who unfortunately is both.

Long distance space travel, even to the nearest star system, would require both of the above, plus a little magic along the way.

With immortality, you surly would make it to any star, and at whatever speed you want. Nevertheless, maintaining your sanity forever floating in outer space, after the Doomsday particle has destroyed your craft, would be unbearable.

If there is a proper definition of a Hell, this is it. Without immortality, it would be more logical to live your simple and "perfect" incorruptible life here and now on planet earth and pass away with some piece of mind.

Since man has not obtained either of these conditions, nor gained some power of magic, interstellar space travel is, for now just a pipe dream. This must also be true for any kind of exobiological 'ETs" as well. If there were only one impossibility, it would forbid any life form getting to another star system.

DANGERS OF INTERSTELLAR PARTICLES

Interstellar ships are vulnerable to the same hazards found in interplanetary travel, such as vacuum, cosmic dust, radiation, and micrometeoroids.

Cosmic dust is a problem that needs solving before any ship takes off. Cosmic dust contributes to the radiation hazard, because the dust particles are actually lumps of high-energy nucleons at relativistic velocities. A serious problem will be the sputtering of a ship's bow or a radiation shield by the relativistic dust particles.

At extremely high speed, interstellar dust and gas can cause considerable damage to a craft. Larger objects (such as macroscopic dust grains) are far less common, but would be much more destructive. The risks of impact of such objects, and methods of mitigating these risks, have not been adequately assessed.

In every cubic kilometer of space, there are an estimated 100,000 dust particles (made up of silicates and ice) weighing only a tenth of a gram. At such a velocity, colliding with even one of these tiny objects could destroy a spaceship if it is not shielded properly.

STELLAR SANDBLAST

Let us say that the possibility of the impact of a particle over the size of 1/100th of a gram is zero. Let say, there is only cosmic dust along the way. These particles cannot penetrate the vehicle skin and are not expected to constitute a serious hazard. Yet, traveling at that speed we would be effected as if being sandblasted for 43 years.

Stellar Sterility

The continual impact of many such small particles would have a gradual pitting, eroding, and sandblasting effect that could become troublesome in the case of exposed windows or optical surfaces, such as a telescope lens. This would also tax the resources of any flight with constant repairs.

As to the bigger particles, scientists say, that *"collisions with a few larger particles, weighing more than one gram, are also expected. These particles could puncture the walls of a spacecraft. However, recent data collected from experimental spacecraft indicate that the chances of a spacecraft being hit by meteors large enough to disable it are very small—much smaller than was anticipated a few years ago. Besides, various kinds of protective devices are available. These include a meteor bumper, or extra skin outside the main spacecraft skin, and several layers of thin-skinned, self-sealing materials. There is also a suitable alarm system whose purpose is to indicate both the occurrence and the position of a puncture"* [19].

Nevertheless, even though traveling at a slower speed would reduce the "intensity" of an impact, it does nothing to protect against larger impacts by particles larger than one ounce and up. It may be true that the chances of being hit by meteors large enough to cause fatal damage is very small, but it is also true that it only takes one to do it. If the meteor is big enough and there is no way to deflect or avoid it, then it means complete disaster.

As to protective devices, no meteor bumper devised is going to bump aside a doomsday meteor; no thin-skinned, self-sealing material is going to fill-in the hole, especially if the hole is half the size of the spacecraft! We may talk about micro-punctures and ways to avoid or repair them, but when traveling at such high speed, there is no protection against the big ones.

All it takes is one miscalculated large one ounce rock. Do not forget, the increase of "speed" necessarily means the increase risk of danger and death. Plus, at 9/10ths the speed of light a $1/10^{th}$ of a gram particle becomes a hydrogen bomb! Notice, 9 x 9.95 tons of TNT is now equal to 90 tons of TNT; and a one gram size particle would effectively produce an explosion equal to 9 x 2,800 tons TNT or 25,200 tons of TNT!

Speed is necessary, even if limited by the speed of light. Due to the tiny atoms scattered throughout space, any ship traveling at extreme speeds, will be impacted with such force that they would tear through even the strongest known metals.

Two options do remain. Humans or machines could constantly patch the damage. This would require impractically large amounts of repair material or the ship is made of elastic material, which self-heals. The good news is NASA has done research into such materials. The bad news is that they do not think them feasible.

RADIATION RISKS

Ultraviolet radiation: UVR is a form of short-wavelength radiation emitted by the sun. On earth, the atmosphere filters out a large percentage of the sun's ultraviolet radiation. In space, however, exposure causes the human skin to burn from 10 to 50 times as fast. Spacecraft shielding is one remedy for this.

X Rays or X radiation is encountered in space, although it is of low intensity and the walls of spacecraft cabins should provide adequate shielding from it. Alternately, secondary X rays generated by the bombardment of cosmic ray and solar flare particles on the spacecraft are of great risk.

Cosmic Rays are atomic nuclei traveling at very high speeds. They carry a positive electric charge and therefore, are deflectable to a certain extent by the earth's magnetic field.

Solar Flares of the sun and other stars also emit large numbers of particles consisting mostly of high-speed, high-energy protons, which can become very intense.

The Van Allen Radiation belt is a large, donut-shaped radiation belt. The high-speed protons of the Van Allen belt have effects similar to those of cosmic rays, and prolonged exposure must be avoided. Short-term exposure to Van Allen radiation is not likely to cause any serious effects. [20]

RADIATION AND LONG DURATIONS

For long duration missions, space radiation presents significant health risks including cancer mortality. Of these different kinds of radiation, some pose only slight problems for space flight. Others, however, constitute major hazards for spacecraft and their occupants. To categorize them, we have solar particle events (SPEs), medium energy protons galactic cosmic ray (GCR), high charge and energy (HZE) nuclei.

As to the protection against radiation on extended interstellar travels, no really "practical" or cost efficient method of protecting astronauts against exposure to cosmic ray and solar-flare particle bombardment is available. No magnetic or electrostatic shielding [alone] is, apparently, applicable for protection against nucleonic radiation of the oncoming flow, because of the presence of the neutral component in interstellar gas.

The most dangerous is interstellar gas, which acts as a flow of nucleonic radiation bombarding a relativistic starship. Radiation flux is extremely high even for moderate relativistic velocities, and therefore, proper windward shielding is a necessity. The presence of a neutral component in interstellar gas excludes the implementation of magnetic shielding alone.

Technically speaking, there are combinations of shielding efficient enough to guard against radiation, if one has an unlimited budget and unlimited resources. For instance, there are two potentially viable shielding systems. First, there is the material and magnetic shielding combination. A thick massive shield made of carbon (or Titanium) would afford relatively good (but partial) protection.

Combined with a magnetic deflection field, shielding from radiation is theoretically feasible: *"A radiation-absorbing windscreen installed in front of a spaceship...of material and magnetic shielding...can provide sufficient protection, however it becomes dramatically thicker with acceleration above 0.3c (3/10th light speed) and reaches several meters."* [21] Secondly, a water and/or ice shielding is another theoretical possibility. *"Shielding by water could also be an option... Placing a water tank (or an ice bulge) in front of a ship is advantageous... It eliminates the damaging embrittlement of solids under intense nucleonic radiation... a water tank of several tens of centimeters [10-15ft] in thickness would be sufficient... however cruising speeds closer to the speed of light would require tens of meters [65+ft] of water shielding... tons of additional load."* [22]

We said, "Technically speaking," which excludes speaking financially, for the construction of a sufficiently powerful magnet (for electro-magnetic shielding) within the limits of spacecraft weight requirements is beyond the capabilities of present-day technology.

Moreover, magnetic shielding by itself is insufficient and only partially shields against radiation bombardment. The addition of extremely thick material shielding in combination with magnetic shielding also adds huge amounts of weight, and thus too costly. This adds to the need for more propulsion fuel. The addition of such a huge amount of water is also at present impractical.

COSMIC RAYS AND BIOLOGICAL ORGANISMS

The Effects of Cosmic Rays and Solar Flares are similar in their effects on living creatures. Secondary particles are particles created by the impact of primary particles on a material. When a high-speed cosmic ray or solar-flare nucleus enters the atmosphere, the wall of a spacecraft, or any other material, it collides with atoms in the material creating particles called gamma rays. Such secondary radiation can cause serious damage to living cells.

The body apparently suffers no ill effects from short exposure to a small number of low-intensity cosmic ray or solar flare particles. However, as the radiation intensity and the time of exposure increase, there are various ill effects. Some of these, such as localized graying of the hair and the growth of tumors, take time to develop, becoming noticeable sometimes weeks, sometimes years, after exposure.

Very intense radiation can have immediate effects, such as blindness or brain lesions. In extreme cases, death may result. This radiation may also cause long-term genetic effects, which may be noticeable only in the children or grandchildren of the person exposed.

COSMIC RAYS ON ELECTRONICS

In addition to their effects on living organisms, intensive radiations have a harmful effect on certain organic materials, such as plastics, electrical insulators, and paints, and on certain electrical components, particularly semiconductors.

Radiation affects computer memory systems, which is a major problem: *"Every high-energy nucleon passing through an electronic component inevitably produces free electrons, i.e. deposits some electric charge in it, often resulting in parasitic signals and causing bits to flip, latch up, or burn out in computer memory. The deposition of charge can "upset" the memory circuits...Damage occurs when nucleons slow down and nearly come to rest at the end of their penetration depth. As a result, they knock silicon atoms out of their proper locations in crystal lattice, creating defects in crystal structure capable of trapping conduction electrons."* [23]

DANGERS AND PROBLEMS OF GRAVITY

The structure of our bodies actually depends on gravity. When humans do not live in normal Earth gravity, our bodies begin to suffer. After a few weeks or months, our bones become brittle and our muscles fatigue, with much more unpleasant long-term effects.

Stellar Sterility

These can be combated somewhat with various exercises and diets, but after years or decades in outer space, the human body becomes permanently damaged.

Even for relatively short flights, eyesight deteriorates so badly that NASA consider it a major boundary needed to be overcome before undertaking manned missions to Mars. Rather than living in weightlessness, acceleration from gravity can be induced by rotating the spaceship quickly. Unfortunately, this requires huge amounts of energy and fuel, and causes nausea in the short-term. The long term effects have not been studied, but are considered poor.

NUTRITION: FOOD, AIR AND WATER

Any humans living on a ship for long periods need life-support. They need to eat, drink, breathe, urinate, excrete, wash, and sleep. Many of these are addressed in space flights already made. However, on longer journeys, the amount of food and water needed becomes too large to take.

The most probable solution is to make the ship into a self-contained ecosystem. Plants could produce air, food, and be used to transform human wastes. Any ecosystem is slightly inefficient compared to earth, but it could still possibly sustain itself long enough to reach the destination – if radiation does not stunt growth and mutate the botanicals. The ship's equipment would gradually decay from the various gases being recycled, but clever maintenance or new materials might circumvent this.

The most efficient system would involve a single plant. Algae have been studied for their potential, with the *"spirspiralingae"* being looked at most closely. It would take care of air, wastes, and food. It is not a complete source of nutrition in itself, and becomes toxic if contaminated or when eaten in large quantities, but genetic engineering could change that in the future. Soylent Green in outer space?

ARGUMENTS AGAINST LIFE BEYOND EARTH

EXOBIOLOGICAL AND GEOCHEMICAL

Comparing the oldest earth rocks (cherts and microfossils) with outer space materials such as lunar regolith samples, meteorites, Mars soil samplings and Martian photography has revealed no evidence of biogenetic materials as of January 2013, and from the following examples, never will in any future date.

THE MOON IS DEAD

Lunar regolith and core sample studies show the Moon to be sterile and in most cases toxic to living organisms.

The Apollo missions collected rocks using a variety of tools, including hammers, rakes, scoops, tongs, and core tubes. The samples were placed inside sample bags and then a special environmental sample container for return to the Earth to protect them from contamination. [24]

Rocks collected from the Moon have been measured by radiometric dating techniques. They range in age from about 3.16 billion years old (basaltic samples), up to about 4.5 billion years old (rocks derived from the highlands). [25]

Lunar "soil" is the fine material of the regolith found on the surface of the Moon. Its properties differ significantly from those of earth. The physical properties of lunar soil are primarily the result of mechanical disintegration of basaltic and anorthositic rock, caused by continuous meteoric impact and bombardment by interstellar charged atomic particles over billions of years. The process is largely one of mechanical erosion in which the particles are ground to finer and finer size over time.

This situation contrasts fundamentally to terrestrial soil formation, mediated by the presence of molecular oxygen (O2), humidity, atmospheric wind, and a robust array of contributing biological processes. The term "soil" is not correct in reference to the Moon because on the Earth, soil is defined as having organic content, whereas the Moon has none. [26]

There are two profound differences in the chemistry of lunar and earth soils. The first is that the Moon is very dry. As a result, structures such as clay, mica, and amphiboles are totally absent from the Moon. The second difference is that lunar regolith and crust are chemically reduced, rather than being significantly oxidized, like the Earth's crust. [27]

A microbial assay of the lunar samples was performed by NASA scientists. Two surface samples were assayed for the presence of indigenous *kicrootganisms. "Neither terrestrial contaminants nor indigenous biota were demonstrated under a variety of growth conditions."*

In fact, opposite to earth "soils" lunar dirt is sterile and lacks all evidence of having any biogenic traces. In fact, because of the higher content of calcium, aluminum, ilmenite, nickel, and scandium, most of the dirt is toxic and lethal to living organisms. Gerard Taylor said *"A microbial toxicity was demonstrated by extracts of Apollo 11 core material which had been in contact with complex nutrient media for four months"* (p. 1939).

Attempts to grow Pseudomonas aeruginosa (ATCC15442) in extracts of Apollo 11 core material that had been in contact with TGY nutrient for four months resulted in the inhibition reported in earlier tests. (See, Taylor et al., 1970a) *"The presence of this extract resulted in the loss of all colony forming units (cfu) within ten hours."* (Taylor., 1971. p. 1941). Studies done on other lunar samples gave no indication of microbial growth. (Taylor., 1971. p. 1944).

The studies done so far demonstrate *"that the fluids extracted from complexes of Apollo 11 lunar sub-surface material, which have incubated in nutrient media for four months, are toxic to all the bacterial species tested"* (Taylor., 1971. p. 1946).

Taylor's summary in "Microbial assay of lunar samples" reports finding the total lack of demonstrable, viable microorganisms (either contaminant or indigenous to the lunar material) in any of the four samples tested. [28]

LUNAR PORHYRINS

Porphyrins have been touted as proof of life. Porphyrins are a group of organic compounds, many naturally occurring. There are two kinds; those produced from biogenetic processes and those from abiotic or geochemical processes.

A geoporphyrin, also known as a petroporphyrin, is a porphyrin of geologic origin. They can occur in crude oil, oil shale, coal, or sedimentary rocks. Abelsonite is possibly the only geoporphyrin mineral, as it is rare for porphyrins to occur in isolation and form crystals. [See Wikipedia under "Porphyrins"]

Stellar Sterility

Does the presence of porphyrins mean that there is or has been life on the Moon? Not at all. There are different kinds. Not all porphyrins are derived from biological processes. Some porphyrins are created by abiotic geochemical processes alone. As a group, porphyrins can form abiogenically, and are present on the Moon and in meteorites. Therefore, we have to be careful about their interpretation.

The 1969 discovery of lunar porphyrins probably said less about the chances for biochemistry there, than about how common their generation may be elsewhere in the universe. In 1978, Dr. Simionescu et al. were able to produce porphyrins under laboratory conditions similar to those of primeval Earth, before the genesis of life.

They summarized the results in the journal Origins of Life: *"Experiments with gas mixtures intended to simulate the primeval atmosphere of the Earth yielded many biologically important chemicals. Investigations into the synthesis of porphyrin-like compounds from methane, ammonia and water vapor were carried out by using high frequency discharges. Microanalyses of porphyrins showed that porphyrin-like pigments were formed in this way. The presence of divalent cations in the reaction system increased the yield of porphyrin-like pigments also involving the direct synthesis of their metal complexes. The ready formation of these compounds in abiotic conditions is significant, suggesting the possibility of their appearance during the early stage of chemical evolution."* [29]

Wherever you have life, you will have "porphryins", but where ever you find porphryins does not necessarily mean you will find life.

Stellar Sterility

The reasons why the Moon is antagonistic to life forms, besides soil toxicity, are numerous. There is no significant amount of water; no air, practically no atmosphere; no ozone for cosmic ray protection; is beyond extreme temperatures; and the angle at which the moon is tilted, its speed of revolution around the sun, speed of its rotation around its own axis are other reasons.

No atmosphere means there are no aerobic organisms. It also means no protection from UV radiation and micrometeoroid bombardment. The Moon lacks a molten iron core to generate a magnetic field to protect its atmosphere in being stripped away by cosmic rays and solar winds. The Moon is not big enough and so does not have enough gravitational pull to hold on to an atmosphere. Without an atmosphere that is thick enough, the Moon cannot get warm enough, so there cannot be liquid water. Without water or moisture, life is practically impossible.

The average temperature on the surface is about 40-45 C lower than it is just below the surface. In the day, the temperature of the Moon averages 107 C, although it rises as high as 123 C. The night cools the surface to an average of -153 C, or -233 C in the permanently shaded south polar basin. A typical non-polar minimum temperature is -181 C (at the Apollo 15 site). [30]

This type of variation in temperature, and the lack of all other hospitable conditions make it impossible for life to exist or function over a long time on the Moon - unless it has a spacesuit.

The Moon turns out to be just as dead as early scientists thought it would be. *"No life, living or fossil, has been found on the Moon. The Moon is even poor in the key chemical elements that are essential for life:*

hydrogen, carbon, nitrogen, and others. (There is more carbon brought to the Moon by the solar wind than there is in the lunar rocks themselves." [31]

MISSING MOON, MARS MICROBES

Have scientists found life from the moon? Russian biologists say they recognize fossils of microorganisms in the lunar soil. They reported new microscopic analysis of samples from the lunar surface, collected from the then Soviet Union's Luna missions in 1970 and 1972.

On September 24th, 1970, for the first time, an unmanned spacecraft delivered a lunar "soil" sample to Earth. The Soviet Union's Luna 16 spacecraft returned from the moon's Sea of Fertility with 101 grams of lunar regolith in a hermetically sealed container. [32] In February 1972, only 120 kilometers from the Luna 16 site, Luna 20 used a drill with a ten-inch, hollow-core bit to collect another regolith sample that was also hermetically sealed on the moon. [33]

The samples contained spherical particles that are *'virtually identical to fossils of known biological species'* in size, shape, distribution and even the way they are deformed during fossilization. They claim these fossils are solid evidence for ancient life on the moon and elsewhere in space. Other scientists disregard these claims and account them as geochemically produced. [34]

One of the problems with the alleged Martian meteorite life forms was that contamination from Earth life is almost inevitable. As to the supposed lunar microbes, this should not be a problem. The Luna samples were hermetically sealed when they were collected. However, the Russians may have

Stellar Sterility

underestimated the abilities of microbes to invade sample containers, and no one knows for sure how effectively sealed they were during the return from the moon almost 30 years.

The origin of life from non-living chemicals (called 'spontaneous generation ', 'abiogenesis or 'chemical evolution') is chemically impossible for many reasons, even under the best conditions. The chance of life originating on the moon is slim to nothing. It has extreme temperatures, little atmosphere to shield radiation, little to no moisture, and the regolith is poisonous and toxic to life forms. If the microbes had turned out to be genuine evidence for life on the moon, this life could not have begun there.

Rather, it may be Earth life transported to the moon from violent meteoritic impact knocking earth material out, with a speed exceeding escape velocity. Apparently, the same way rocks from Mars get thrown out to Earth, rocks from Earth can get knocked out to Mars and the moon. Furthermore, moon rocks can also be ejected to Earth.

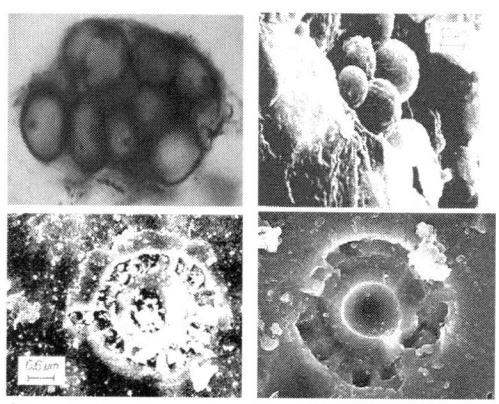

(Top Left) Luna 20: Fossils similar to modern coccoidal bacteria Siderococcus or Sulfolobus, Lithified by metallic iron. (Upper scale bar = 1.2 micrometers). (Top Right) Ancient Fossil Bacteria: Cyanobacteria

Stellar Sterility

from the www.ucmp.berkeley.edu/precambrian/bittersprings.html Bitter Springs chert, Central Australia. Site date: Late Proterozoic, 850 myo.

A colonial chroococcalean form.

(Bottom LEFT) Luna 16: A silicated fossil found in lunar regolith similar to modern spiral filamentous microorganisms such as *Phormidium frigidum*. (Bottom RIGHT) A 10-μm-diameter micro-crater on lunar glass, from Donald E. Brownlee et al. [abstract], p 1767 v 304, Science, 18 June 2004

The abiotic theory is supported by the fact that the moon fossils were 'virtually identical to fossils of known biological species' and had an 'unmistakable resemblance to modern spiral filamentous microorganisms like Phormidium frigidum, found in growing stromatolite, in Shark Bay, Australia.

Evolutionists frequently use common structures to 'prove' a common ancestry (although a common designer would explain them better), so it is difficult to believe that almost identical structures evolved independently on different places with vastly different environments.

What we may have here are naturally geochemically created artifacts, morphologically similar to biological artifacts, much the same way a micro-meteorite craters look like fossils?

MARS IS DEAD

MARINER 4 PROBE

Mariner Crater, as seen by Mariner 4 in 1965, suggested that Mars is too dry for any kind of life. Streamlined islands seen by Viking orbiter show that large floods occurred on Mars. Therefore, some scientists theorize that traces of life may be hiding deep in the Martian soil. The probability is very slim. Life needs more than just water to exist.

Mariner-4 performed the first successful flyby of the planet Mars, returning the first pictures of the Martian surface in 1965. The photographs showed an arid Mars without rivers, oceans, or any signs of life. Further, it revealed that the surface was *"covered in craters, indicating a lack of plate tectonics and weathering of any kind for the last 4 billion years."* The probe showed that *"liquid water could not exist on the planet's surface"* within the last 4-billion years. The probe also found that Mars has no global magnetic field that would protect the planet from potentially life-threatening cosmic rays.

This means, whatever could be on the surface would not live long. This tells us that life has NOT existed on the surface of Mars for the last 4 billion years, because favorable conditions [i.e. flowing water, favorable life-supporting atmosphere, nutrients] for the evolution of multicell organisms ceased four billion years ago. [35]

If we are to look for multicellular organisms, we must dig a lot deeper into the Lunar and Martian soils - like 4 billion years deeper. Now, if we go back in earth history 4 billion years, we do not find "multicellular" organisms. What we find are smaller and more primitive bacteria. Consequently, after Mariner 4, the search for multicellular life on Mars changed to a search for bacteria-like living organisms, as the environment was clearly too harsh for them. (36)

VIKING MISSIONS

What of the VIKING Probes? Did they reveal any chance evidence of life on Mars? From all pre-Viking findings, there is no reason to design and send probes to look for anything other than microscopic life forms. Thus, the primary mission of the Viking probes of the mid-1970s was to detect microorganisms in Martian soil. The tests were to look for "microbial" life. Of the four experiments, only one returned a positive result, showing increased $14CO_2$ production on first exposure of soil to water and nutrients.

The Viking experiments are considered by the majority of scientists as inconclusive. Most scientists agree on two points from the Viking missions: 1.) That radiolabeled $14CO_2$ evolved in the Labeled Release Experiment, and 2.) that the GC-MS detected no organic molecules. However, there are vastly different interpretations of what those results imply.

Consequently, an almost general consensus discarded the Labeled Release data as evidence of life, because the gas chromatograph and mass spectrometer, designed to identify natural organic matter, did not detect organic molecules.

PHOENIX LANDER, 2008

The Phoenix mission landed a robotic spacecraft in the polar region of Mars on May 25, 2008. One of the mission's two primary objectives was to search for a "habitable zone" in the Martian regolith where microbial life could exist. There was an electrochemistry experiment, which analyses the ions in the regolith and the amount and type of antioxidants on Mars.

Phoenix's preliminary data revealed that Mars soil contains 'perchlorate,' and thus may not be as life-friendly as thought earlier. The pH and salinity level are viewed as benign from the standpoint of biology. The analyzers also indicated the presence of bound water ("an extremely thin layer of water surrounding mineral surfaces") and CO_2. (37)

However, there was no trace of life found. This is one more strike against exobiological life forms teaming on other worlds. What about the latest Mars Science Curiosity Rover?

MARS CURIOSITY ROVER

The Mars Science Laboratory mission is a NASA spacecraft launched on November 26, 2011 that deployed the Curiosity rover, a robot carrying instruments designed to look for biological activity. The Curiosity rover landed on Mars on Aeolis Palus in

Gale Crater, near Aeolis Mons (Mount Sharp) on August 6. It has not found any biological evidence yet.

Curiosity has found evidence that water did flow on Mars (4 billion years ago?), at least in one of the areas on Mars. The camera recorded a rock formation "outcrop" at "Hottah" (named after Hottah Lake in Canada's Northwest Territories) displaying layers of what we call "gravel-bearing conglomerate" rock.

Martian conglomerate shelf – round pebbles!

The above site lies between the north rim of Gale Crater and the base of Mount Sharp, a mountain inside the crater. The imagery shows material "washed" down from the rim, streaked by many apparent channels, sitting uphill of the new finds. The rounded shape of some stones in the conglomerate indicates long-distance transport from above the rim, where a channel named Peace Vallis feeds into the alluvial fan. (38)

Stellar Sterility

A conglomerate is a rock consisting of individual clasts within a finer-grained matrix cemented together. Conglomerates are sedimentary rocks consisting of rounded fragments and are differentiated from breccias, which consist of angular clasts. Both conglomerates and breccias are characterized by clasts larger than sand.

This rock was once part of a flowing river or stream, where the water flow polished and smoothed out the sharp edges of the stones, rocks and boulders. This creates rounded fragments from large boulder to sand grain sizes.

Comparison of Mars and Earth conglomerates. Mars shows water worn pebbles!

Though all of the basic geological processes are present, such as flowing water, erosion and "gravel-bearing conglomerate" rock, this does not necessarily mean that carbon life forms are present. In fact, the only thing found is a sharp contrast between earth's plentiful supply of biological materials and the complete absence of it beyond earth's atmosphere.

No matter how much astro-biological digging is done, all previous experiments indicate that all future experiments will only reveal negative results. Are we alone in this solar system? So far, no evidence has revealed that we are not.

EXOBIOLOGICAL ARGUMENTS

PANSPERMIA - BUGS FROM OUTERSPACE

The search for proof that life exists beyond earth does not stop with planetology. Strong evidence in support of the Panspermia Theory came in 2001 when two Italian scientists from the University of Naples, molecular biologist Giuseppe Geraci and geologist Bruno D'Argenio, discovered bacteria in a meteorite that is over 4.5 billion years old.

The "spores" had found shelter in the crystal structure of the minerals that were present in the formation. The two scientists "claimed" that they had managed to resurrect the bacteria in a standard medium. The spores survived even after a treatment with alcohol at very high temperatures. The most interesting finding was that the DNA of the bacteria from the meteorite was not similar to any life form on Earth – according to their opinion.

After an in-depth survey of bacterial species, it was later determined to be an unknown strain of "existing" bacteria, but no further studies have been made to explore the origin of the found organisms.

RED RAINS OF KERALA AND SPACE BACTERIA

Other scientists who cannot send probes to other planets probe the earth's geology and skies for evidence. Some scientists suggest new biology of the ET red rain

Stellar Sterility

Kerala is indicative of the panspermia theory – the belief that life forms traveled through space and seeded earth.

In 2006, Dr. Godfrey Louis (physicist at Mahatma Gandhi University) published a paper in 'Astrophysics and Space Science Journal' in which he hypothesizes that the samples taken from the mysterious blood-colored showers that fell across the Indian state of Kerala in the summer of 2001 contain microbes from outer space.

Louis supposedly isolated strange, thick-walled, red-tinted cell-like structures about 10 microns in size. Dozens of experiments suggested that the particles lacked DNA. They also reproduced even in water superheated to nearly 600 degrees F. (The known upper limit for life in water is about 250 degrees F).

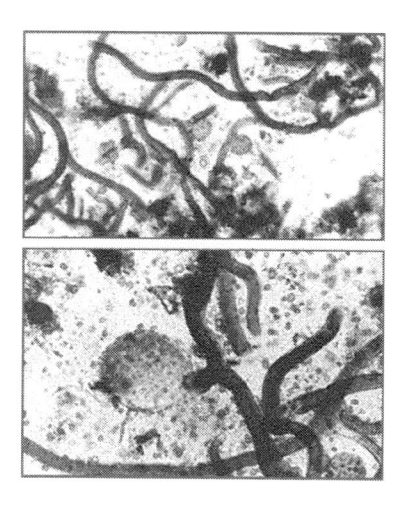

His explanation was that the particles were ET bacteria and that the microbes hitched a ride on a comet or meteorite that later broke apart in the upper atmosphere, and mixed with the rain clouds above

India. If the red rain particles are biological cells, then the phenomena can be a case for cometary panspermia.

Milton Wainwright, who is part of another British team now studying Louis's samples to confirm whether the cells truly lack DNA, said one preliminary DNA test has come back positive. He said, *"Life as we know it must contain DNA, or it's not life."* And, *"even if this organism proves to be an anomaly, the absence of DNA wouldn't necessarily mean its ET."* (39).

The Tropical Botanic Garden and Research Institute put out a research report noting that, almost all the trees, rocks and even lamp posts in the region were full of lichens. Samples were collected and observed under the microscope and confirmed the identity as Trentepohlia with orange-red pigment haematochromes, inside the cells."

It continues: When these samples were grown in culture, "Trentepohlia was seen in the cultures along with Sporangia with spores. It was therefore concluded that the algal spores found in the colored rainwater from Changanacherry [Kerala] were of local origin." So much for alien micro-blobs from Beta Reticula! (40)

ETB's or OSB's

In 2001, a team of international researchers found what could be the first proof of life beyond our planet, in the form of clumps of ET bacteria (ETB's) in the Earth's upper atmosphere.

What they believed they found was Earth atmosphere bacteria originating from extraterrestrial outer space bugs. The 'bugs' from space (OSB's) are similar to bacteria on Earth, but the living cells found in samples of air from the edge of the planet's atmosphere are too far away to have come from Earth.

Stellar Sterility

Professor Chandra Wickramasinghe, an astronomer at Cardiff University in Wale said, "*There is now unambiguous evidence for the presence of clumps of living cells in air samples from as high as 41 kilometers (25 miles), well above the local tropopause, above which no air from lower down would normally be transported.*" Professor David Lloyd, a microbiologist at Cardiff University suggests that, "*The most likely possibility is that the bacteria have arrived from another planet.*"

The professor has long argued that comets and space dust probably brought the seeds of life to Earth. He and the eminent astronomer Sir Fred Hoyle put forward the so-called Panspermia Theory.

Other scientists working in the field of astrobiology - the study of life in the Universe - are unconvinced by the Cardiff evidence. UK scientists interested in ET life said they would need more hard evidence. Dr. Alan Penny (Astronomer from Rutherford Appleton Laboratory), told the BBC that "*there were unresolved questions about contamination in the experiments.*" He said "*I think Chandra is being a bit optimistic.*" Professor John Zarnecki, from the Open University, said there were experimental errors that are potential problems: He asks, "*How do we know there is no contamination? Let's see the work published. Let's see it peer reviewed.*" (41). So much for Bacterianauts.

METEROIC EVIDENCE
THE MARTIAN METEORITES

NAKHLA

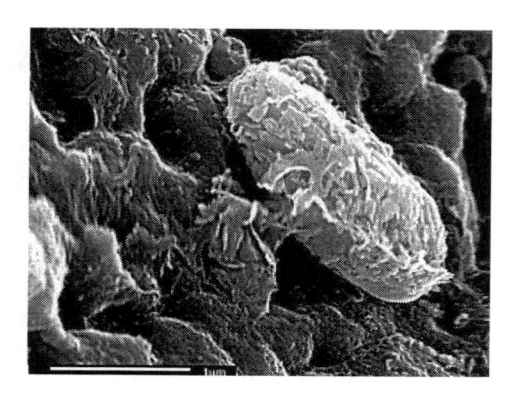

How about digging alien DNA out of Martian meteorites? The Nakhla meteorite (member of the Nakhlite group of achondrites) is composed mostly of clinopyroxene with minor amounts of feldspar, sulfides and oxides. The Martian meteorite is 1.3 billion years old and contains "clay- filled" cracks of Martian origin.

Microscope images reveal rounded micrometer sized structures embedded within the clay-filled cracks. The features consist of 0.5-2 micron-sized spheres with irregular surfaces in distinct cluster-like distributions within the clay. These spheres and ovoid are similar to the fossils of terrestrial coccoidal bacteria. However, it is important to emphasize that spherical and/or ovoid "morphology" alone are not indicative of biogenic activity.

ALH84001

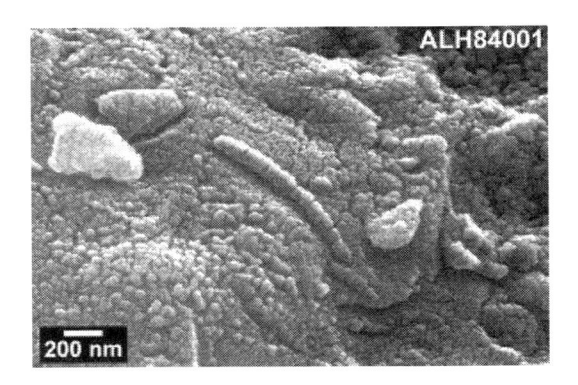

In 1996 a team of scientists found a Martian meteorite in Antarctica. Ejected from the surface of Mars it came to rest on Earth, bearing odd, microscopic structures that resemble fossilized remnants of bacteria. High-resolution scanning with electron microscopes revealed tiny "ovoid" which may actually be fossil remnants of bacteria. Moreover, the so-called microfossils are 100 times smaller than any bacteria microfossils found on Earth! They call these morphological shapes potential "nano bacteria" trace fossils.

SUPPOSED EVIDENCE OF LIFE

Exobiologists say that the indication of life hinges on three important pieces of evidence. 1.) Is the abundance of polycyclic aromatic hydrocarbons (PAHs)? Some overly hopeful scientists say they found these on the fractured surfaces. [These are a family of complex organic molecules formed by non-biological chemical reactions.] 2.) The mixture of PAHs found on ALH84001 is very different from that found on dust grains and other meteorites, suggesting the possibility of a biological origin. 3.) Thousands of different types of PAHs are all over the Earth, but those in ALH84001 do not appear to be contaminants, which have leaked into the meteorite. They must have leaked in while still on Mars.   However, PAH's can be formed from abiotic (non-biological) processes. Apparently, the Martian biogenic explanations can be explained by abiotic processes. They can be reproduced artificially and chemically, without any biological influence. For example:

The proof of Martian life in the meteorites ALH84001, Nakhla or Shergotty is not satisfactory, say the scientists. The visual (morphological) shape of a supposed exobiological "alien" fossil should not be the only proof for biological origin. There are abiotic processes that can duplicate bacterial and nano bacterial looking fossils.

As far as the Martian meteorites are concerned, Mr. Gibson said, "*The observations are intriguing but two fundamental aspects need to be addressed [i.e. before scientists can conclude life has existed separately outside the earth.]. [1] The ability to distinguish between no biogenic and biogenic bacteriomorphs, and [2] the extent of terrestrial contamination in these Martian meteorites.*" (42)

MORPHOLOGICAL (SHAPE) ARGUMENT

Geologist, Dr. Mary Sue Bell caught the attention of planetary scientists and deflated hopes of finding little green men. Her research centered on ALH 84001, the famous Martian meteorite. Scientists claimed that magnetite crystals on the meteor could be fossilized remains of tiny bacteria.

Stellar Sterility

However, Bell's re-creation of those magnetite crystals in a lab dealt a major blow to that theory. Dr. Bell used a powerful canon-like instrument at the Johnson Space Center to simulate a meteorite impact. *"Bell fired a small projectile into a target of siderite – an iron carbonate rock – at a speed of nearly 1.5 kilometers per second and at a pressure of 49 gigapascals."*

This simulates the force of the meteor that struck Mars, sending Martian rocks flying to Earth. The shock waves passing through the siderite in the lab caused instantaneous chemical reactions that created magnetite crystals identical to those found on the Martian meteorite, Bell said *"There's only one thing we know for sure about this meteorite – that it was shocked by impact."* Bell continues: *"Using the one process we know occurred, we've created magnetite crystals indistinguishable from the crystals in the (Martian) meteorite."* Bell concludes that, *"The shock from the impact is a more likely explanation for the magnetite crystals on the meteorite than any biogenic process."* (43)

Recent analysis of these Martian meteorites believed to have originated from Mars has shown that the sulfur isotopes they contain were produced by atmospheric chemical reactions, not by bacteria. (44)

One might say that if it *"looks like a dog, barks like a dog, has four legs like a dog, has hair and wags its tail like a dog, then it must be a dog."* Normally this is true. However, in certain situations, it might not be true that it is a dog. At Disneyland, we might very well find out that it is an audio-animatronic. This is the case so far with Mars.

CARBONATES ARGUMENT

The carbonates argument proposes that since carbonates are created only by living organisms and are found on such planets as Mars life must therefore exist on other worlds, especially Mars. But, carbonates can be formed without interference from biological organisms

Researchers have now discovered a new way to form carbonates on Earth without interference from biological organisms. They suggest this process likely takes place on Mars as well, and could indicate that the meteorite is a "Martian" example of this process.

The carbonates seen in ALH84001 possessed unusually high levels of the isotope oxygen-17. Chemist Robina Shaheen at the University of California at San Diego discovered anomalously high levels of oxygen-17 in carbonates found on dust grains, aerosols and dirt on Earth as well. This hinted that a chemical process common to both planets might be at work.

Shaheen's analysis of the carbonates in ALH84001 suggests the aerosols formed in the ancient Martian atmosphere. NASA's Phoenix lander recently detected carbonates linked with particulates in the dusty atmosphere of Mars. "*We think it might be this same mechanism that is operating,*" she said. (45)

HEMATITE CRYSTAL ARGUMENT

Martian "Blueberries" of Valles Marineris transported and deposited at Meridiani Planum, and Aram Chaos. Some say,*"Unless these balls rolled up hill, this indicates a biological source."*

There are scientists that believe that Mars once had life and that the Martian blue-ball looking hematite crystals "blueberries" are proof of that. One such pro-life Mars scientist is Dr. Rhawn Joseph, who said, *"On earth, hematite is a waste product produced by a common species of bacteria, 'shewanella.' In the absence of oxygen the bacteria metabolizes minerals such as iron, and excretes hematite. When exposed to large amounts of water, a spherical ball begins to form and pops out onto the surface after considerable water erosion. However, these round Martian balls are found on both the crater floor, which may have once been filled with water and on the soil above the outcrop of the crater.*

MARS HAS LITTLE BLUE BALLS

Dr. Joseph also suggests that MARS HAD CRABS! Dr. Joseph said, "*The only other creatures on Earth which creates similar spherical balls and which lives around large bodies of water, are crabs. Crabs are ancient and can live even at the bottom of the frigid ocean beneath thousands of pounds of pressure. However, they first crawled forth from the sea hundreds of millions of years ago--a time period when oceans may have splashed across the surface of Mars. Thus it is now apparent that Mars had been a very wet planet, which long ago was cris crossed with oceans, lakes, rivers and streams. And where there is water, there is life.*"

However, most reputable scientists say that water, by itself is not the ONLY necessity for the production and sustaining of life. (46) In other words, water by itself does not necessarily mean there is life. It appears that some believe that because where there is life there must be water or moisture, so also, where there is water there must be life. This is what the media and movie industry pawn off on the unsuspecting audience.

For a better understanding, let us think about this simple analogy: Whatever has water has life; Mars has water. Therefore, Mars has life. So, let's keep looking and we will eventually find life. This sounds a lot like the famous logical fallacy based on false premises: Whatever has wings can fly. Pigs have wings. Therefore, pigs can fly. The false premise is that Pigs have wings.

Stellar Sterility

But what about Mar's blue balls? OK. Let's try this syllogism: Earth has blue balls. Whatever has blue balls has life. Mars has blue balls. Therefore, Mars has life? Can you pick out the false premises?

Dr. Joseph said,

1.) *There is no evidence life began on Earth. 2.) Life was present on this planet from the very beginning. 3.) Life on Earth has a cosmic ancestry. 4.) The preponderance of evidence demonstrates life on Earth, came from other planets. 5.) Our ancient ancestors journeyed here, from the stars."* (47)

This is an interesting set of statements. Dr. Joseph's claim as mention in No. 3 is not proven and is inconclusive. No. 4 is an outright lie. No. 5 is a false assumption. No. 1. There is no evidence to prove that life did not start on earth! Yet, we must give Dr. Joseph one credit out of the five mistakes for being right that life was present from the beginning. (Oh, what a smart man!)

Dr. Joseph is correct about his No.1 statement, that there is no evidence proving that life began ("evolved") on earth. It just popped up! He said *"Life was present on this planet from the very beginning."* This substantiates exactly what the Genesis account said in the first Book of Moses. Yet, Dr. Joseph prefers to call the account a "religious fantasy".

Enhanced close-up of Blueberry

Speaking of fantasies and moving along from alien microbes and space crabs, Hematite can be created abiotically - without biological influence. There are two possibilities for the aqueous formation of hematite on early Mars: First, the abiotic oxidation of ferrous iron, accompanied by a redox equivalent loss of hydrogen to space, and secondly biological oxidation. (48) The second "biological oxidation" explanation has not been proved and probably never will at the rate of robotic experimental failures.

Stellar Sterility

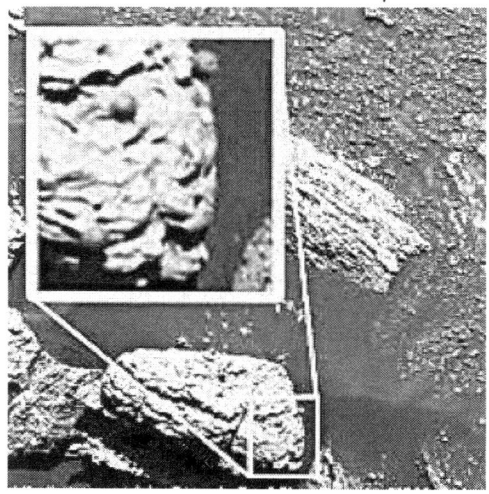

Close-up of Martian Blueberries

The majority of scientists say that the hematite "blueberry" crystals are the products of abiotic non-biological processes. *"The spherules originally formed at Valles Marineris due to the interaction of volcanic deposits and acidic hydrothermal fluids. During the wash out flows from Valles Marineris the blueberries were transported and deposited at Meridiani Planum, and Aram Chaos as alluvial/fluvial sedimentary deposits along with other sulfates and rock fragments. Diagenesis of the hematite-rich spherules may have also been a possible mechanism following sediment transport and emplacement. The hypothesis is consistent with available relevant information to date and provides an insight into the understanding of Martian surficial processes."* (49)

THE "PAH" ARGUMENT

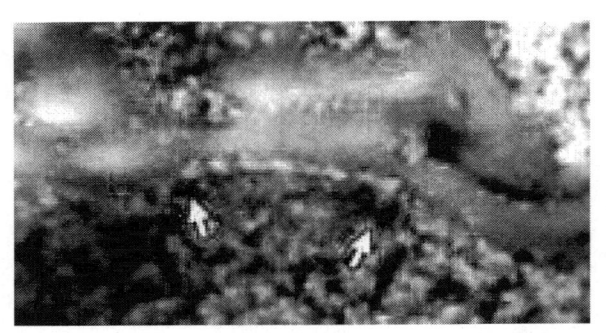

Magnetic Crystals: Bacterial Magnetosomes

Pro-alien life scientists, studying Martian meteorite fossil content, (looking for extraterrestrial life forms) push the idea that only biological processes produce "polycyclic aromatic hydrocarbons" on Mars and would have us believe that life "does" exist on Mars, or at least did at one time in the past. Therefore, if we keep looking we will find it.

Pro-alien life scientist Dr. Joseph said, without mentioning alternatives, that *"The positive discovery of Organic molecules called polycyclic aromatic hydrocarbons (PAHs) (with a carbon-hydrogen bond)...[are]... form by biological processes and are the byproduct of cellular decay. PAHs form when bacteria die and begin to decompose. These PAHs, however, were not from Earth, but from Mars. On Earth, PAHs found in fossil molecules are derived from biological activity associated with plankton and early plant life. Although PAHs form at very high temperatures from burnt tobacco or petroleum exhaust, these are also biological substances. However, the Martian PAHs are different from those found in Earth's atmosphere and are unlike 99% of all PAHs found on this planet. They are inside, but not on or near the outside of the meteor, which rules out contamination. Furthermore, the PAHS were found adjacent to the carbonate globules, which were also biological in origin. Thus, it could be assumed that the Martian PAHs may have been produced by microbes, bacteria or even plankton and plants. When these creatures died, they left PAHs. However, they also left the mineralized fossils of their bodies."* (50)

Wow! This is a fantastic chain of one-sided statements. What the pro-alien life Exobiologists do not want to admit is that, *"polycyclic aromatic hydrocarbons"* (PAHs) can be created without interference from biological organisms. Mikhail ZolotovEverett Shock said, that they can be manufactured by pure geothermal and chemical processes, even on Mars for *"PAHs and aliphatic hydrocarbons on Mars and Earth could be formed without the contribution of biogenic carbon. Some PAHs could be formed because of pyrolysis of other hydrocarbons formed earlier by the FT type synthesis or other processes."* (51) Mr. Shock further said, *"PAHs are not formed by biological means on Earth anyway. They occur in petroleum due to transformation of ringed organic compounds like steroids through loss of functional groups around the periphery. They can be converted from organic matter by the high-energy conditions and drastic chemistry that occurs around deep-sea hydrothermal vents. PAHs are thermodynamically favored over other types of molecules. The resonance energy is a gain. That is why they occur throughout interstellar space, and probably are seen in comets, the atmosphere of Saturn's moon Titan, and on Mars. PAHs are not directly biogenic. I said it almost twenty years ago and now those doing the laboratory work concur."* (52)

Mr. Rich Deem said, *"In principle, it would be impossible for life forms to have been formed on something as small as a meteorite. Since virtually all carbonaceous chondrites formed 4.55 billion years ago... It would also be impossible for any such bacteria to fossilize inside such a meteorite. However... astrobiologists have a philosophical bias of belief that all life was seeded in the solar system and the earth through some form of panspermia. The panspermia idea has gained supporters primarily because attempts to generate a realistic origin of life mechanism on primordial planet earth seems increasingly unlikely. Although I favor an ET source for life's origins, it involves a Supernatural mechanism."* (53)

Dr. Bates supports this same understanding: *"If bacteria are found elsewhere in the solar system, it will be hailed as proof that life can 'just evolve'. However, we have previously predicted in print that in such an unlikely event, the organisms will have earth-type DNA, etc., consistent with having originated from here as contaminants—either carried by recent man-made probes, or riding fragments of rock blasted from Earth by meteorite impacts."* (54)

There are other more plausible explanations for the development of "fossil-like" inclusions in Martian meteorites and surface sample photos. Geological processes can create many biological looking shapes.

Allan H. Treiman said *"Purported biogenic features of the ALH84001 Martian meteorite (the carbonate globules, their submicron magnetite grains, and organic matter) have reasonable inorganic origins, and a comprehensive hypothesis is offered here. The carbonate globules were deposited from hydrothermal water, without biological mediation. Thereafter, ALH84001 was affected by an impact shock event, which raised its temperature nearly instantaneously to 500-700K, and induced iron-rich carbonate in the globules to decompose to magnetite and other minerals. The rapidity of the temperature increase caused magnetite grains to nucleate in abundance; hence individual crystals were very small."* (For full quote, see footnote 55)

METHANE ARGUMENT

The biological verses geochemical production of Martian Methane is inconclusive to date. Methane has been detected on Mars and is the subject of controversy at NASA.

For over 40 years, astronomers have found various hints of methane on Mars. These reports have always generated a lot of excitement because they seem to provide some clue to the habitability of our planetary neighbor. But, recently scientists have measured large amounts and wonder how that much methane is being produced. Michael J. Mumma of the Goddard Space Flight Center is sure the methane is actually there, being supplied, and somehow consumed, at a surprisingly high rate, which is another topic of controversy.

Methane is a gas that we consider a bi-product of predominately biological (decaying) processes. Yet, it can and is produced by geochemical ones. *"Mumma and his colleagues have estimated that the largest plumes on Mars would require a source that releases methane at rates comparable to the world's largest hydrocarbon seep, which is in California. Although thoughts turned to little methane-belching Martians, geochemical processes could potentially produce these levels of methane on Mars." At the moment, NASA said "The jury is still out as to which of the various candidates is the most likely source."* (56).

The short lifespan (300-600 years) of methane (from a biogenetic origin) would suggest the presence of life on Mars. Methane breaks up in the presence of ultraviolet solar radiation, which constantly bombards the Martian surface. The continual production (seasonally?) suggests an active source. The methane that is observed today cannot have been produced 4.5 billion years ago, when the planets formed. This means that "something" is producing fresh methane within a 600 year period. So what can explain the presence of this gas on the Red Planet? Some scientists suppose this might be a biological source.

The biological origin of Martian methane could be explained by the existence of living micro-organisms, called methanogens, existing deep under the surface, and producing methane as a result of their metabolism. Another explanation is biotic. Either long-extinct microbes, which disappeared millions of years ago, have left the methane frozen in the Martian upper subsurface, and this gas is being released into the atmosphere today as temperatures and pressure near the surface change, or some very resistant methane-producing organisms still survive.

Others explain it as a product of geochemical processes. It could be produced, for example, by the oxidation of iron, similar to what occurs in terrestrial hot springs, or in active volcanoes. This gas could have been trapped in solid forms of water, or 'cages' that can preserve methane of ancient origin for a long time. These structures are known as 'clathrate hydrates'.

A geochemical process called serpentinisation could also produce the abiotic methane.

Serpentinization is a geological low-temperature metamorphic (and hydrothermal) process involving heat, water, and changes in pressure. It occurs when olivine, a mineral present on Mars, reacts with water, forming another mineral called serpentine, in the presence of carbon dioxide and some catalysts.

When certain catalysts are also present, the hydrogen combines with the carbon to form methane. On Mars it is possible to find all these primary elements: olivine, carbon dioxide and some catalysts, but the chemical reaction needs liquid water to occur. This implies that, if the Martian methane comes from serpentinisation, it could be related to subsurface hydrothermal activity. (57)

From all the tests made and the ones being made on Mars by the rovers and experiments, and the negative results in finding microbes, it appears that the methane is most probably being produced by abiotic geochemical processes.

MARS CURIOSITY ROVER FINDS METHANE

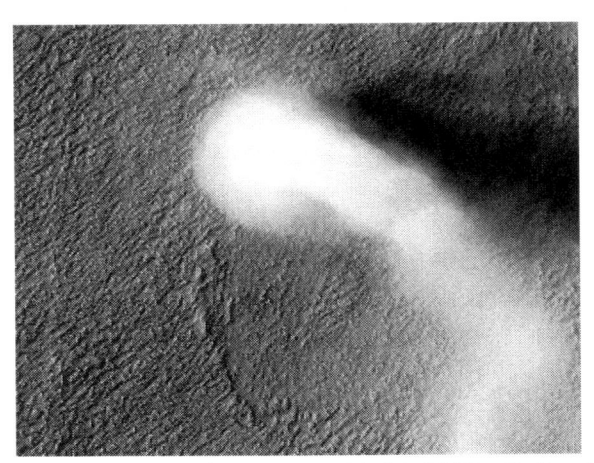

A Mars dust devil. (Credit: NASA/JPL-Caltech/UA)

Curiosity's on-board Sample Analysis at Mars (SAM) instrument recently detected the presence of methane gas on Mars. Methane can be geological in origin. Or, the gas is biotic, or caused by living organisms. According to an article published by Science Daily in May 2012, researchers were able to show that methane escapes from a meteorite *"if it is irradiated with ultraviolet light under Martian conditions."* Therefore, the team concluded that the methane detected in the Martian atmosphere is a result of high-energy UV radiation triggering the release of methane from meteorites.

METHANE DUST DEVILS

Another possible explanation for methane on Mars is dust devils. A team of researchers, led by Arturo Robledo-Martinez from the Universidad Autónoma Metropolitana, Azcapotzalco, Mexico proposed that *"the discharges, caused by electrification of dust devils and sand storms, ionize gaseous CO_2 and water molecules and their byproducts recombine to produce methane."* (58)

But, recently, NASA scientists released new evidence to the contrary. Mars is not as methane producing as previously thought.

The MSL/Curiosity team just announced that *"So far we have no definitive detection of methane."* They've limited it to less than 5 parts per billion at a 95% confidence level... The most publicized of the papers claiming there's methane on Mars is one by Mike Mumma and colleagues. In addition to detecting methane concentrations well over 10 parts per billion, they also claim to see a seasonal and regional *variability in the methane concentrations... Curiosity has a laser on board tuned to the wavelengths needed to detect methane. In theory, it should be able to detect methane concentrations down to at least 1 part per billion... the Curiosity team has limited methane to be less than 5 parts per billion. This is well below the 10's of ppb's Mumma claim to observe."* (59).

As for as what the percentage of methane is the scientists are not exactly sure. It sure is not above 5 parts per billion. They said, *"At this time, we don't have a positive detection of methane on Mars,"* said Sushil Atreya of the University of Michigan, a SAM co-investigator. *"But that could change over time, depending on how methane is produced and how it is destroyed on Mars."* (60)

Another dead end! There is methane on Mars. No there is not methane on Mars. Or, there is methane on Mars, but no one knows yet how much. One thing seems to come out of the above opinions and that is, there is less methane on Mars than previously measured. Another temporary conclusion is no one is sure what is producing the methane. Yet, we can bet and most probably win the turkey by saying that it is coming from abiotic sources.

The scientists who study Mars and the methane mystery are undecided as to which of these processes might be producing it. If biological, then life exists on Mars. If it is being produced geochemically, then this is another indication that life may not exist.

The evidence of life on Mars continues to stay inconclusive. In each case above, there are vital and very necessary problems that have not been solved. So, the question of whether life has an ET origin and exists outside our earth's atmosphere remains open according to secular humanist astrobiologists. The "Microbial life in Martian Meteorite" theory is inconclusive, and so are all the other potential astro-biological processes.

THE WATER ARGUMENT
"If there is water, there is life."

Where ever there is water, there is life. Man has found water on Mars. Therefore, there must be life on Mars. Uh? Man found water (frozen) on the Moon. But, no one has found life there yet. So, what is wrong with the following syllogism? "If there is water, there is life. Mars has water, thus Mars has life." It is not true that where ever there is water there is life. Geodes have water in them and there is no life in geodes unless it has been contaminated in the process of being cracked open.

Well, so far there has been no evidence found to support life on Mars. Mars does have water, frozen in the permafrost, but so far, no one has detected any life forms or any fossil remains of life forms. Why the apparent absence of life? Because, if there is to be life on Mars it would have to have more than just water to live and survive - a lot more!

It is a fact that just because you may have water does not mean you will have life. The premise of the above syllogism is actually false. We may say, again, syllogistically: "Where ever you have water does not necessarily mean you have life. Mars has water, thus, Mars does not necessarily have life."

To have life, or to have had life at one time, Mars must have the right soil composition, the proper atmosphere and practically everything that earth has to support any life. Needless to say, the above premise that wherever water is found, life will also be found or at least the evidence that it once existed is false.

Stellar Sterility

The presence of water is not evidence in favor of the existence of life in any place of the universe. Water is surly the main factor that contributes to life, but it is not the solitary factor involved in the creation and sustaining of "protobionts." We could find water, organic compounds and enough energy at a given place, but not find one single form of life! We should wonder why. Maybe it takes God to blow "life" into things. Besides, we have more proof that life begets life, than life deriving from 'stone-cold' lifeless matter. Has anyone seen a rock divide into two pieces, grow bigger and then get up and walk (or swim) off? Evolutionists say they did at one time in the far distant past.

Why have we not found life on Mars plentiful? Mars, as NASA said, is an unfriendly planet for life. The conditions on Mars cannot support any form of life. The living things inhabiting our planet are so sensitive to environmental changes that many of them are threatened by extinction due to the present changes of the climate of the Earth. If life is in danger on this planet that is "hospitable" for living beings, how could they even being to live and survive on a world as adverse to the conditions of life as Mars? It is easier to turn a sow's ear into a silk purse than to turn the Martian soil and find an isopod.

Why is Mars dead? First of all, Mars is a planet with scarce water and it has no protective ozone layer. Its gravity is weaker than Earth's gravity and its magnetic field is almost zero. According to evolutionary theories, it takes a lot more water to produce life than a 10 ml test tube filled to the top. The theoretical synthesis of a single protobionts and many organic molecules need huge amounts of water. In case a protobionts had emerged on Mars, how would it be maintained alive with such small quantities of water?

If all water vapor in the Martian atmosphere were condensed, it would make a pond smaller than a hundredth of one centimeter deep.

If a living form existed on Mars, it must have flown away in some UFO. It surly is not there today! Our hope on finding traces of life on Mars have dropped to almost zero. Scientists have concluded, by the analysis of Martian meteorites that Mars has been an icy planet for almost four billion years. The approximate emergence of bio systems in our Solar System happened about 3.5 billion years ago, when Mars already was a frozen inhospitable deep freeze.

Forget "War of the World" and "Mars Attacks", the Martians are not coming and never will. The only aliens invading a planetary body will be those made of metal and rubber stamped with the letters N.A.S.A.

MOON / MARS SOILS: ADVERSE TO LIFE FORMS

The composition of Martian soil is extremely adverse to the prospects of life forming, evolving and being preserved in and on it. The oxidizing agents in the soil impedes the consolidation of complex organic compounds. Because of this adverse composition, it is more than likely that complex organic materials will not be found in the Martian soil. (61)

The effects of cosmic rays upon the Martian surface are totally adverse to life, as they are deadly to living organisms. The high quantity of inorganic super-oxides and peroxides in the Martian soil also provokes the decomposing of the organic molecules inhibiting the biological and the chemical synthesis and stability of organic compounds - up to 10 cm deep. The compounds in the atmosphere of Mars are equally aggressive for living beings, no matter which form of life may land on it (without protection). (62)

THE "EXTREMOPHILE" ARGUMENT

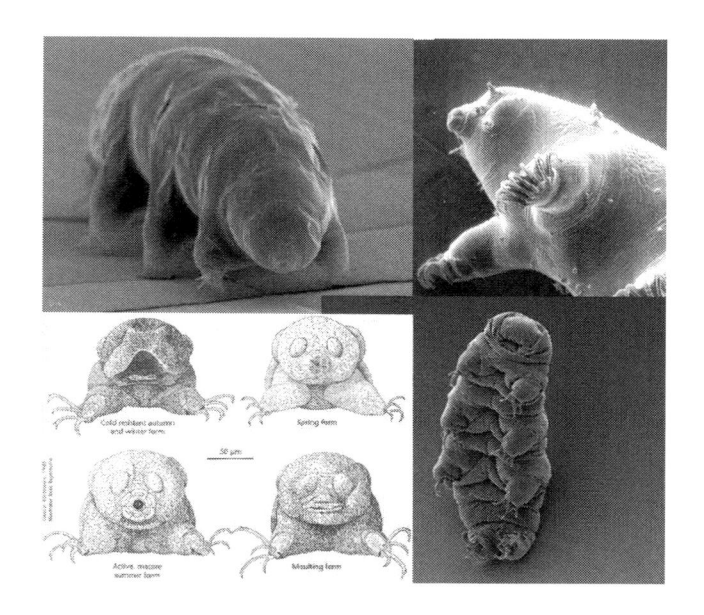

Tardigrades (known as waterbears or moss piglets) are water-dwelling, segmented micro-animals, with eight legs. They were first described by the German pastor J.A.E. Goeze in 1773. The name Tardigrada (meaning "slow stepper") was given three years later by the Italian biologist Lazzaro Spallanzani. – [Wikipedia]

Gads! What is an Extremophile? The Tardigrade is considered the king of the extremophiles. There are other weird looking ones like The Pompeii Worm. A team of French scientists first discovered the ten-centimeter-long Pompeii worm at hydrothermal vents in the Galapagos Islands in the early 1980s.

Stellar Sterility

Some of these microscopic organisms look like clear gummi bears come to life (hence their more common name, "Water Bears") and have proven to be more durable than Twinkies. These Tardigrades (See above picture) have been discovered all over the world, and in the most amazing places, from the peaks of the Himalayas to the sea floor, from temperatures approaching absolute zero to temperatures over 300 degrees.

Like the Chernobyl fungus, another extreme condition survivor microbe, these wonderful Water Bears can withstand doses of gamma rays lethal to humans without flinching. Tardigrades can also withstand the extreme pressure of a vacuum.

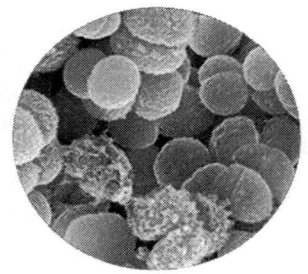

Deinococcus radiodurans (extremeophile)

NASA astrobiologist Richard Hoover is leading the hunt for more extremophiles, hoping to prove that some of these little fellas are not of our world, but interstellar hitch-hikers that came here millions of years ago on meteors.

The existence of organisms like Deinococcus radiodurans and the Tardigrades gives weight to the argument that some of these extremophile life forms are actually aliens among us.

Stellar Sterility

These creatures are thought to be able to exist in the vacuum of space and withstand high levels of radiation. Some exobiologists believe that these abilities are evolutionary traits that enabled them to arrive here, on Earth, from somewhere else in the galaxy. These scientists consider that if our Earth supports life, as a consequence any place in the universe can support it. And that would be very true if life was transplanted to that environment. We have found bacterium that is resistant under extreme conditions.

Therefore, evolutionary scientists assume from this that we will be able to find them in other places of the universe. Why? Because, IF they just popped up here (out of nowhere?), then they can pop up anywhere.

This is not scientific reasoning, especially in light of the fact evolutionists cannot show how "life" popped up here on earth. They can only point out that it popped up, but cannot explain it. Ultimately, they have to derive a self-replicating Mycoplasma mycoides JCVI-syn1.0 somehow without any external intelligent influence – from primordial rock to Providencia with no procreator!

So far, earth continues to be the unique center of all life forms in the universe, until proven otherwise. Furthermore, as prolific as life is on earth it is a wonderful mystery why it is not popping up all over our solar system. We should, according to the above rhetoric, find life popping up everywhere in our solar system - on all the planets, moons and asteroids.

We have found "extremophiles" living in Earth's most extreme environments. Earth has all the exact conditions to support life. Earth supports a wide variety of living organisms only because it is designed to support them. The scientists say, if we find this

Stellar Sterility

elsewhere, we will find life too. Yet, one slight change in the "conditions" can end with the extinction of whole species, if not the whole planet. Our geological records show this to be true with many extinct species of plant, fish and animal life.

It is the contention of this book that we cannot obtain universal conclusions from specific premises when those arguments apply to a specific place and time. Not all phenomena, which are true in a known time and/or place would be valid for other unknown eras and/or other places.

The possibility of finding biological systems on other planets is potentially non-existent, because the conditions "given" on Earth have been essential to the preservation of biological systems. Consequently, the conditions on other planets would have to be almost exactly like those of Earth's and from the very beginning too.

Now this is no guarantee that life must exist if the conditions are exactly like earth's conditions. Any variation in the given conditions, more than a fraction, would place earth in the same condition as Mars, if not the Moon.

The conditions have to be exact for life to even exist, not to mention evolve; and we do not see this with the Moon, Mars or any other planet in our system. The real point is that life comes from life. It does not and cannot be proved to pop out of lifeless matter. Obviously, it must either be planted here by transport or placed here by some intelligent life form. The alternative to these, when proven impossible is what is not acceptable by conventional science — life was put here by design from a supernatural source.

It's "possible" that bacteria exists in our solar system, but it is exceedingly unlikely. As far as

Stellar Sterility

intelligent life is concerned, the entire focus of the cosmos is mankind on Earth; the living forms on Earth's beautifully balanced biosphere are part of our created life support system. Man does seem to be the center of the universe when it comes to rational consciousness.

If bacteria are found elsewhere in the solar system, it will be hailed as proof that simple primitive life can just pop up out of nothing and with very little programming evolve into super complex multicellular highly intelligent life-forms. This is theory at its best and if it is really possible then humans will grow feathers, become chickens, lay eggs and flap their way to Mars.

However, it is more than likely that the organisms will have earth-type DNA, consistent with having originated from here as contaminants - either carried by recent man-made probes, or riding fragments of rock blasted from Earth by meteorite impacts. As far as other star systems, they are out of the question - because mankind will never reach them.

The absence of evidence for life may not be evidence of absence, but it surly increases the improbability when "absence" is continually observed. Nevertheless, if there is non-human intelligent life out there then, *"Where is everybody?"* asks Nobel-Prize-winning physicist Enrico Fermi. SETI, the Search for ET Intelligence, has failed to obtain a single "intelligent" signal from outer space in over 50 years. (63)

In April 2000, 600 scientists met at the First Astrobiology Science Conference, held at NASA's Ames Research Centre, California, to evaluate the evidence of ET life in the universe.

The general consensus was encapsulated by British paleontologist Simon Conway: *"I don't think there is anything out there at all except ourselves."* Dan Cleese, a Mars program scientist Pasadena's Jet Propulsion Laboratory, said that it is time to *"tone down expectations."* (64) It is a shame that this was not televised. But why not? NASA ratings and support for the space programs would plummet!

Obviously, all these scientists are disappointed in the "absence of evidence" for ET life forms and should feel that way, because there isn't anything out there. Furthermore, it has been known for sometimes that the probability of life on Mars is very unlikely - all the conditions for life are missing. In fact all the conditions are opposed to life as compared to earth.

MARS CANNOT SUSTAIN LIFE

What is the condition of Mars compared to Earth? The Martian climate is highly unstable and impossible for continued existence of life. Mars has very little water and if liquid, would only amount to be 2 inches deep, sulfuric and acidic. This is equal to about 4% of the earth's South Pole. It has extremely small layered "frozen" polar-regions, no ozone and is not protected from cosmic radiation. It has about 1/4 gravity of earth, has superficial temperatures average 55% below zero and the atmosphere is 95.3 % saturated with carbon dioxide. The atmospheric pressure is adversely low and is therefore adverse to any life on the surface. The soil is mostly sulfuric and is inhospitable to life forms. It has a weak geomagnetic field and no protection from solar electromagnetic storms. So far the evidence of life beyond earth adds up to this — nothing!

The lack of evidence for a "physical" alien presence in our solar system is indicative of a lack of alien

Stellar Sterility

presence in the rest of the universe. All efforts to detect life beyond earth have failed so far and it is believed by some that this will continue, until humanity wakes up to the truth of a supernatural Creator.

The search began with the moon, where astronauts walked during six lunar landings from 1969 through 1972. After it was concluded that the moon was a sterile, lifeless place, the search moved to the other planets and their moons. Once Mars began to disappoint the alien hunters at NASA and elsewhere, the search shifted to more recent theories such as the icy and watery moons of Jupiter and Saturn.

Viking probes to Mars in 1976 performed experiments designed to detect life, including microscopic organisms, with negative results. Two unmanned Voyager craft, whose destinations include Jupiter, Saturn, and Uranus—have taken thousands of pictures of the outer solar system. They reveal harsh, non-livable conditions everywhere.

Searches of deepest space have been carried out by radio telescopes, instruments that are able to beam messages of greeting toward any planets that might be circling distant stars. Radio telescopes also "listen" for any space messages that may be coming in earth's direction. During the past few decades, scientists have searched dozens of nearby stars for intelligible radio signals. The results are once again completely negative. At this point, it appears that life as we know it is unique to planet earth.

Apparently, the lack of evidence points to a monogenetic geocentric origin of life - planet earth. This conclusion has been very upsetting to evolutionists, who believe that life began spontaneously on earth and that the same thing probably happened elsewhere in the universe. This theory, which is based on the assumptions of evolutionary polygenesis and panspermia, has no supporting substantial evidence. In fact, all the data points to the contrary.

The lack of "physical" evidence indicates that aliens have not visited our solar system nor visited our planet. If they have, then where is the evidence? We have found a lack of evidence of ET life on earth, on the Moon, on Venus and Mercury, on Mars, on all the other planets, as well as the moons and a lack of ET communications detectable with electronics.

For decades, speculation about ET life has been boosted by tales of flying saucers and encounters with aliens. It is now being fueled from a more serious source. In August 1996 NASA researchers claimed to have found evidence for simple life forms in a meteorite allegedly from Mars.

Since then, this 'proof' of life in the 'Mars rock' has very much lost favor among the scientific community. In spite of this, the two kilogram rock found in Antarctica has ignited a new surge of 'Mars fever'. In the next 20 years, the Americans, Europeans, Japanese and Russians plan around 20 projects to explore our neighboring planet, some 78 million kilometers away at its closest approach to us.

Meanwhile, fantastic belief in ET intelligences continues to grow with an almost religious fervor, while ironically the evidence continues to maintain a constant zero. It could be very true that there is a lot of wasted empty space out there. Is it really empty or is it for our observational enjoyment? Many people look at existence pragmatically. Others see it as an artist or tourist.

ARGUMENTS AGAINST PHYSICAL ALIENS

Is there life in other solar systems and galaxies? Is there any life out there anywhere? If not, we might think *"If we are alone, then that sure is a lot of wasted space out there"* (Quote: Movies, "Contact"). The concept of "wasted space" is an opinion based on utilitarian thinking: That if something is not used or occupied by humans, it must be a waste, observations not included. Paintings are "observed" in museums. They are not touched or played with or occupied by humans. Are famous paintings waste or wasting space? Stars and planets are no different than paintings on a wall. They serve the purpose of being observed and enjoyed by humans.

Now, the probability of life in the universe is slim to none. With the billions of years the universe has existed, there should be (by now) civilizations out there advanced enough to reach us. And if there is such advanced alien races here, then there should be plenty of evidence. Unfortunately, there is no evidence to prove aliens are presence in our solar system.

Nevertheless, a series of so-called 'flying saucers' caused a great stir in southern England before being revealed as hoaxes, complete with sound effects from built-in batteries and loudspeakers.

Harvard University psychiatry professor John E. Mack recently attracted worldwide attention with his collection of cases of people claiming they were 'abducted by aliens'.

There was also the release of a film alleging to be of an autopsy on an alien from a crash in New Mexico close to the U.S. Air Force Base at Roswell. The blurry footage, which most have dismissed as an obvious and crude forgery, was nevertheless the main attraction at the 1995 UFO World Congress in Düsseldorf, Germany.

Then of course, there was the 'alien invasion' film Independence Day, which grossed more in its opening week than any previous film in history.

A recent poll in Germany revealed that 17% of the population believe in visits by alien craft, while 31% believe there is intelligent life in other galaxies.

In all the attempts of science to prove an exobiological existence, not a trace of 'little green men', or indeed any life, has been found on any of the planets which our probes have been able to explore. So far, man has not found any alien stuff on earth. There has been nothing found on the moon and nothing yet found on Mars. Many think that because life is found on earth then it should be found on other planets, because there are other places out there where the conditions of life are similar to Earth.

According to the known laws of physics, astrophysics, carbon based biology and many other factors, conditions have to be "just right" for life to exist on any planet in the (physical) universe. Earth seems to be the "only" pattern by which life can exist and live. Outside this pattern, life cannot live very long, if at all. Hence, this "pattern" seems to be what is used (imitated) for astronauts to survive outside the earth's environment.

Despite whatever exceptions there may be, the rules for now are: The habitat must be at the right distance from its sun, so as to be neither too hot nor too cold. The habitat must have more than just "water" as a foundation for life to exist. Just having liquid water is completely insufficient, despite the excitement reigning when such was detected as possibly being on the surface of Jupiter's moon, Europa – Hollywood even made a movie called "The Europa Report" to push the idea of advanced life elsewhere.

A habitable planet must have the right gases, atmosphere, gravity, radiation shielding, and every known element on the chart to function, plus the information programming "software" to tell it what to do, when to do it and how to do it. A computer is a good example of truth about life: Soldering all the elements on the chart together will not give you a response. It needs energy. But, energy is not all. A computer can have all the material elements and energy, but without intelligent programming, it will never do anything.

Materialists cannot understand that life must derive from life and intelligence must come from intelligence just as the information in computer software must be put there (programmed) by intelligent humans.

ALIEN COSMOLOGY

In all the attempts of science to prove an exobiological existence, not a trace of 'little green men', or indeed any life, has been found on any of the planets which our probes have been able to explore. It is because an abiotic origin of biotic life cannot be proved.

The same is true with alien messages of the origin of the universe and life from nothing or from some unintelligent mysterious singularity. The most commonly held cosmology is that the universe was once a gravitational singularity, which expanded extremely rapidly from its hot and dense state. While this expansion is well-modeled by the Big Bang theory, the origins of the singularity remains one of the unsolved problems in physics. (See Wikipedia under "Cosmogony.")

The problem is life cannot form spontaneously without intelligent, creative input. Yet, many people swallow alien lies that the Universe popped out of nothing and from a non-intelligent source. Others nurse the intoxicating beverage of an eternal Universe – very tempting indeed.

Without intelligent, creative input, lifeless chemicals cannot form themselves into living things. Without this unfounded evolutionary speculation as a foundation, Ufology would not have its present grip on the public imagination.

LITTLE GREEN OLIVE-EYED IDOLS

There should be no protest ever against creative thinking, art, sculpting, painting, and image making for entertainment purposes, as we know very well that these things are creations of the artistic mind and fictional. Creativity and inventiveness is what drives society forward in progress.

Nevertheless, to create and pawn off the fictional as real is to lie and deceive, and usually for some good reason – to distract from the truth. It is perfectly fine to enjoy a fictional Mickey Mouse or Sherlock Holmes, but to believe in "aliens" that have no more substance than a Mickey Mouse is false religion, especially when it detracts and distracts from true religion.

It is a very important point that the records do show Christians are the least likely, if at all, to be abducted by so called aliens. We must ask why this is, deal with it, and come to the one and only conclusion: the alien phenomenon is demonic and those who are protected by the good spirit of Christ are immune.

REAL PURPOSE OF THE STARS

Many people ask for what purpose were the heavens made if not for life forms to exist in. The reasons stars were made are given to us in several places in the Bible, not only in the well-known Psalm 19, but especially in the Creation account. In **Genesis 1:14** we read: *'And God said, Let there be lights in the firmament of the heaven to divide the day from the night; and let them be for signs, and for seasons, and for days, and years.'*

We see from this that the stars are there for humankind on earth. Add to this the sequence of creation (on the first day the earth, and only on the fourth day all of the stars), and it is easy to see the thrust of the biblical testimony, that the purpose of creation is uniquely centered on this earth. Everything out there was created for our joy, fun, study and exploration in Edenic times, but now is beyond our reach until the resurrection. We could have multiplied, filled the earth and subdued it as well as the Universe as immortals, but now it seems we all are doing time on planet earth and the only way off is in a box and not in a metal spaceship. The heavens were made for our enjoyment, observation and short distance exploration, education and entertainment. But as far as a second home, forget it!

ALIEN ARTIFACTS NO PROOF OF EVOLUTION

The discovery of ET technology would certainly be one of the most significant findings in human history; for even if this technology were non-functional, it would give us some certainty that life and intelligence has developed elsewhere. With so many places for a small observational probe to hide in our own backyard, we may as well keep our eyes open. (71)

The use of the word "developed" betrays the motive for the search for ET life. Notice the difference in the phrase "exists elsewhere." Apparently, the hope to prove chemical evolution is more important than proving ET life. Ironically, even if alien life was found, the debate would still be at a stalemate. Just because life would be found elsewhere, does not prove it developed. They could just as well have been created. Finding "alien life" does not prove evolutionary theory. It also does not prove creationism either. It only proves the "existence" of alien life and nothing more. It does not answer where "life" came from and "how" it came about originally.

Nevertheless, lack of alien life and ET artifacts favors more the theory that they do not exist. Actually, with zero evidence for something one might justly say that "it" does not exist - that is, until proven otherwise. People believe and even suggest that Unicorns might have existed, but in all of 5000 years plus, and within all the fossil records, not one has been found. As to aliens and alien artifacts, until something substantial is found it remains alien to archaeology and as fictional as unicorns.

ASTROARCHAEOLOGY NOT A VALID SCIENCE

The alien artifact is a popular theory among pro-alien enthusiasts that believe ET's have visited our solar system. They believe that there is evidence of alien technology and ancient architecture on the moon, similar to the pyramids in Egypt, the Nazca lines in Peru and Stonehenge in England. They suspect that ET's may be responsible for humankind's technological development as well as origin similar to the plot of the movie Prometheus. If these theories hold true, perhaps the best evidence awaits us on the moon, since it is the closet planetary partner to earth.

This exciting, but damaging farce is becoming extremely popular, particularly on the internet blogs and forums, where people can share their concepts and views at an alarming rate. It is a shame that most do not take the time to check original sources, and it pains me to think that maybe they enjoy the myth rather than the truth.

Hundreds of thousands of lunar photographs are being examined for telltale signs of alien artifacts on the moon by these tall-tale moon monument dupes to determine whether it has been visited. Colonial caravans of careless cosmic creatures might have left scientific instruments, heaps of rubbish or evidence of mining on the dusty lunar surface that should have been detected by our orbiting space crafts.

We shall be taking a look at some of these claims in the upcoming photographs and we shall see that all efforts have failed miserably.

Some scientists are focusing their scientific attention on NASA's Lunar Reconnaissance Orbiter Camera (LROC), which has mapped a quarter of the moon's surface in high resolution since mid-2009 in hopes of finding such evidence of contact. The LROC has captured over 340,000 images and is expected to reach one million by the time it has mapped the whole lunar surface. They say, that a manual search of these photos is going to be a never ending job.

Scientists at Arizona State University are proposing that these high resolution photographs of the moon may reveal signs of alien intervention. ASU's Dr. Paul Davies and Robert Wagner, both pro-alien artifact hunters, submitted a paper, "Searching for alien artifacts on the moon", to Acta Astronautica, the official journal of the International Academy of Astronautics in hopes of furthering the cataloging process of the photos as well as the finding of any alien presence. In the abstract they state that the search for aliens by SETI has a low probability of success, and should broaden its search from simply listening for alien signals to looking for astro-archaeological evidence.

The ASU professors argue that such a project is worthwhile and could be accomplished with a small budget. They also contend that with new computer software, geologists and amateur artifact hunting enthusiasts could help speed up the process of cataloging the photographs and surface features at the same time as searching for alien evidence.

They say that one way to scan all of the images could involve writing software to search for strange-looking features, such as the sharp lines of solar panels, or the dust-covered contours of quarries or domed buildings. It can be argued that since NASA already has a program that counts craters, then it would be easy to convert the software to count geometric shapes. Another approach could be sending photos to volunteers over the internet for study, examination and cataloging and surface features. (72)

Though this could lead to disagreements over what constituted an unusual and potentially alien feature, not to mention the geological arguments, it would facilitate the mapping of the LROC lunar features.

It is not a bad idea to sucker alien presence buffs, fans, and even (pro-alien) Exobiologists to do a lot of the work. It could even become a very helpful new department of NASA, lower the cost to taxpayers and advance scientific knowledge.

Nevertheless, the danger involved though, would be the potential misrepresentation of the lunar surface data, the fostering of a practically hopeless search for aliens and alien remains, and a distraction from human responsibility in favor of alien dependency. Another possible disadvantage could be the dissuasion of the public from time-proven religion and the cultural integrity that it maintains through the promotion of human contact, unlike the internet.

Since the subject of alien artifacts is becoming an important topic in the search for an alien presence, it is also important to point out the necessity to show all other alternatives, theories and possibilities to avoid deceptions that might creep in to true science, religion and culture.

Though myths, legends, ghost stories, and even fun talk of friendly aliens is good for human creative thinking, they can become infectious problems to science and religion when not kept around the camp fire. It can and does seem to boost the cost of exploratory devices such as those being sent to Mars.

Hence, the purpose of this book: To defend true science and religion against the potential infiltration of pseudo-scientific theories and misinformation by exposing the falsehood of the alien presence fiction, and thus the false belief of life beyond earth.

It is hoped that after reading this book it will become apparent that no aliens have traveled from other star systems to our world, that there is no "physical" activity of aliens in our solar system and that there are no alien artifacts to prove it. Furthermore, and consequently speaking, it may become apparent that the Universe is sterile of any biological life forms beyond the confines of planet earth,

Stellar Sterility

just as scriptures testify to and just as our planetary probes are proving.

Now speaking of evidence to support the alien presence, let us analyze the available materials for alien artifacts starting with the Moon and see whether we can find any proof that aliens have been visiting us.

PREFACE TO THE SECOND SECTION

In the first section we discussed George Leonard's supposed evidences for "Somebody Else Is on the Moon" and the alien presence theory. George was one of the first to popularize the theory (mythology) that aliens have visited our solar system and have occupied the moon and our other planets, starting a chain reaction that has evolved to great proportions. To do this, people like George Leonard must promote unjustified "conspiratorial" theories to pawn such nonsense off on the public at the expense of the credibility of NASA. This results in slandering and maligning the character of our space programs for hiding the so-called facts when there are no facts being hidden. This volume (volume-2) is the continued attempt in exposing the horrific problem of Fred Steckling and Richard Hoagland in seeing meaningful patterns or connections and objects (alien artifacts) in random or meaningless data.

Following the follies of Mr. Leonard comes Fred Steckling with his "We Discovered Alien Bases on the Moon" and Richard C. Hoagland with his best seller "The Monuments of Mars." Is it true that we are "not" alone in the Universe? Is there any truth to alien artifacts in backing up the claim of alien presence? We shall deal with this in the following volume.

The view that aliens inhabit our solar system and have left artifacts strewn about our planetary neighbors has spread like wild-fire into every corner of society and government. The alien presence "religion" is now a global reality and is considered true by more than half the world's population. It intrigues the rich, the poor, the educated and the illiterate as as well as politicians and religious people. In the last 60 years it has practically taken over the world, is universal, non-sectarian, indiscriminate and its membership exceeds that of all the religions of the world combined. In fact, most religions allow it to be believed because it is a religious belief itself with not a shred of evidence to back it up. So, what harm is it to believe?

The prevenience of this new "aliens are here" religion, and particularly the belief that life can be found on Mars is demonstrated with one internet poll revealing that 39.97% of the voters believed Mars is teaming with microbes. Another 40.85% (majority) supported Mars having had life a long time ago but not at the present, while 10.76% believed aliens live under the surface hidden from view, while 8.93% considered Mars stone dead, sterile and having no life now nor any in the past. Obviously, the poll taken showed almost 91% believing in some kind of extraterrestrial life at some time in the past, if not in the present.(1)

Everyone and their dog, from politicians to religious leaders seem to be playing party to the alien presence heresy. It is as if "aliens" are basically harmless, helpful and ultimately beneficial to humanity, albeit their obsessive compulsive habits of abduction and *palpatio per anum.* This must be true since they haven't destroyed us yet. They must have some vested interest in us. We must be valuable to some extent.

Little do people know of the "depth" of deception they are being sucked into. Like cattle to the slaughter, mankind is complacently falling head over heals into gleefully accepting the alien presence religion, which is leading to a complete ruination of our traditional foundations. Our foundations may be littered more than the moon has supposed artifacts with trashy history, but we are still alive today! Just because our past is filled with sin, corruption and error does not mean it is expendable and replaceable. At least we know what we have, where we've been and what lessons we have learned from it. Whereas, with aliens, we really have no idea what we are getting. By the fact that aliens are so illusive and because of the unjust abductions and experiments we have suffered, it should tell us something about the comfort we have gained from our hard earned past, and how much distrust we should have for this new paradigm from outer space. It may not be very logical, but it may very well be true, that by displacing the very traditional foundations of religions we do have, we open the door to our very destruction. It seems that the more we dispose of our traditions the more they have have influence. This tells us much about them. For example: Within all the UFO literature on our book shelves, is there one account where we may read that aliens have given us any credence for our hard earned religious beliefs?

PROPONENTS OF "WE ARE NOT ALONE"

Mr. Leonard greatly contributed to the modern belief that extraterrestrial physical life exists. Nevertheless, his arguments from morphological shapes, extreme highlights and shadows, and creative interpretations of surface features just do not hold up in scientific court rooms. Peer pressure needs hard evidence and George did not bring it to the table. Skeptics complain that all we see is a play of shadow and light upon funny shaped surface features, and that there are no such evidences. We saw this geophysical farce demonstrated in the first volume of "Is Anyone

Else on the Moon? -- The Search for Alien Artifacts."

Despite the heavy skepticism, Fred Steckling, Jack Swaney, Richard Hoagland, and many others, continue promoting the Leonard fiction of morphological madness. Aliens on the moon, lunar bases, extraterrestrial vegetation and other exotic signs of E.T. life are pushed to great extremes. Gene Roddenberry could not have done better. Skepticism has reclined if not retired and very few take it serious now.

The most prevalent contender for the existence of interstellar invaders is Mr Richard C. Hoagland. He has pushed the E.T. garbage to such an extreme that even government agencies, such as the National Aeronautics and Space Administration (NASA) have been persuaded to re-photograph supposed alien structures such as the Face on Mars. Many lunar and Martian features, claimed to be of extraterrestrial origin were re-photographed, scanned and analyzed along with the run-of-the-mill surface features. The desperate fetish to find fossils and artifacts in the photographic data sent back by our spacecraft has continued until every internet site, blog and forum, including NASA web sites, promotes moon and Martian artifact hunting.

ASTROBIOLOGY, LEAVEN IN THE NASA LUMP

Without much criticism from right-wing traditionalists (and for good lucrative reasons too - most of them have jumped in on the band wagon), astrobiology (and its derivative astropalaeobiology) has grown into the largest pseudo-religious cult on the face of the planet, and it is rapidly spreading to the moon, Mars and to all the other planets of our solar system. Traditional religious beliefs are diminishing in favor of the new belief that man is not alone and that one day it will be proved life exists beyond earth. There will then be no need for the prophets of old or the old middle-eastern Savior -- humanity will have found a new one - even if it is only a microbe.

LACK OF ASTROBIOLOGICAL EVIDENCE

It is a given thing today that government figures and a large amount of scientists think aliens exist (or at least extraterrestrial microbes) and that one day it will be proved they inhabit our solar system. Almost every space probe mission houses alien life detection equipment, despite the fact that for decades now and after many planetary missions, not one proof has been found.

Unfortunately for them the evidence so far is zero. The budgets are now being cut in favor of more reasonable and simplistic scientific experiments. Planetary probes are not looking for gigantic exo-paleontological evidences such as large alien structures, astrodomes, ruins of buildings or UFO airports. They are instead poking expensive microbial drills and other tools into the dirt, rocks and whatever else that might get in the way looking for biosignatures.

Over half the planet watches the incoming planetary data in hopes of finding alien artifacts, a piece of extraterrestrial junk or maybe catching an alien passer-by dropping their half-eaten jelly donut. Teams of scientists and engineers are now proposing a 2020 Mars rover mission that will borrow the design of the 1-ton Curiosity and seek out biosignatures on Mars (2).

One civilian enthusiast is actually sewing NASA

for hiding supposed facts about what appears to him to be a jelly-donut or at least a fast growing Martian mushroom type plant - some say it is a humongous lichen. Others argue that a Martian must have past by and dropped the donut-looking "thing" either by mistake or on purpose, maybe to just taunt the human's space probe (3). The traditional NASA conservatives say it is a rock dislodged and flipped over by one of the wheels of the Rover. Could it be that a Martian tossed the sulfuric mushroom at the little alien rover thinking it might be hungry?

In the following study of alien artifacts we will look at Fred Steckling's "We Discovered Alien Bases on the Moon" and Jack Swaney's "Objects on the Moon" along with a fun and polemical exposé of Richard Hoagland's "The Monuments of Mars."

Our conclusion will be the same as volume one that both the moon and Mars are "stone" dead and adverse to any life forms; that they cannot naturally maintain life nor at anytime in the past have ever supported life forms. In a greater scope of cosmology, such dire lack of evidence infers greatly that even the universe itself is sterile and void of life, being just as adverse to life forms as our closest planetary neighbors - Mars, Mercury, Venus, Jupiter, Saturn and all the moons, etc. But, before we jump off into the alien artifact collection of imaginative artifices, let us look first at the evidence against extraterrestrial aliens and their habitats as existing outside the earth's atmosphere.

TWO OPPOSING PARADIGMS

Contemporary views upon the subject of the origin of life can be divided into two evolutionary branches of belief - Monogenetic Evolution (M.E.) and Polygenetic Evolution (P.E.) with the pro-life exobiologists on one

Stellar Sterility

side and the traditional geo-physicists and cartographers on the other.

Monogenists or single origin Evolutionists believe life and matter began from a single point or location and spread out from there. In other words, both matter and life had a big bang, where the former gave birth to the latter.

Alternately, polygenists or multiple origin Evolutionists believe that life popped up universally in many locations throughout the Universe, disconnected from one another and totally antonymous, where life is unique and completely independent in origins. Therefore, E.T. life must exist just as it does here on earth and should not have the same genetics.

M.E. postulates that all genetics ultimately originate from a singularity or a single big bang with earth being the most likely ontological point in time and space. Thus, the universal is most probably completely sterile of any life, since life has not seemingly evolved beyond or outside of earth. If it is out there beyond earth then it has evolved beyond earth and will share similar genetics as earth based life forms.

P.E. Alternately says that extraterrestrial life has just not been found it yet and when it is found all history books can be rewritten. If P.E. is correct, other planetary bodies will eventually demonstrate life of some sort just as earth does. The search for extraterrestrial extremophiles is a big thing now, since the finding of these little creatures in the most harsh and inhospitable terrestrial environments. P.E.'ers think that if certain bacterial forms can exist in deadly locations here, then they can exist on Mars or at least deep below the surface of the planets.

IF M.E. is correct then no matter what E.T. rock is

Stellar Sterility

found it will show no life at all or what life it does hide will have some connection to earthly origins.

True P.E. must demonstrate (eventually) that whatever life is found is completely dissimilar to earth life. It must prove conclusively and formally a restricted polygenesis.

The argument from the silence of evidence, evidence which is necessary to support the super abundant polygenetic origins theory rather favors greatly the truth of a monogenetic evolutionary origin. It seems that the closer the evidence is to earth the more complex life becomes. It does not get simpler.

This invalidates polygeneticism. For example (and this may be a weak argument), it should be as easy to find evidence of life in Martian rocks and in other planetary materials as earth rocks. But this is not the case. In fact, it is just the opposite. The further away from the center of life the weaker the evidence, if not a total lack of evidence. The further away from earth we get the more inhospitable space becomes to life.

All the data and the lack of evidence points to a monogenetic geo-centric origin more so than of a P.E. Origin. We have a total lack of any bio-signatures and a constant indication for a monogenetic geo-centric evolutionary origin.

So far, all evidence points to life beginning on earth. The Martian rocks and all other space materials so far are found to be sterile. There are some few chemical similarities and morphological similarities, but life seems to have "popped up" here extremely abundantly and no where else.

M.E. teaches that what life we do find can have and will most probably have a connection to earth life. There can be no other origins for life with life forms popping up and spawning elsewhere. Some NASA's scientists hope that if we prove this then we would prove P.E. To this date, it has not been done. P.E. Continues to be a hopeful potential and nothing else. P.E.'s greatest task is to first prove irrefutably the doctrine of absolute dissimilarity that there are multiple points in space where life began totally separate from earth base life forms. Obviously, this is practically impossible.

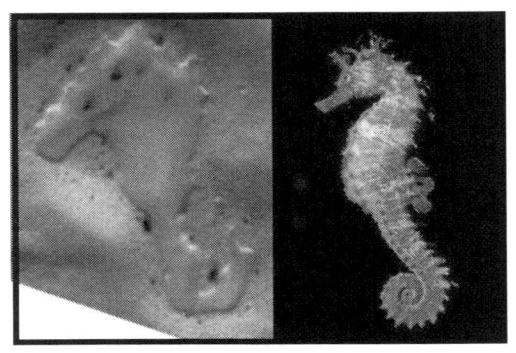

Mars Global Surveyor
MOC narrow-angle image M04-02091

STERILITY BEYOND EARTH

It seems that no matter where we send a probe to find life it returns a negative. We suggest a comical analogy, that, man will never find an odd-toed ungulate outside of earth's biosphere and therefore, he will never find an astro-horse on Mars, not even an Eohippus, a seahorse, nor any other kind of four-footed animal for that matter. But, the pro-alien life goat-ropers, for the moment, are continuing to "horse around" looking for the illusive (missing) extraterrestrial *Equidae, Equus ferus caballus*. They have failed to find one on the Moon and are now failing

Stellar Sterility

to find any on Mars. Why? Because, both the moon and Mars are stone dead! All lunar and Martian probes, from the Russian Luna probes through the American Lunar Ranger, Apollo and Martian landers have found no traces of life.

For example, Mariner 4 in 1965 found that Mars is too dry for any kind of life. According to NASA, the Mariner 4 mission forced most exobiologists to accept that life would not be obviously found on Mars. "The New York Times" remarked that Mars is "*probably a dead planet.*" However, there was a large degree of uncertainty, because the mission had only imaged a part of the planet and had spent less than half an hour doing its work there. Thus, a new view began to emerge that life could still be present on Mars perhaps lurking in "micro-environments" such as in volcanoes, or in hot-springs somewhere. This view appears closer to our understanding of proposed Martian life today. (3a)

The Viking missions found that large floods occurred on Mars, but detected no life forms or traces of early life. Lawrence Bergreen, project scientist on the Viking Missions, said "*The mission was hailed as a great success, of course, but in terms of the search for life on Mars, it turned out to be a great disappointment. To our shock, there was nothing organic on Mars, let alone what we would call life.*" Eventually, he says, they all thought, "*Well, there isn't any life on Mars. It's dead, deader than the moon.*" (3b) In his testimony, he said their hopes end with them picking up their marbles, so to speak, and going home. Again, "*The Viking landers found strong oxidants in the Martian soil,*" says planetary scientist Mark Bullock of the Southwest Research Institute in Boulder, Colorado. Mr. Bullock concludes that "*Not only would these be hostile to life, but they would also consume any fossil remains of life."* (3c) In other words, the minute any life popped up, it would be destroyed immediately!

The negative evidence continues with the Phoenix mission landing a robotic spacecraft in the polar region of Mars on May 25, 2008. Phoenix's preliminary data revealed that Martian soil contains *'perchlorate'* and thus may not be as life-friendly as thought earlier. The pH and salinity level are viewed as benign from the standpoint of biology. There was no trace of life found.

The Mars Science Laboratory mission vainly deployed the Mars rover Curiosity, a robot carrying instruments designed to look for biological activity. The Curiosity landed on Mars on Aeolis Palus in Gale Crater, near Aeolis Mons (Mount Sharp) on August 6, 2012. It has not found any biological evidence yet. It has found Martian spherules or hematite blueberries, sedimentary strata, carbonaceous material, opal, ice, lots of salts and plenty of sulfides. It has found practically everything, including kitchen sinks according to some artifact hunters, but not one microbe!

No matter how much astrobiological digging is done, the sum of all extant experiments indicate that all future experiments will only reveal negative results. Are we alone in this solar system? So far, no evidence has revealed that we are not. Not one rock turned over has proved otherwise.

Nevertheless, to maintain public support, cash flow and to hope for a P.E. proof, NASA scientists continue to probe the planets. From digging on the moon to drilling on Mars, to flying by Europa to potentially probing Uranus, and sending oceanographic submersibles to other frozen moons - the fallacious search continues for E.T. life from one failed experiment to another.

NASA GOES UNDERGROUND

It is becoming evident that it is impossible for life to exist superficially upon the surfaces of other planets other than earth. So, the groundhogs are now planning to dig below the surfaces in search for their illusive extremophile truffles. If life is not found on the surface of the planets then it must therefore be hiding underground or in sub-surface permafrosts or liquid oceans. Through Hell or high waters, sulfuric or hydrochloric ecological niches, the search continues.

Planetary scientists are now theorizing that life may exist under the cold icy surfaces of the moons of our planetary neighbors. Like teeth to be found under a pillow, the fairytale hope of finding extraterrestrial life forms is saying we will surly discover life under the ice caps of Jupiter's moons. For example, the reddish brown color of some portions of icy Europa may be caused by eruptions spewing out living (red, brown, pink?) bacteria onto the surface, which then are instantly flash frozen into what we are seeing as the reddish ting. New plans are being made to land a device to explore the underwater world thought to be below the surface of Europa, since the surface is too deadly to sustain life forms. [4]

Seems now that Mars is becoming like the moon, not too hopeful of offering any proof of extraterrestrial life, unless one digs a little deeper. What if extraterrestrial life existed in the past and has since vanished? Well, if it ain't on the surface it must be hidden below the surface, either alive or in fossil form. Amateur web sites, professional scientists, NASA and other microbial mummy hunters theorize that if there is any evidence of E.T. life it will surly be found as fossilized remains. Seems they are giving up on the false hope of finding living evidence.

As far as E.T. fossils, they say we need a way to identify possible alien remains in the search for and the analysis of fossilized extraterrestrial evidence. So, astrobiologists have spawned a new science called astropalaeobiology -- the search for extraterrestrial (microbial) fossils. This is much different from astrobiology as one only needs to dig, drill, grind, polish and then look with microscope devices for physical evidence rather than living organisms - for example, the Martian meteorites. Needless to say, the search for evidence continues with the new addition of astropalaeontology. As long as the Sun shines and it doesn't rain, the space probe equipment will never run out of energy and it can continue to search and drill holes till the cows come home or they find one on Mars!

EXAMPLES OF TERRESTRIAL ENVIRONMENTS

Of course, the theory that our planetary neighbors hide living organisms is logical, no matter the freezing cold, deadly radiation, extreme heat, poisonous and alkaline environments, and the rarity of liquid water. Contemporary reasoning seems to go like this: Microbes have flourished on earth for more than 3.5 billion years and in the most deadly environments too. They thrive in the most inhospitable earth conditions such as beneath ice sheets in Antarctica. Therefore, since Mars is an extremely cold planet similar to Antarctica, having large ice deposits, it is possible that similar microbial organisms might be found there. (4a)

EARTH ANALOGOUS TO OTHER PLANETS

Palaeobotanists find micro-fossil evidence in early

Devonian sedimentary (chert) deposits of the prehistoric hotbed ecosystems of Rhynie and Windyfield Scottland. They have found living extremeophiles in geochemically active extreme conditions that are detrimental to most life on Earth. [5] For example, they find polyextremophile thermophilic and barophilic organisms living inside hot rocks deep under Earth's surface. [6] There are also the pH tolerant radioresistant xerophiles, psychrophiles and oligotrophes living at the summit of a mountain in the Atacama Desert. [7] Also found are microbes living in liquid asphalt in Pitch Lake at La Brea in southwest Trinidad. [8] Other "things" are found living in ice 3,700 metres (12,100 ft) deep at Lake Vostok in Antarctica [9], as well as weird little Frankensteinish things in boiling water, sulfuric acid and in the water core of nuclear reactors, not to mention other stuff living in salt crystals, toxic waste and a whole range of other extreme habitats that were previously thought to be deadly to life.[10] So, why not in the extreme conditions on Mars and other planets? Though all of these habitats usually rest outside the "habiltible zone" such as where earth is, they might, under certain conditions offer habitable niches.

Is it possible for life to exist on some of these bodies? It is important to remember that the habitable zone, by definition, is the region where a planet could potentially have surface temperatures that would support liquid water, but a more accurate name for this zone might be the "zone of liquid water," as scientists now believe that habitability can occur outside this zone in certain conditions.

For example, NASA believes astro-varmints may survive in subsurface groundwater, which may still exist on Mars today as well as in subsurface oceans like the one believed to exist on Jupiter's moon Europa. Recent images taken by NASA's Cassini Spacecraft show what may be liquid water reservoirs that erupt in Yellowstone-like geysers on Saturn's moon Enceladus. Some believe that, even if the waters are found to be super acidic, maybe the little varmints their evolved into sulfuric and other acidic resistant forms.

POSSIBLE ASTROBIOLOGICAL ENVIRONMENTS

The above terrestrial examples have provided data for extrapolations to the likelihood of microorganisms surviving frozen in extraterrestrial habitats. Therefore, in search of unearthly life forms, living microorganisms are the most likely candidates for a biota of an extraterrestrial habitat. Thus, by analogy with earthly extremophiles, potential niches or biological oases have been postulated for Mars [11] and other planetary systems. They may potentially live on the planets Venus and Mars, along with several natural satellites orbiting Jupiter and Saturn, and even comets are suspected to possess niche environments in which life might exist. A subsurface marine environment on Jupiter's moon Europa might be the most suitable habitat in the Solar System, outside Earth, for multicellular organisms. [12]

It follows that Mars and other water and ice worlds may harbor endolithic communities or biotic oases where planetary parasites, astro-amoebas and exo-skeletal micro-cryocrabs may exist, such as in deep subsurface permafrost, subsurface oceans and geothermal vents. Even the subsurface water ocean of Europa may harbor life, especially at hypothesized hydrothermal vents at the ocean floor. Microbes (extremophiles) could also exist in the stable cloud layers 50 km (31 mi) above the surface of Venus [13].

Stellar Sterility

The desparate attempt to find alien life goes so far as peeking into cosmic dust water droplets, potential asteroidial ecological niches and other far-off, out of the way, unlikely places.

Hell, if earth based bacteria (Streptococcus mitis) can survive in the lunar Surveyor 3 camera for 2 years, until Apollo-12 Astronauts found them and brought them back home (13a), then they can live anywhere! Right? Sorry to say, the answer is, "No." There is a big difference in survival times between 2 years and billions of years when considering radiation.

The Wikipedia article "Life on Mars" tells us that life on Mars, if it ever existed, would have only lasted at the most 1/2 million years under present conditions: "Currently, ionizing radiation on Mars is typically two orders of magnitude (or 100 times) higher than on Earth. (13b) Even the hardiest cells known could not possibly survive the cosmic radiation near the surface of Mars for that long. (13c) After mapping cosmic radiation levels at various depths on Mars, researchers have concluded that any life within the first several meters of the planet's surface would be killed by lethal doses of cosmic radiation. The team calculated that the cumulative damage to DNA and RNA by cosmic radiation would limit retrieving viable dormant cells on Mars to depths greater than 7.5 metres below the planet's surface." And, "Even the most radiation-tolerant Earthly bacteria would survive in dormant spore state only 18,000 years at the surface; at 2 meters —the greatest depth at which the ExoMars rover will be capable of reaching— survival time would be 90,000 to half million years, depending on the type of rock." (13d)

In other words, if life ever began on Mars, it died 1/2 million years (at the latest) after the Martian

Hesperian Period (named after Hesperia Planum): between 3.0 to 3.7 billion years ago. This fits exactly to the time period that life supposedly originated on earth 3.6 billion years ago, with the appearance of simple cells (prokaryotes). (13e) The oldest known fossilized prokaryotes were (are believed to have been) laid down approximately 3.5 billion years ago, only about 1 billion years after the formation of the Earth's crust 4.6 billions years ago. The origin of these simple cells or prokaryotes are said to have evolved out of (we cannot explain) self-replicating molecules, which (somehow?) formed proto-cells or "primitive" cells - assemblages of essential components or vesicles (somehow?) encapsulated and (someway?) cooperating, and (coincidentally) interacting within an enclosed membrane (that we do not know how developed). How long this took no ones as yet knows. Did Mars have time to produce its own versions of prokaryotes? If so, it had only one billion years to do it and no proof has been found yet to affirm this.

Some scientists say, that Mars could have had and might still have ecological niches where lichens live and thrive. The German Aerospace Center's Institute of Planetary Research in Berlin ran an experiment proving lichens can live in a simulated Martian environment surviving temperatures as low as -51 degrees C and enduring a radiation bombardment during a 34-day experiment. But the lichen, a symbiotic mass of fungi and algae, also proved it could adapt physiologically to living a normal life in such harsh Martian conditions. Of course, this is only true as long as the lichen lived under "protected" conditions shielded from much of the radiation within "micro-niches" such as cracks in the Martian soil or rocks. (13f)

Unfortunately, lichens probably do not exist on Mars. They are a much later development in the supposed evolutionary chain and would not have developed after the beginning of the extreme harsh conditions of the Hesperian Period and the extinction of the Martian prokaryotes, from which they would have derived. No prokaryotes means no lichens!

The above logic of the space microbe hunters runs as follows: Earth has extremeophiles and simple cells that live in extreme and even deadly envirnments. Mars has extreme and even deadly environments. Therefore, Mars might very well have extremeophiles!

MARS HAS COCKROACHES

If we allow ourselves to believe this, then Cockroaches live on Mars. It was said cockroaches were found to have survived the bombing of Hiroshima, Japan. This was questioned later and because no one was really sure if the Japanese cockroaches really went through the bombing, a test was performed. An experiment was performed by a group of scientists (13g) where cockroaches were subjected to high doses of radiation to test whether it was true that cockroaches could survive an atomic blast.

Stellar Sterility

Now, since the bomb on Hiroshima emitted radioactive gamma rays at a strength of around 10,000 rads, the roaches were radiated for a 30 day period. The results showed that half the roaches exposed to 1,000 rads were still kicking, and a remarkable 10 percent of the 10,000 rad group was alive. The results confirmed that cockroaches can survive a nuclear explosion.

Well, what we have here is proof that cockroaches might very well be living on Mars! According to the above logic, if cockroaches can survive an atomic bomb radiation blast on earth, they can survive the radiation bombarding the surface of Mars. We have shown that Earth has radioactive resistant cockroaches. Mars has radiation too. Therefore, Mars could have radioactive resistant cockroaches. Seems reasonable doesn't it?

What about the radiation-resistant cocci "Conan the Bacterium" belonging to the genus Deinococcus radiodurans from sewage sludges and animal feeds? Six strains of radiation-resistant gram-positive cocci were isolated from sewage sludges and animal feeds in Japan after gamma-irradiation of more than 1.0 Mrad. All six strains were able to growon nutrient agar slants, and some strains were also able to grow on glutamate agar slants. [13h] How about Bdelloid rotifers - two species shrugged off as much as 1,000 Grays and were still active two weeks after exposure.[13i]

A Brazilian study in a hill in the state of Minas Gerais which has high natural radiation levels from uranium deposits, has also shown many radioresistant insects, worms and plants. [13j] Turtles can swallow 15Gy. Goldfish can swim in 20Gy and Escherichia coli is ablivious to 60Gy. German cockroaches can twitch threw about 64Gy, while Shellfish are hardened against 200Gy. The Fruit fly flies threw some200+Gy,

Stellar Sterility

but the Amoeba has it beat at 1000Gy. The blue ribbon winners of higher Gray levels are Braconidae up to 1800Gy and Milnesium tardigradum at up to lethal doses of 5000Gy. The top world champs of chomping on high Rads are Deinococcus radiodurans (previously talked about) and the global champ, Thermococcus gammatolerans (T.G.). Conan the Bacterium was not the most radiation resistant organism that we know of anymore. T.G. was discovered in 2003 near hydrothermal vents in the Guaymas Basin and it can withstand 30 thousand Gy! [13k]

ASSESSING THE PROBABILITY OF THE EXISTENCE OF COCKROACHES IN OUR GALAXY

$N = R^* \times Fp \times Ne \times Fl \times Fi \times Fc \times L$

Using the above "Drake" formula, where:

N = the number of cockroaches in our galaxy with which we might hope to be able to find.

R* = the average rate of star formation in our galaxy = 1.025 Msun/yr. [Total star formation rate = 0.68 to 1.45 Msun/yr. = Averaged "mean" = .60 + 1.45 = 2.05 div by 2 = 1.025 Msun/yr. See, http://arxiv.org/abs/1001.3672]

Fp = the fraction of those sunlike stars that have planets. [= 20% See, http://www.extremetech.com/extreme/170404-kepler-2 0-of-sun-like-stars-have-habitable-planets-alien-life-dr ake-equation-finally-has-a-leg-to-stand-on]

Ne = is the average number of planets that can potentially support life per star that has planets. [= 60 billion. See, http://www.space.com/21800-alien-planets-60-billion-h

Stellar Sterility

abitable-exoplanets.html]

Fl = the fraction of the above that actually go on to develop simple life forms at some point. [= pick random estimate = .01% of 60 billion = 6,000,000.ooo]

Fi = the fraction of the above that actually go on to develop cockroaches. [= random estimate = .01% of 60 mill = 600]

Fc = the fraction of cockroaches that develop a large enough population that releases detectable biosigns (gases, fumes, stinches) of their existence into space. [= random estimate = again, .01% of 600 = .06]

L = the length of time such developed cockroach populations release detectable signals into space. [= 3.6 billion years]

A population of organisms grows at a rate that depends on how many organisms are currently in the population. For instance, if there 100 cockroaches:

Well, suppose each mother cockroach can have 10 baby cockroaches per month. For simplicity, let's also say that once a cockroach has babies, it then dies. If you started with 100 cockroaches, after a month you would have 1,000 cockroaches (each of the 100 would have 10 babies, and 100 * 10 = 1,000). If you started with 1,000 cockroaches, after a month you would have 10,000 (each of the 1,000 had 10 babies, and 1,000 * 10 = 10,000).

Starting with 1,000 cockroaches, you end up with 10 times as big a population at the end of the month as you would be starting from just 100 cockroaches. But despite the different population sizes, both populations of cockroaches grew at the same rate. That rate was 10 new cockroaches for each old cockroach during the month. This is known as the "intrinsic growth rate" of

Stellar Sterility

the cockroaches, and this kind of unlimited growth is called "exponential growth."

In equations, exponential growth looks like this:

Population (time + 1) = growth_rate * Population (time)

This just says that the number of individuals one time unit in the future (one month, in our example) will be equal to the number of individuals now times the growth rate of the population. If the adults don't die before the kids grow up, then the formula becomes slightly more complicated because we have to add today's adults into the next time's population:

Population (time + 1)
= Population(time) + growth_rate * Population(time)

This is the same as the first formula, but here the population at the next time point is made up of both the babies born and todays adults.

Exponential growth can make a population grow very quickly. Let's start with the 100 cockroaches and grow them for twelve months. We will see that in one year, there will be 1e+26 roaches.

Number of Months
    Starting population
        Ending population
1   100   1,000
2   1,000   10,000
3   10,000   100,000

4	100,000	1,000,000
5	1,000,000	10,000,000
6	10,000,000	100,000,000
7	100,000,000	100,000,000,000(b = billion)
8	100(b)	100 trillion
9	100t	100[15] or 1e+17
10	100[15]	100[18] 0r 1e+20
11	100[18]	100[21] or 1e+23
12	100[21]	100[24] or 1e+26

Multiply the further over a continued 3.6 billion years, or the estimated time it took "mankind" to evolve and this will give us an estimated total population of cockroaches.

$N = R^* \times Fp \times Ne \times Fl \times Fi \times Fc \times L$

CALCULATING THE ABOVE:
R*		1.025 Msun/yr
x Fp	x	20%
x Ne	x	60,000,000,000
x Fl	x	6,000,000,000
x Fi	x	600
x Fc	x	.06
x L	x	3.6 billion years

TOTAL = 9.56448e+21 = N

JUMPING TO JUPITER: THE SEARCH CONTINUES

There are many extreme environments that might host life forms as well as many possible extraterrestrial environments such as the moons of Jupiter and Saturn, Neptune, the asteroid belt and other planetoids.

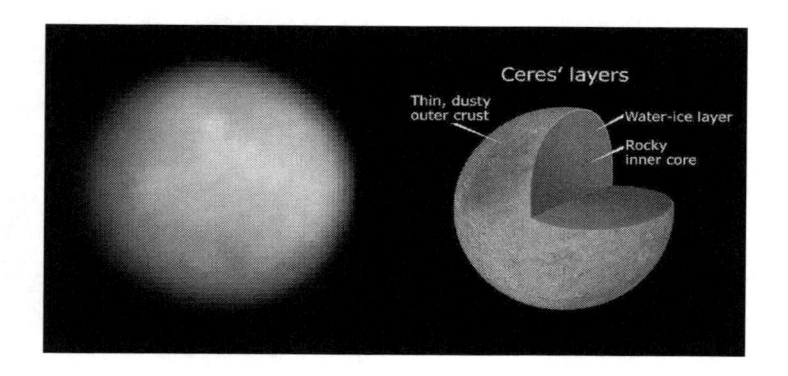

CERES

Ceres is the largest asteroid and the only dwarf planet in the inner Solar System (asteroid belt). It was discovered in 1801 by Giuseppe Piazzi. It is named after Cerēs, the Roman goddess of growing plants, the harvest and motherly love. Ceres is recently confirmed to have water vapor in its atmosphere. Frost on the surface has also have been detected.[14] The presence of water and the temperatures on Ceres make it possible that life may exist there.[15] The Dawn space probe is scheduled to enter orbit around Ceres in spring 2015.[16]

Some speculate that Ceres may offer extreme habitats for life. Others suggests it's too damn cold. But, Ceres is gushing water vapor from its unusual ice-covered surface. This raises the question of whether

Stellar Sterility

it might be hospitable to life. The spewing water vapor caused by the Sun warming parts of the icy surface provides some proof that Ceres may have some kind of atmosphere. There may be liquid water under the frozen surface of Ceres and that may be the origin of the vapor shooting out of the geysers or icy volcanoes.

Scientists think Ceres holds rock in its interior and is wrapped in a mantle of ice that, if melted, would amount to more fresh water than is contained on Earth. Ceres is one of the few places in the solar system aside from Earth where water has been located. A big question is what this water vapor means regarding the possibility of life.

There's a lot more ingredients than just water that's required for life. And whether Ceres has those other ingredients, it's too early to say.(16a) So far no signs of life and no biosignatures have been found.

MORE MOON MICROBES

The heated subsurface oceans of water that are speculated to exist deep under the crusts of the four outer moons of Jupiter -- Ganymede, Callisto, Europa and Io are also likely locations for life forms, at least according to the polygenists.

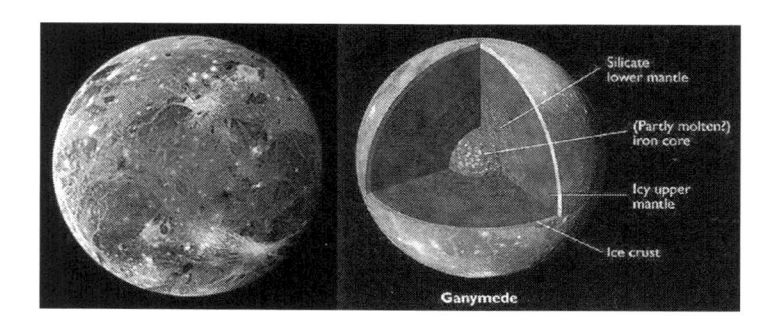

GANYMEDE

Ganymede, one of the moons of Jupiter, is believed to be a likely candidate for life. It houses a saltwater ocean sandwiched between layers of ice some 200 km below the surface.[17] This is a very tempting environment for life forms to exist. Ganymede is, like Europa, a large, ice-covered moon. It too has a subsurface ocean, which could potentially host life. Whether it exists in pockets or as a continuous band around the moon, are questions to answer. Ganymede is more complex than all other moons. It has a magnetic field and probably a liquid-iron core powerful enough to generate an aurora, like Earth's. It also has a tenuous oxygen atmosphere formed by the breakdown of water ice on the surface. This moon appeals to geologists, astrobiologists, magnetophysicists and atmospheric scientists because it is clearly a very rich environment. [17a] No life or signs of life have been found to date.

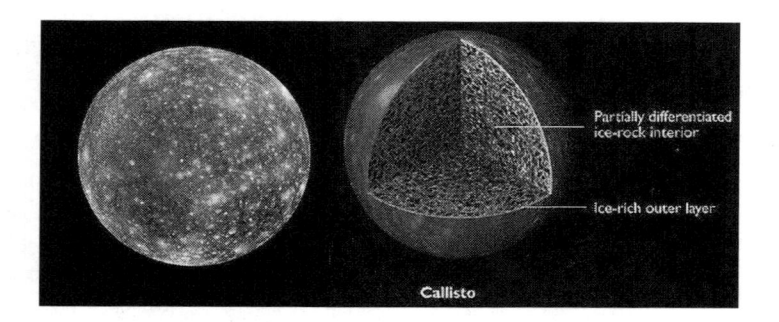

Callisto

CALLISTO

Callisto, another moon of Jupiter, also offers some hope of finding life forms. Its composition is about half water ice and half rocky material. it consists of water, magnesium, iron-bearing hydrated silicates, carbon dioxide, sulfur dioxide, possibly ammonia and other organic compounds. The presence of an inner ocean leaves open the possibility that it could sustain life. Its low radiation levels present Callisto as being considered a suitable place for future exploration. [18] Again, so far, no life or bio-signs have been found.

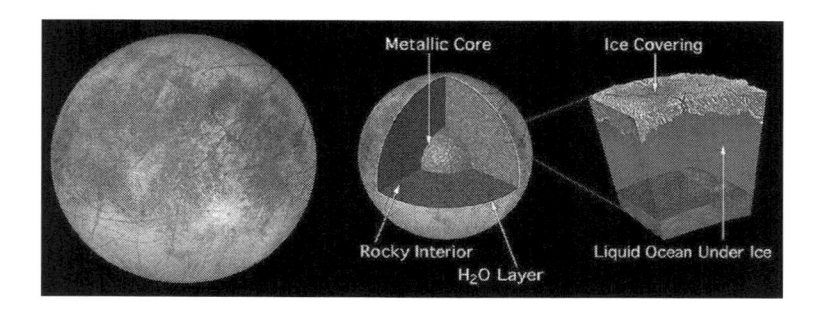

EUROPA

Europa, one of the more hopeful moons of Jupiter, on the other hand is mostly frozen water. It absorbs infrared radiation differently than normal ice and thus displays a funny red color. This is not normal and something is causing this oddity. Researchers think this is because something is binding the water molecules together. For example, salts of magnesium sulphate would make the molecules vibrate at different frequencies. Another explanation would be bacteria. This red tinge of Europa could be caused by frozen bits of bacteria. This could explain Europa's mysterious infrared signal.

Bacteria cannot survive on Europa's surface, but they could live in the liquid water inside Europa's icy crust. Bacteria could be blasted out to the surface in some kind of eruption and flash frozen, thus causing the reddish color. [18a] Nice moon, but again, scientists have not yet found any living microbes and may never find any.

Stellar Sterility

Europa could be flat dead, say some scientists! Chemicals called oxidants found on the surface of Europa might jeopardize chances of life evolving there. The level of acidity in its ocean cannot be friendly to life. Acidity messes with delicate membrane development and the building of large-scale organic polymers.

These destructive oxidants are rare in the solar system because of the abundance of reductant chemicals such as hydrogen and carbon that react with these oxidants to form oxides such as water and carbon dioxide. This is how planet earth works. Europa happens to be different. It does not have these reductant elements and so is super rich in strong oxidants such as oxygen and hydrogen peroxide, which are created by the irradiation of its icy crust by high-energy particles from Jupiter. Also, the oxidants on Europa's surface are likely carried downward in potentially substantial quantities by the same churning that causes water to rise from below. This possibly causes a reaction with the sulfides and other compounds in its interior ocean generating sulfuric and other acids. If this has occurred for just about half of Europa's lifetime, not only would such a process rob the ocean of life-supporting oxidants, but it would become relatively corrosive, with about the same pH level (2.6) of a soft drink or what we call a soda-pop. [19] Instead of oceans full of water, just imagine swimming for long periods of time in an ocean of Pepsi Cola!

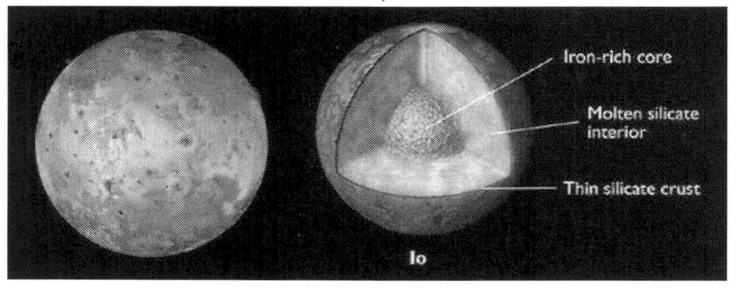

IO

Io is the innermost of Jupiter's large satellites and the most volcanically active body in the solar system. It is considered a possible candidate as a hot spot for extreme extraterrestrial (theoretical) sulfur compound based life forms. Conditions on Io might have made it a friendlier habitat in the distant past for carbon based life forms as we know them. But, life on the surface as it is now is biologically impossible.

It is categorically dead right away because of it's extreme hydrogen sulfide environment. No organic molecules have been detected on Io's surface. And no wonder, any organic compounds that once existed on the surface or that may today still emanate from the subsurface would be quickly destroyed by Jupiter's radiation. Jupiter's radiation stripped all water from Io's surface leaving it deadly to life forms.

This does not mean they do not exist underground. Whatever developing life there might have been could have easily retreated underground, where water might still be abundant, and geothermal activity and sulfur compounds could provide microbes with sufficient energy to survive.

If probes could dig down further into the rocks of Io, they might possibly find life-forms. Microbes are common deep under ground and in lava tubes on Earth.

Stellar Sterility

If lava tubes exist on Io, they could also serve as an especially favorable extraterrestrial environment for life by protecting organisms from Jupiter's radiation and providing thermal insulation, trapping moisture and providing nutrients such as sulfurous compounds. [19a] Life is absent from the surface. Future missions will tell for sure if Io has any life.

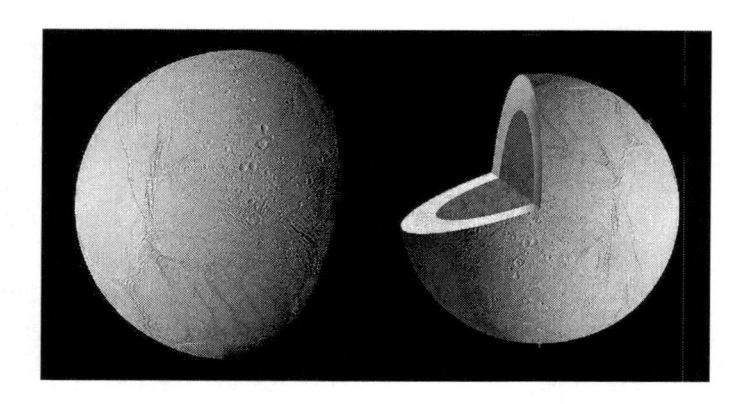

ENCELADUS

Enceladus is the sixth-largest of the moons of Saturn. It seems to have liquid water under its icy surface and water-containing plumes or geysers at the south pole shooting large jets of water vapor, other volatiles and some solid particles into space. This is very similar to those of Jupiter's moons Europa and Io and Neptune's moon Triton with its nitrogen "geysers."

The outgassing of these plumes obviously originates from a body of subsurface liquid water, and the unique chemistry found in the outgassing plumes suggests the possibility that Enceladus may be important as the most habitable spot beyond Earth in the Solar System for life as we know it.[20] Until a lander tests the surface and interior waters, we are left to speculation as to the presence of life.

TITAN

Titan is the sixth moon of Saturn and the second largest moon in the solar system, after Jupiter's moon Ganymede. Titan has a diameter 50% larger than Earth's natural satellite, the Moon, and is 80% more massive. It is the only natural satellite known to have a dense atmosphere,[21] and the only object other than Earth with a stable body of surface liquid.[22] It is primarily composed of water ice and rocky material and has a dense, opaque atmosphere preventing any view of its surface. The Cassini–Huygens mission in 2004, discovered liquid hydrocarbon lakes in Titan's polar regions.

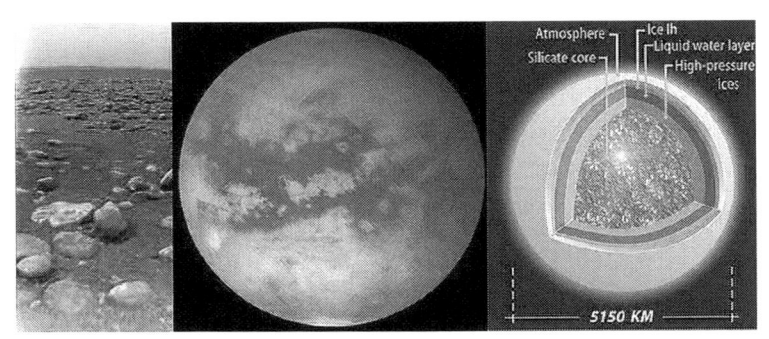

Image of Titan's surface. Huygens probe, January 14, 2005

The geologically young surface is generally smooth, with few impact craters, mountains and several possible cryovolcanoes.[23] The atmosphere of Titan is largely composed of nitrogen, methane, ethane clouds and nitrogen-rich organic smog. The climate creates surface features similar to those of Earth, such as dunes, rivers, lakes, seas (made of liquid methane and ethane), and deltas.

Titan is thought to be a prebiotic environment rich in complex organic chemistry[24] with a possible subsurface liquid ocean serving as a potential biotic environment.[25] The Cassini–Huygens mission showed an environment on Titan that is similar, in some ways, to ones theorized for the primordial Earth.[26]

One model suggests an ammonia–water solution deep beneath a water-ice crust with conditions that could support life. Detection of microbial life on Titan would depend on its biogenic effects. That the atmospheric methane and nitrogen might be of biological origin has been examined.[27] Life could exist in the lakes of liquid methane on Titan, just as organisms on Earth live in water[28] if one speculates that life on Titan uses instead a liquid hydrocarbon, such as methane or ethane, rather than water as a liquid solvent.[29] Such creatures would inhale H2 in place of O2, metabolize it with acetylene instead of glucose, and exhale methane instead of carbon dioxide.

There are formidable obstacles to life on Titan. At a vast distance from the Sun, Titan is frigid, its atmosphere lacks CO2 and water exists only in solid form. Because of these difficulties, Titan is less likely to be a habitat for life. NASA notes in its news article on the June 2010 findings: *"To date, methane-based life forms are only hypothetical. Scientists have not yet detected this form of life anywhere"*.[30]

NEWS AND ENTERTAINMENT MISINFORMATION

Needless to say, these planetary systems make for exciting speculations about life on other worlds. This frenzy of false hope and misinformation of unearthly life forms is demonstrated in such sci fi movies as Operation Ganymede (1977), 2010, The Year We Make

Stellar Sterility

Contact, Buddy Holly is Alive And Well on Ganymede, Night Caller From Outer Space (Horror) and Blood Moon (2008). The movie Outland is also set around Io, another moon of Jupiter, where mining operations are being done, while "The Europa Report" depicts astronauts finding life forms under the ice sheet and barely escaping being a Gany-meal.

This theme of life on other planets is continued in new TV series and even dates back as early as black and white silent films: Escape from Jupiter (1994); Le voyage sur Jupiter (1909 short). The list of Hollywood exo-planetary prokaryotic life forms goes on and beyond the moons of Jupiter. The classic sci fi movie Green Slime depicts a large astroid (from where, we do not know) headed toward earth oozing with a photosynthetic green slimy microbial substance that enjoys electrocuting people.

Apparently, from the above short lists of both government and entertainment industries, the idea of E.T. life is a given truth based upon an undying belief that life exists beyond earth and man must find it - and at all costs, too! There is no need to emphasize this fact that, as Rovers rampage across the Martian surface, many children suffer malnutrition even in our own country. But, never mind the gaunt faces of impoverished children hiding behind the black-budget curtains of the alien presence theory, the show must go on! The search for the Holy Green-Slime Grail must be sought!

In volume-1 "Is Anyone Else on the Moon" we looked at the Moon and Mars as potential places for life to dwell. As anticipated, we found the moon to be dead, Mars to be sterile and both lacking alien artifacts, even though the yarn-spinners keep producing blurry and fuzzy objects depicting alien junk, motor parts, screws,

Stellar Sterility

square-things and even faces and big-foot looking creatures staring at the Rover's camera. From the above synopsis of our other planets and their moons we see that life is still yet to be found. Most every planetary system we have studied has demonstrated a lack of life.

The problem is that none of these extraterrestrial planets or moons has the proper "soil" like environment as that of earth. I propose this as the problem. If you are going to have life, you first have to have the proper environment or "soil." None of these planets or moons have the proper "soil" or environment to produce or maintain life. It might be predicted that in the future, just as it has been in the past, no life will be found. No soil, no seed growth. If there ever were microbes on these cosmic orbs, just as soon as they evolved and began to thrive, they all would have died immediately,

With such a solar system wide lack of microbial life, bio-signatures and even fossil remains, one should get the message that exobiological life forms do not exist. This should tell us something about the potential reality of alien cultures, their developing on other habitable exo-planets and their coming from other planets to visit humans and, best of all, traveling trillions of miles only to kidnap humans and probe their bums!

But these points do not stop people from searching for advanced space perverts and microscopic exobiological extremophiles. After Leonard hatched his "Somebody Else Is on the Moon" book, others followed with more hyped up pseudoscientific pro-alien studies, unthwarted by the conventional scientific warnings that there are probably no space beings.

In this volume we will critique Mr Fred Steckling's

bizarre discourse on alien bases on the moon, a few other E.T. trash and garbage hunters and then end with the laughable myth-making pareidolean apophenia (seeing things that aren't there) theories of Richard Hoagland.

FOOTNOTES:

[1] http://www.space.com/video

[2] http://www.space.com/24834-strange-mars-meteorite-life-evidence-debate.html

[3] http://news.discovery.com/space/mystery-rock-appears-in-front-of-mars-rover-140117.htm.

And, http://www.space.com/24356-mars-rock-mystery-opportunity-rover-photos.html

[3a] Mariner 4: "First Spacecraft to Mars." by Elizabeth Howell, SPACE.com Contributor. See, www.space.com/18787-mariner-4.html

[3b] Bergreen, Lawrence. (2000) "Voyage to Mars: NASA's Search for Life Beyond Earth." Riverhead Books, N.Y. P.188-189.

[3c] :Phoenix discovery may be bad for Mars life." Maggie McKee, space editor. NewScientist.com.

http://www.newscientist.com/blog/space/2008/08/phoenix-discovery-may-be-bad-for-mars.html

[4] NewScientist.com under search "Europe."

[4a] "Habitability and Biology" by The Phoenix Mars Mission. See, http://phoenix.lpl.arizona.edu/mars143.php

[5] Rampelotto, P. H. (2010). Resistance of microorganisms to extreme environmental conditions and its contribution to Astrobiology. Sustainability, 2, 1602-1623.

Rothschild, L.J.; Mancinelli, R.L. Life in extreme environments. Nature 2001, 409, 1092-1101

[6] Oskin, Becky (14 March 2013). "Intraterrestrials: Life Thrives in Ocean Floor". LiveScience. Retrieved 17 March 2013. Also, Nanjundiah, V. (2000). "The smallest form of life yet?". Journal of Biosciences 25 (1): 9–10.

[7] Atacama Desert, a plateau in South America, covering a 1,000-kilometre (600 mi) strip of land on the Pacific coast, west of the Andes mountains.

[8] Astrobiology, Vol. 11, p. 241-258)

[9] "Detection, recovery, isolation, and characterization of bacteria in

glacial ice and Lake Vostok accretion ice". Ohio State University. 2002.

[10] Carey, Bjorn (7 February 2005). "Wild Things: The Most Extreme Creatures". Live Science.

Cavicchioli, R. (Fall 2002). "Extremophiles and the search for extraterrestrial life". Astrobiology 2 (3): 281–92.

The BIOPAN experiment MARSTOX II of the FOTON M-3 mission July 2008.

Surviving the Final Frontier. 25 November 2002.

[11] Horneck, Gerda. "The microbial world and the case for Mars"- Planetary and Space Science 48 (2000) 1053-1063

[12] Wikipedia, under search "Extraterrestrial Life."

[13]   Venusian Cloud Colonies :: Astrobiology Magazine Geoffrey A. Landis Astrobiology: The Case for Venus. Cockell, C. S. (December 1999). "Life on Venus". Planetary and Space Science 47 (12): 1487–1501.

[13a] "Earth microbes on the Moon". Science.nasa.gov.

[A common type of bacterium, *Streptococcus mitis*, accidentally contaminated the Surveyor's camera prior to launch, and the bacteria survived dormant in the harsh lunar environment for two and one-half years, supposedly then to be detected when Apollo 12 brought the Surveyor's camera back to the Earth. This claim has been cited by some as providing credence to the idea of interplanetary panspermia, but more importantly, it led NASA to adopt strict abiotic procedures for space probes to prevent contamination of the planet Mars and other astronomical bodies that are suspected of having conditions possibly suitable for life. See Wikipedia, under "Surveyor 3"]

[13b] Keating, A.; Goncalves, P. (November 2012). "The impact of Mars geological evolution in high energy ionizing radiation environment through time". Planetary and Space Science – Eslevier 72 (1): 70–77.

[13c] Than, Ker (January 29, 2007). "Study: Surface of Mars Devoid of Life". Space.com. "After mapping cosmic radiation levels at various depths on Mars, researchers have concluded that any life within the first several yards of the planet's surface would be killed by lethal doses of cosmic radiation." And,

Dartnell, Lewis R.; Storrie-Storrie-Lombardi, Michael C.; Muller, Jan-Peter; Griffiths, Andrew. D.; Coates, Andrew J.; Ward, John M. (2011). "Implications of Cosmic Radiation on the Martian Surface for Microbial Survival and Detection of Fluorescent Biosignatures"

[13d] Dartnell, L. R.; Desorgher, L.; Ward, J. M.; Coates, A. J. (2007). *"Modelling the surface and subsurface Martian radiation environment: Implications for astrobiology"*. Geophysical Research Letters 34 (2): L02207. Bibcode:2007GeoRL..34.2207D. doi:10.1029/2006GL027494. *"Bacteria or spores held dormant by freezing conditions cannot metabolise and become inactivated by accumulating radiation damage. We find that at 2 m depth, the reach of the ExoMars drill, a population of radioresistant cells would need to have reanimated within the last 450,000*

years to still be viable. Recovery of viable cells cryopreserved within the putative Cerberus pack-ice requires a drill depth of at least 7.5 m." And,

Lovet, Richard A. (February 2, 2007). "Mars Life May Be Too Deep to Find, Experts Conclude". National Geographic News. "*That's because any bacteria that may once have lived on the surface have long since been exterminated by cosmic radiation sleeting through the thin Martian atmosphere.*"

[13e]http://en.wikipedia.org/wiki/Timeline_of_evolutionary_history_of_life

[13f] Lichens on Mars, Astrobiology Magazine: http://www.astrobio.net/exclusive/5932/lichen-on-mars

[13g] "Cockroaches Survive Nuclear Explosion" http://www.discovery.com/tv-shows/mythbusters/mythbusters-database/cockroaches-survive-nuclear-explosion.htm

(13h) http://ci.nii.ac.jp/naid/130000027414/.

[13i] http://blogs.discovermagazine.com/notrocketscience/2008/03/24/bdelloid-rotifers-the-worlds-most-radiation-resistant-animals/#.U0naWaKwUY0

[13j] http://piecubed.co.uk/conan-bacterium-radiation/

[13k] http://en.wikipedia.org/wiki/Radioresistance

Cordeiro, AR; Marques, EK; Veiga-Neto, AJ (1973). "Radioresistance of a natural population of Drosophila willistoni

living in a radioactive environment.". Mutation research 19 (3): 325–9

Moustacchi, E (1965). "Induction by physical and chemical agents of mutations for radioresistance in

Saccharomyces cerevisiae". Mutation research 2 (5): 403–12

[14].http://www.panspermia-theory.com/extremophiles/extremophiles-and-panspermia

[15] Küppers, M.; O'Rourke, L.; Bockelée-Morvan, D.; Zakharov, V.; Lee, S.; Von Allmen, P.; Carry, B.; Teyssier, D.; Marston, A.; Müller, T.; Crovisier, J.; Barucci, M. A.; Moreno, R. (2014-01-23).

"Localized sources of water vapour on the dwarf planet (1) Ceres". Nature 505 (7484): 525–527.

Campins, H.; Comfort, C. M. (2014-01-23). "Solar system: Evaporating asteroid". Nature 505 (7484): 487–488.

A'Hearn, Michael F.; Feldman, Paul D. (1992). "Water vaporization on Ceres". Icarus 98 (1): 54–60.

O'Neill, Ian (5 March 2009). "Life on Ceres: Could the Dwarf Planet be the Root of Panspermia". Universe Today. Retrieved 30 January 2012.

Catling, David C. (2013). Astrobiology: A Very Short Introduction. Oxford: Oxford University Press. p. 99.

Is there life on Ceres? Dwarf planet spews water vapor into space.

(22 January 2014).

[16] "Dawn has departed the giant asteroid Vesta". NASA/JPL. 2012-09-05.

[16a] "Water vapor plumes raise question about life on dwarf planet Ceres." By Will Dunham, WASHINGTON Wed Jan 22, 2014.
Ref: http://www.reuters.com/article/2014/01/23/us-space-ceres-idUSBREA0M02J20140123

[17] Solar System's largest moon likely has a hidden ocean". Jet Propulsion Laboratory. NASA. 2000-12-16.

[17a] NewScientist.com "Why Jupiter's moon Ganymede is an exciting destination" 15:30 3 May 2012. Justin Mullins, contributor.

[18] Space.com "Callisto: Facts about Jupiter's Dead Moon" Kim Zimmermann

[18a] NewScientist.com "Bacterial explanation for Europa's rosy glow" 10:34 11 December 2001 by Nicola Jones

[19] "Jupiter Moon's Ocean May Be Too Acidic for Life."
Charles Q. Choi, Astrobiology Magazine Contributor. March 01, 2012

[19a] "Is there Life on Europa, Io or Ganymede?" New Mission Set to Jupiter's Moons.
http://www.dailygalaxy.com/my_weblog/2012/05/life-on-europa-io-or-ganymede-new-mission-to-jupiters-moons-.html

[20] Lovett, Richard A. (31 May 2011). "Enceladus named sweetest spot for alien life". Nature (Nature). doi:10.1038/news.2011.337.

[21] "News Features: The Story of Saturn". Cassini–Huygens Mission to Saturn & Titan. NASA & JPL.

[22] Stofan, E. R.; Elachi, C.; Lunine, J. I.; Lorenz, R. D.; Stiles, B.; Mitchell, K. L.; Ostro, S.; Soderblom, L. et al. (2007). "The lakes of Titan". Nature 445 (1): 61–64.

[23] "NASA Titan - Surface". NASA. Retrieved 2013-02-14. And, G Mitri (2007). "Hydrocarbon lakes on Titan".

[24] Staff (April 3, 2013). "NASA team investigates complex chemistry at Titan". Phys.Org.

[25] Titan is thought by some scientists to be a possible host for microbial extraterrestrial life. Grasset, O.; Sotin, C.; Deschamps, F. (2000). "On the internal structure and dynamic of Titan". Planetary and Space Science 48 (7–8): 617–636. And,

Fortes, A. D. (2000). "Exobiological implications of a possible ammonia-water ocean inside Titan". Icarus 146 (2): 444–452. Also, McKay, Chris (2010). "Have We Discovered Evidence For Life On Titan". SpaceDaily. Retrieved 2010-06-10. Space.com. March 23, 2010.

[26] Raulin, F. (2005). "Exo-astrobiological aspects of Europa and Titan: From observations to speculations". Space Science Review 116

(1–2): 471–487.

[27] Fortes, A. D. (2000). "Exobiological implications of a possible ammonia-water ocean inside Titan". Icarus 146 (2): 444–452.

[28] McKay, C. P.; Smith, H. D. (2005). "Possibilities for methanogenic life in liquid methane on the surface of Titan". Icarus 178 (1): 274–276.

[29] Committee on the Limits of Organic Life in Planetary Systems, Committee on the Origins and Evolution of Life, National Research Council; The Limits of Organic Life in Planetary Systems; The National Academies Press, 2007; page 74.

[30] "What is Consuming Hydrogen and Acetylene on Titan?". NASA/JPL. 2010.

SECTION – 2
IS ANYONE ELSE IS ON THE MOON! NO!
A Critical Analysis of
George Leonard's
"Is Someone Else on the Moon"

"One has not only an ability to perceive the world, but an ability to alter one's perception of it; more simply, one can change things by the manner in which one looks at them"
- HENRI BERGSON, Matter and Memory

Mr George Leonard titled his book, "Somebody Else Is on The Moon." Is this statement true? Was somebody else actually on the moon? Is somebody else on the moon right now? Are there alien artifacts, construction sites, machinery and garbage dumps on the moon?

It seems that everyone researching and analyzing lunar landscapes and archaeologically digging for extra-terrestrial devices is finding all kinds of junk! In the last thirty years, this cult of alien artifact hunting has grown to pandemic proportions!

To find out the truth we must start somewhere and Mr George Leonard is the best cult leader to begin with, since his book is the first and most extravagantly, albeit verbose, extra-terrestrial travesty of selenology theories available to us. In fact, it is a master piece of macro-cosmological alien horse apple mythology ever written.

In the search for alien trailer-trash on the moon, it really does not matter where the curious ET private eye starts. Those who are artifact hunters are probably familiar with George Leonard's work "Somebody Else is on the Moon." If not, you're not a real Gung-ho alien hunter!

When it comes to lunar locations of such evidence, it also does not matter. According to George, anywhere on the moon is good enough. The moon's surface is loaded with garbage, strewn with trash and heaped with piles of junk. An astronaut alien garbage recycler would go crazy.

Mr Leonard can find ET litter anywhere on the Moon and suggests that anyone can. A re-evaluation of Mr Leonard's observations will help in the decision as to whether there really are aliens on the moon or not. Let us take his suggestion and study the Moon for ourselves. Pull up your socks and grab your Schnapps, you're going for a wild ride in search for – nothing!

HOUSTON, HE HAS A PROBLEM!

"All truth passes through three stage:
First, truth is ridiculed and then it is violently opposed. Finally, after much time and debate,
it is accepted as being self-evident" - Anon

ET excavators, like magicians, love to take lunar photos and pull phantom artifacts out of shadows and highlighted dirt piles, crater rims and even transform boulders into lunar rabbits. The art of "deception" is to make something appear that is not there and distort what is there into something that is is not. We may conclude that all falsity passes through several stages of conflict and perversion before it is finally accepted as truth.

For example, a selenology blob may be whatever a person imagines it to be. Next, after displaying such foolishness to the public, some reasonable scientific people oppose it. Next, after science disregards it, scoffers follow mocking it. Finally, after serious violent opposition, ridicule and mockery retires back to the laboratories, the falsity (eventually) prevails from the lack of serious complaint.

From here, the lunacy spreads like a virus throughout the majority of society such as through grocery store pulp, rag-mag publications all the way to the top into government agencies. The once mocked and scoffed at foolery gains attention in the better magazines and book publishers. Money begins to flow and "talk" and the BS begins to walk its way up the cultural as well as scientific ladder - even into national space agencies. Wisdom is, once again proven true, that the majority prefers to believe the lie, the falsehood, the deception, the myth, the legend, and the fabulous rather than the truth. Hence, the lie gains credibility from a lack of discrediting, while the truth is lost. Thus, a great woe comes upon society in saying what is false is true, and that which is true is false. Furthermore, false hopes take root and usurp true hopes.

Alien artifact hunters have the certain ability to perceive the world in an uncanny and deceptive way. By altering their rational perceptions of things, more simply, they can filter vague natural lunar objects through the degenerate powers of their warped imagination and alter the manner in which they look at things, and with enough persuasion, make other lame folk do the same – see artificial alien creations. Without a solid grasp of lunar processes and considerations for poor quality images, the dim-witted reader swallows the lie of alien artifacts rather than see the reasonable "shadow" that should be creating doubt.

Mr Leonard said intelligent people want to hear what is on the cutting edge of current truth about the moon, not swallow a bunch of pseudo-scientific NASA status qua garbage (See p.16). However, according to Mr Leonard, the cutting edge of scientific truth in lunar research consists of collecting and swallowing exobiological and astrophysical pseudo-scientific "garbage" left by supposed Extraterrestrials.

According to George, the burden of proof rests upon anyone trying to prove an alien presence on the moon. Apparently, Leonard self-contracted the burden, whereas his colleagues preferred not to get their hands lunar-soiled. Therefore, in defence of ET existence and

Stellar Sterility

moon monuments, he wrote a book and a fancy one to boot!

In his book of 238 pages, Mr Leonard graces his readers in burdening himself to belabouring the subject of alien artifacts supposedly found as proof for an alien presence. He graces his readers as well, in that he burdened himself not to crank out 2380 pages of Selenic sewage. Further blessings abound that on his Moon there are mountainous heaps of alien leftovers and he cherry picked the best for the book.

Questioning and having healthy skepticism, as well as testing proofs are virtues in the field of science, said Mr Leonard. For, he admits all sorts of rascals might invade and lower the standards. The standards he speaks of are the standards of scientific proof. Otherwise, fictions and falsehoods creep into science and if this happen, how can we be sure that breakthroughs were real? (p. 17).

Therefore, we should be on the lookout for possible rascals infecting the realm of astrophysical sciences with trash, imprudent nonsense and alien presence artifact refuse. Yes, let us credible scientists explore and belabor these same artifacts and see what kind of burden is developed - in refraining from impertinent laughter - as Georges' breakthroughs are re-examined.

No matter his set of virtues, his conclusions are beyond incredible. He has a complete disregard for the very virtues he has set as precedence. Consequently, if the reader is to avoid the same consequence, the same virtues must be considered when analyzing the data, but without preconceptions. In this way, by using these standards of scientific evaluation, rather than disregarding them, the reader may waffle through Leonard's laughable lectures with immunity. As light exposes darkness, true science will expose Leonard's play upon shadows and highlights revealing their true contents.

Mr Leonard claims that the photos taken by the lunar missions scream with evidence that the moon has life on it. He begins bellowing loudly that, *"The moon is occupied by an intelligent race or races (and) is firmly in the possession of these occupants. Evidence of their presence is everywhere: on the surface, on the near side and on the hidden side, in the craters, and in the highlands*" (p.23).

Good Lord and lunar land sakes alive, alien rubbish is all over the moon! Therefore, it will not be difficult to blindly pin the tail on the lunar donkey and hit a gold mine of moon-dweller artifacts. Yet, with proper scientific insight, such alien rubbish will be rather only found in his book.

Without further ado, let us cast aside, as George said, all *"misconceptions and preconceptions shoveled into us*" (p. 26) and by George, and by Jove keep our minds open. We should also brace ourselves as we slam on our booster rocket "brakes" of reasoning, when we happen-stance upon alien engineering of a macroscopic scale, dwarfing anything seen on earth. Furthermore, let us be prepared to have shoveled down our throats fantastic new explanations for old mysteries on the moon, such as the artificial origin of some craters, the causeway explanation of the white crater rays (i.e. Tycho) and seriously doubt the old orthodox explanations (p.27). Let us volitionally suffer the hilarity of fantastic selenology fabrications with the full understanding that it is all alien and pure rubbish that we are looking at.

Is there some reason why Mr Leonard wants us to cast aside our old "orthodox" concepts and scientific explanations – basically, all sanity and decent research?

Stellar Sterility

There are more reasons we must do this. The first and most important reason is we can become boon-doggled, easily swallow fabulous alien interpretations, and therefore rest assured in our new religious heresy.

Another good reason, if we are privy to the myths of Mr Leonard, we can learn to create our own ET farce and parley (publish) the alien corn upon others, thus spreading the boon-doggle — and make money to boot.

Consequently, though most may waste time and money looking for lunar phantasms, fragments of asteroid Munchhausen or the Eye of Harmony, we may *"neither flux nor wither"* in profiting from our pulp fictions. Apparently, this is the case with George Leonard, for in the above noir tableau we find the photo facts do not fit the tall tail and the book is more a Daleks' search for Finn McCool's giant causeway than a science lesson.

A TRIP TO THE MOON

Let us take off in an imaginary rocket of scientific re-evaluation and fly to randomly picked spots, the first being the Sea of Tranquility. Here we will begin our reconnaissance mission into the imaginative world of George Leonard's alien artifacts, lunar life forms, and moon monuments. This is the best place to start with lunar sciences as this is where the first astronauts landed, played with feathers, hammered out lunar rocks, experimented with gadgets, and swung golf clubs.

For Mr Leonard, there must be some hidden reason other than scientific to pick this as a first choice. Of course, it is simple: Alien artifacts are there! Therefore, fasten your seat belt and stuff a feather in your hat, because our spaceship is entering the lunar atmosphere and it is settling into a stable orbit around the Moon.

1
THE SEA OF TRANQILLITY

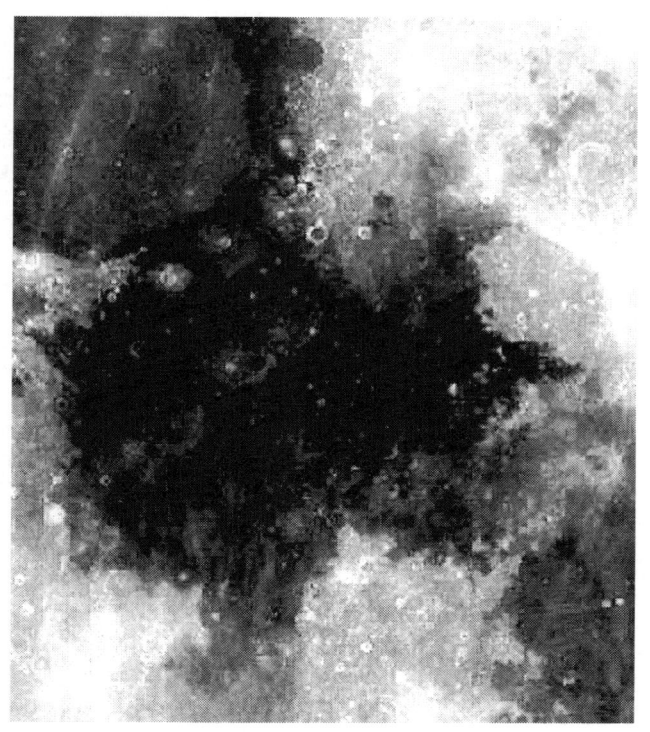

Mare Tranquillitatis

Sea of Tranquility: Named in 1651 by the astronomers Francesco Grimaldi and Giovanni Battista Riccioli in their lunar map 'Almagestum novum.'

As we direct our ship to the appointed location, we begin to see what looks like the floor of a very large ancient domed city. Now, what we are looking for are alien artifacts mentioned by Mr Leonard and other extreme astrobiologists, if not the aliens themselves.

Stellar Sterility

This is also the crash site of the Ranger-8 photographic satellite (Feb. 1965), as well as the landing area of the Apollo-11 Mission (July 20, 1969). It is also a very hefty site of alien spacecraft and constructions.

To the adventurous and scientist, the view from our spaceship is a once in a lifetime grandiose experience that cannot be duplicated on earth. It is the most breath-taking sight ever to behold short of entering heaven. Tranquility Base, here we come - again!

The panorama we see is a large area, a lunar "mare" (large "sea") that sits within the Tranquillitatis basin on the Moon. What we notice is mare material within the basin that consists of basalt and some surrounding mountains and of course, if we look hard enough, alien artifacts, domed areas, crystal lattices and roadways.

The basin has irregular margins and lacks a defined multiple-ringed structure. The irregular topography in and near the basin results from the intersection of the Tranquillitatis, Nectaris, Crisium, Fecunditatis, Serenitatis basins and the two thorough going rings of the Procellarum basin.

Concept drawing of Lunar Dome

Here we notice Palus Somni, on the northeastern rim of the mare, is filled with the basalt that spilled over from Tranquillitatis. This Mare has a slight bluish tint relative to the rest of the moon and stands out quite well when color is processed and extracted from multiple photographs. The color is likely due to higher metal content in the basaltic soil or rocks and not from aliens spray-painting graffiti or from camouflaging their lunar dwellings.

Well, shucks! Unfortunately, what we see is a disappointment, when it comes to bustling alien galactic mutant merchants scurrying about buying and selling their wares. Sorry, there are no domed cities, buildings, tollbooths, trolleybus, artificial structures, glass castles, roads, billboards or sewer pipes. All looks destroyed and what we have entered is not some Beulah land, but a hellish desolate world beaten to a pulp after billions of years of meteoric bombardment.

What are visible are the results of meteoric, geological, and morphological processes. The aliens must have relocated the artifacts to another location. Therefore, let us keep looking, for according to George, we will stumble over something. The "garbage" is everywhere and there is no place you cannot find alien evidence.

ALIEN HABITATION ON THE MOON?

Our mission to the moon went well and notes were taken as to George Leonard's claims, and the following preliminary science report made. [See "Lunar Photography and Selenographic Data Re-evaluation Preliminary Report" NASA SPAM-RSM-101]

Stellar Sterility

So, where did the aliens build their landing platforms, domed houses, airports, and storage facilities? Did they move them off the moon? Did they go underground? We did not find one simple crescent wrench! Not even a threaded bolt was uncovered. Furthermore, as for one single "nut," there was no need to look any further beyond George Leonard.

NASA SP-206
"Lunar Orbiter Photographic Atlas of the Moon"

2
OVAL BLOWN BOULDERS

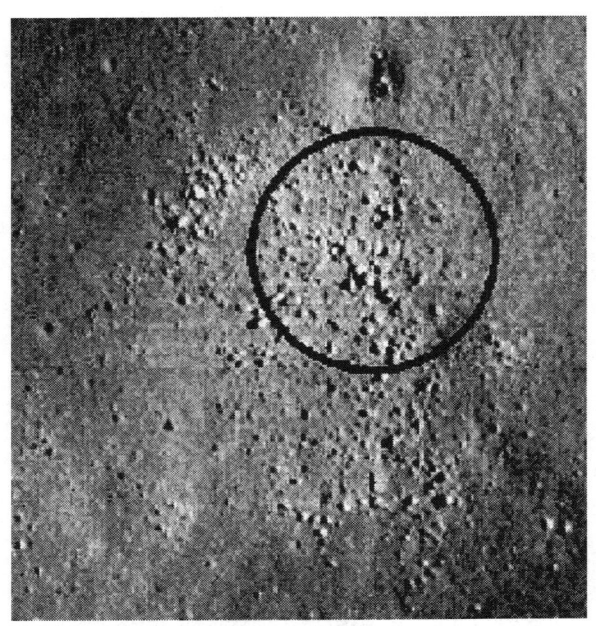

Lunar Orbiter II-042-H1
"The Rock Pile"
Rukl's "Moon Viewed by Lunar Orbiter (p.49)

    Fret not. Trust in George. We might find some ET's. For, George directs us to a special location in the Lunar Orbiter-II photo 042-H1, which he calls "the Rock Pile." He claims that UFOs are stored in a small crater among these rocks. Get your magnifying glass ready, for it is a small little area in the top right corner of the frame. In our reconnaissance of the area we found what we needed, reprocessed the photo from the original negative, enlarged it and analyzed the data, and then listed the following conclusions.

Stellar Sterility

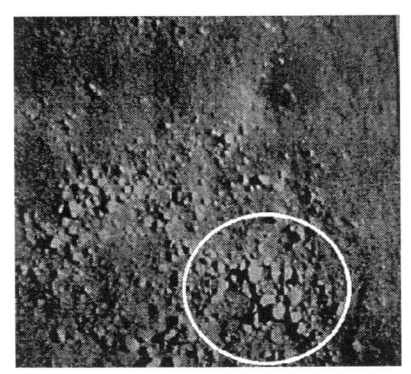

Kopal's Atlas "Megalithic Rings"

For those without access to the NSSDC archives, flip open the Lunar Orbiter Photographic Atlas of the Moon to LO-II-042-H1, and compare it to our new processed picture. You will see what George calls "*a manufactured vehicle.*" He sketches what he sees on page 101, because the picture does not reveal what he said is there, even if you use a microscope. To take his word on it, he said, "*It is gleaming among other little round alien objects.*" He claims the little round ovals look artificial and thus manufactured by aliens. One of the machines has "*evenly spaced rear struts*" and a "*perfect peak.*" It has a beautiful "*molded point*" adorning the front, and has "*cilia like appendages, resembling those of a centipede... absolutely impossible for this kind of irregularity*" (p. 23. See Plate-2, photo 66-H-1612). Well, maybe it is a centipede and an alien one too.

Wow! This should convince and excite anyone who is bewitched into seeing in this "bad" reproduction of the rock pile an oval shaped alien artifice. However, according to George's criteria, it does not matter if you see this or not. There are plenty of other locations, in

Stellar Sterility

fact millions of places to look. Every NASA lunar photo is loaded with exoplasmic belching anti-gravity propulsion devices and mindless meandering digging machines.

However, at this time in 1977, George cannot know that the aliens are actually mining Helium-3, and not just pitching regolith in and out of craters. The gadgets are not digging for minerals or drilling for water, but are full of gas. George also does't know that he too is full of hot air in declaring, that this is *"one more piece of evidence"* in support of a long list *"of enormous machinery and devices that pushed the Moon around and knock down the rims of craters"* (p.23-24). It is a long list of hot air, if you ask any serious scientist.

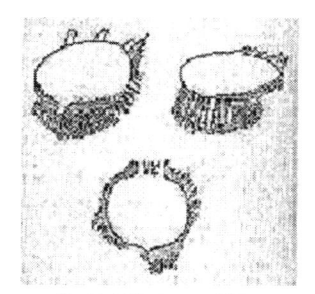

LO-II-042-H1
George's sketch of 66-H-1612. "Oval" shaped vehicles (p. 101)

Yet, George continues to point out that, *"The three 'struts' at one end of the object, with the peak at the opposite end, are so perfect as to demand the conclusion that this is artificial."* He continues to fantasize that, *"it is a vehicle of some kind"* (p.101). In addition, he should demand us to see it, since we cannot determine what it is other than a boulder.

Stellar Sterility

It is true that there are such fuzzy oval shapes. Nevertheless, shadow and light can play weird tricks on the eyes, and if the imagination is predisposed to looking for alien machines then the eyes will see alien machines. No matter what the skeptic said, the results can be convening if you are boondoggled by the alien fantasy.

Our re-evaluation of the area shows something quite different from oval shaped alien 'machines', for we see in these enlarged "detailed" photos that they are only piles of rocks, many of which form circular ring-like patterns on the lunar surface. The rings are probably the result of ancient craters undergoing deterioration and erosion. These particular boulders are actually "the Megalithic rings" and are composed of boulders, approximately 1-10 m in size. They are located in the plains of Mare Tranquillitatis (34.42 degrees E, 2.81 degrees N) as photographed by Lunar Orbiter 2, Nov. 7, 1966.

Rukl's Atlas shows an enlarged photo-enhanced shot of the area processed off the original negative (1b). Kopal's atlas also has a good enlargement of the site (1c). Most of these boulders are large and many appear to have rounded outlines. Of course, boulders can be "oval" on the Moon and can be round looking, though maybe not as smooth as earth boulders. Unlike planet earth that can smooth rocks through water erosion, lunar geophysical processes smooth the sharp edges of rocks by intense heat as they are ejected during meteorite impact or by micrometeorite bombardment over eons of time. Just because the Moon is a terrestrial type planetoid does not mean fish are jumping in and out of craters as George would like us to believe.

Stellar Sterility

The following is a close up of the Megalithic Rings from Lunar Orbiter-II, but unfortunately, there is yet no LROC photos yet found of the area. Maybe you might be able to locate the frame?

3
TOOLS IN TRANQUILLITATUS

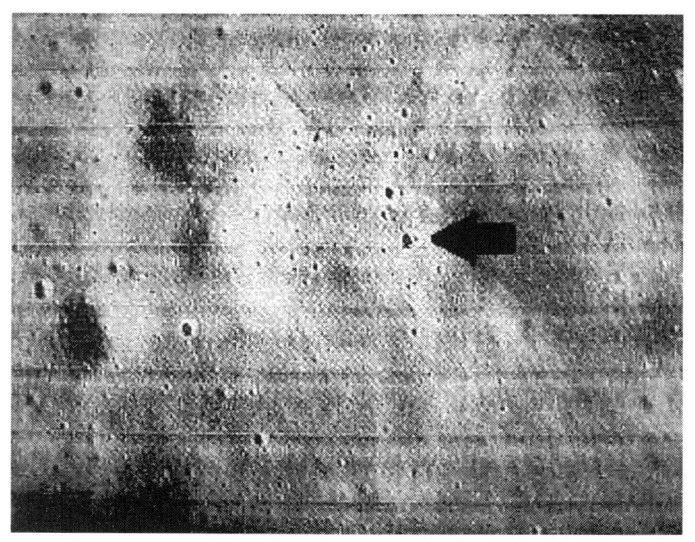

NASA 67-H-304
A sharply defined, machine-tooled object is indicated in the area of Crater Maskelyne-F. (p. 177)

Also in the Sea of Tranquility, south of Maskelyne-F, are other hot spots of supposed space alien activities. Mr Leonard shows another crater with *"Two magnificent machined-tooled specimens"* (See Plate 25, NASA photo 67-H-304). It is a small crater and he said, *"These creatures do love [small] craters."* Mr Leonard likes small craters too and for real good reasons. So much stuff can be hidden, stored and stuffed in "small" craters almost unnoticed.

For instance, located in the dark mare area near a broad dome of some kind is a small crater about one hundred yards in diameter (300 feet). This is one of many, he said, of the smaller more anomalous type craters to consider. Of course, what we are to see stuffed in the crater is a very *"precision made"* mechanical object with unnatural geometry that has *"a perfect square etched in blackness"* (p.177). He suggests that they are entrances to vast underground vaults, maybe caverns filled with ET provisions or even an entire metropolis.

After wasting hours scanning dozens of Lunar Orbiter frames looking for this small 'anomalous' crater, the results were just that – it was truly anomalous and impossible to find. It is a sad shame that it cannot be exposed and debunked directly, but it is not impossible to pitch it in the alien presence garbage can as more debris bunk. When compared to the billions of other empty craters it becomes highly unlikely that it houses anything else other than what every other crater contains – dirt!

Nevertheless, making an imaginative exception in this case, we are to believe it houses a *"metallic specimen"* that appears like some blanket or "lintel" - a decorative architectural element or a combined ornamented structural item like what is found over portals, doors, windows and fireplaces, but in this case a crater. It appears to be crossing over the crater, moving over the top, similar to fabric drawn like a "covering" sealing it up – a crater blanket.

Is it a canopy of some sort covering an entrance or it is an optical illusion? Since it cannot be re-locate, we cannot be sure what the lunar devil it is. George's photo is degraded and blurry, and no conclusive decision is possible. We can take his word for it or write it off as another optical illusion derived from blurry selenography.

Stellar Sterility

Craters can have other weird things hidden in them besides machine tools and left over busted equipment. Accordingly, they can house alien spaceships and habitat of all kinds. These particular craters are extra small, hard to find, yet big enough to hide a spacecraft. They are also so small, yet big enough to hide fantasies! For instance, there can be diggers, dozers, dirt-spewing buggies, and most anything else we might imagine that aliens use to process the lunar surface. Craters are great places for protecting and hiding valuable items, and even excellent spots to stuff imaginary ones too.

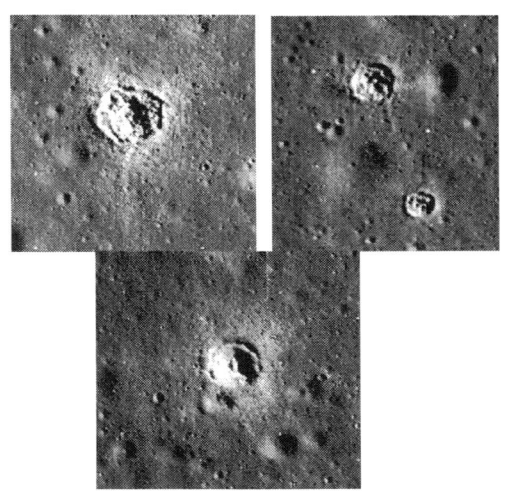

LO-III-194-H2 More craters with devices and debris.

Stellar Sterility

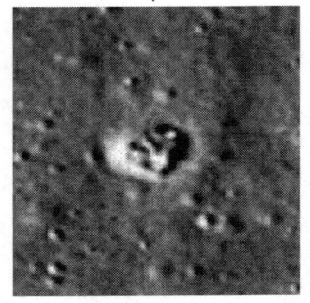

Possible "Play-boy Bunny Rabbit" face

4
FRA MAURO HIDEAWAY-LANDS

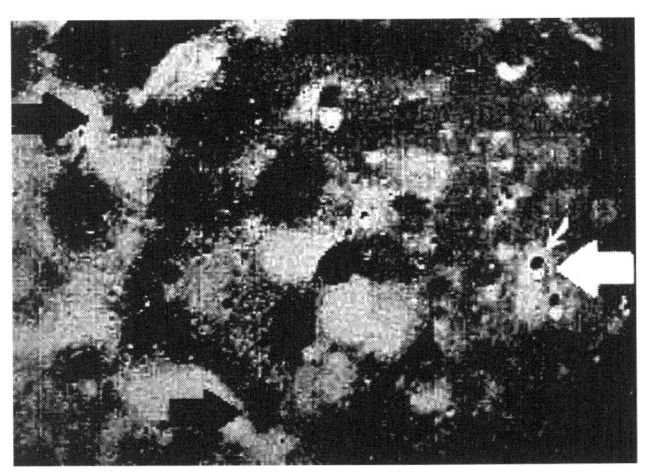

PLATE 26, 70-H-1629
Diggers, Dozers and Dirt-flingers

The Fra Mauro Highlands is a lunar formation on the near side of the moon that served as the landing site for the Apollo 14 mission in 1971. It is a widespread hilly geological area covering large portions of the lunar surface and is composed of ejecta from the impact, which formed Imbrium. It is another fantastic place to look for hidden astro-trash.

The area is primarily composed of relatively low ridges and hills, between which exist undulating valleys. Much of what covers the ejecta blanket of the Imbrium impact is debris from younger impacts and material churned up by possible moon quakes. Fra Mauro is a great hideaway for alien ejecta materials and moon-quack created craters of the same suspicion

Stellar Sterility

as the previous ones. Can we find any hidden anomalous alien artifacts!

Picking a crater at random and with very little discrimination is no gamble for Mr Leonard, when rooting for alien truffles. Alternately, neither is it much of a gamble for us to randomly pick one with little discrimination when exposing lunar artifices.

In Lunar Orbiter frame 70-H-1629, we have what George supposes to be some kind of "machined T-bar" shaped item hiding in an unnamed crater. He said the other craters also have oddities and activities in them. Yet, under closer inspection, these explanations themselves raise even more suspicions as to authenticity. Not one crater ever gives up any artifacts!

Mr Truffle Hunter's T-bar is most assuredly an angular rock, highlighted and shadowed in a way to look like an artificial device. Again, the crater is too small to relocate. Nevertheless, like every other crater that can be located, and there are thousands of them, it is as sure as the Sun rises that it will reveal no treasures.

Another one on the list for a one eyed and half sense inspection is a "tiny" (almost invisible) crater with what appears to be a "control-wheel" (frame LO-II-182-H1). It is very striking and looks like the head of a shank on a large screw.

George asks if this is some kind of plumbing. He speculates that it might be *"an underground community."* He claims that NASA feels *"if we're too blind to see these oddities"* then we must have a grave problem. He obviously shares the same feeling. However, we do not share those feelings. It is not that we are so blind that we cannot see. It is that we cannot

Stellar Sterility

see anything even with 20/20 vision. They are not there and it is plumb foolishness to see otherwise.

Now, sane people will all agree they are probably geological oddities, but not everyone is sane. Some few will believe they are oddities manufactured by aliens, despite all evidence favoring otherwise.

LO-II-182-H1 / 67-H-041 (Plate 27)

Plate 24, 70-H-1630 Control Wheels

A close inspection reveals exactly the geological explanation as well as Leonard's imaginative ability to

Stellar Sterility

mold mare-mud into mare hills and paint ethereal lunar landscapes. The evidence of alien technology around the Moon is all "painted" into the imagination of his readers. Once painted or we should say "tainted," the human imagination will begin to see ET Trash everywhere it looks.

As we walk further through the black & white pages of lunar atlases in search for Leonard-e-facts, we find uncountable craters storing uncountable other extra-terrestrial crafted constructions.

As we strain our eyes, we are supposed to see different kinds of mechanical gadgets and devices. The crater in the above photo is supposedly housing some kind of "control wheel" (See Leonard, Plate 24, frame 70-H-1630).

It is a cute little crater, just like all the millions of other "cute" little craters: Natural, geologically created, full of lunar dust, riddled with rocks, and saturated with the burning heat of the Sun. However, with proper enlargement and enhancement, we see more rocks, boulders and collapsed rims with highlights and shadows.

Do you actually see any "super-rigs," "plumbing," "control wheels," "dirt-diggers," "dozers," "do-what's," "thing-a-ma-jigs," "ding floppies," and "whatcha-call-its" in them? He can fool some people all the time, all the people sometimes, but not all people all the time; and this time he is not fooling anyone anymore.

In NASA frame 66-H-1611 (above), which covers the area of a small crater west of Mare Tranquillitatis, we are to see two bumps or round objects and a "T-bar plumbing" (Leonard pl.32). Almost any small crater you pick (at random) will reveal these cute little alien bumps, pipes and bars, because almost every little

Stellar Sterility

crater on the moon has crater debris, rocks and boulders. So, what are they?

Did aliens wastefully leave billions of tons of their hard work behind or did natural processes leave billions of practically worthless rocks contorted into all sorts of morphological shapes? It is unimaginable that an alien culture hundreds of light years from home would be that wasteful and foolish as to leave behind precious re-usable scrap metal, junk, and hard to get resources.

Now, speaking of pipes and bars, if we get wasted enough we will see anything – not only plumbing but, kitchen sinks as well.

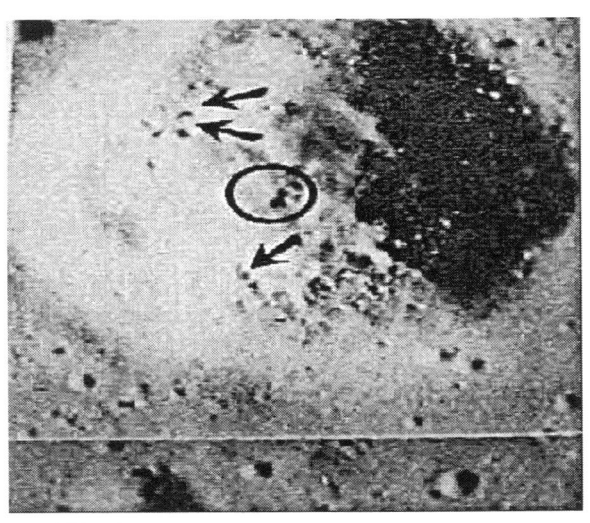

66-H-1611

5
TACKLING TYCHO TECHNOLOGY

Crater Tycho LO-V-125-M

Crater Tycho is another great hiding place for alien machines and construction sites. In fact, Crater Tycho is the ultimate hiding place, a real Disneyland of crater floor disturbances and extra-terrestrial tidbits of technology.

Tycho is located south and a little east from the great Mare area we see with our eyes on a Moonlight night. It is about 85 kilometers across and is clearly visible on our Moon's surface. The freshness of the crater and the rays of material radiating from it suggest that it is a young crater; there has been little time to erode it. Tycho appears close to the southern

Stellar Sterility

polar region of the Moon and surrounded by a bright ejecta blanket of rays extending across the lunar surface.

Leonard said he has found all kinds of mechanical relics lying around the crater floor *"in-between the boulders, inside the crater, outside and along the rim"* – he means everywhere. These gleaming white alien made structural hemispheres, domes, and used equipment are not real. They are irregular mounds with small craters at the peak and round mound shapes strewn about from location to location.

He said, these objects are 1/8 to 3/4 of a mile in diameter, or 400 yards, and usually never differ in size. To Leonard, this similarity in size is evidence to artificiality and the junk found is proof that Tycho is a literal alien trash dump of artifacts. There is so much alien refuse that anyone can stumble over a pile of regolithic material and find something alien!

Crater Tycho Photomosaic taken by Surveyor 7.
The central horizon hills are eight miles away from the spacecraft.

Talk about peak our interest. One location supposedly has about 30 weird objects clustered together inside it. Let us zoom in on some of these locations and see if we can find anything other than rocks, mono-mineralic fragments, various kinds of glasses including agglutinate particles, volcanic and impact spherules, high titanium basalts rich in iron, olivine, pyroxene, plagioclase, and high abundances of ferro-titanic oxide ilmenite. [1a]

Tycho crater's central peak complex about 9.3 miles (15 km) wide. (LROC)

TYCHO CRATER CENTRAL PEAK
View of the summit area of Tycho crater's central peak. Boulder in the background is 400 feet (120 m) wide.
(Credit: NASA Goddard/Arizona State University)

Stellar Sterility

In the above Lunar Reconnaissance Orbiter photos, we see what looks like a "T" and "Golf Ball" on top of the central peak of Crater Tycho. This is something Leonard could never see even in the finest Apollo pictures. He did not have access to the quality of the Lunar Reconnaissance Orbiter photographs. The best he had access to was Apollo. Though very high in resolution, the Lunar Orbiter and Apollo photography does not come close to the digital resolution of the LROC pictures [1].

Further rooting for Tycho technology, George takes us to another location (69-H-1206) where we are supposed to see "*perfectly round gleaming white hemispheres*," that are flung all about the crater floor. They never vary in size, never change much in shape, but are "*symmetrical and scalloped on the straight edge*" sides. They actually rather look like chunks of pie, bitten and chewed on one side by alien pie-holes who owned them.

Nevertheless, Leonard gives us his straight answers and would have us believe they are just down to earth type structures similar to "yurts." They look exactly like the round circular yurt tents built by the Mongolian nomads of Mongolia (p.61). He said, that "*There is, however, no evidence for saying with assurance more than that they are certainly artificial*" (p.61). Begging to differ, what is artificial is certainly this kind of interpretation.

This is not all that is yurting about in Tycho: "*The floor of Tycho shows so much ground disturbance that it is virtually impossible to sort out the* [alien] *tracks from the lava flow and general ground wrinkling*" (p.103-104). What? Let us rephrase this for Leonard: The floor of Tycho is so disturbed and contorted, and so

Stellar Sterility

impossible to distinguish between geological ground distortions and artificial "tracks," that it is very possible to misinterpret ground wrinkling and lava flows for alien activity, and no one would know the difference – with such blurry pictures. However, with better resolution pictures we can know with certainty exactly what they are. Let us look at some higher resolution photos of Tycho and solve this eyesore for the reader.

Leonard's drawings on page 61 describe what he sees – a vast collection of artificial whatnots. Nevertheless, he is kind enough to give us the traditional scientific (orthodox) geological explanation for them; that these mounds are the results of the upward movement of magma, which has warped the overlaying rock. An analysis of the new LROC photos reveal a positive proof of exactly what he gives as the orthodox interpretation – ground disturbances and lava flows.

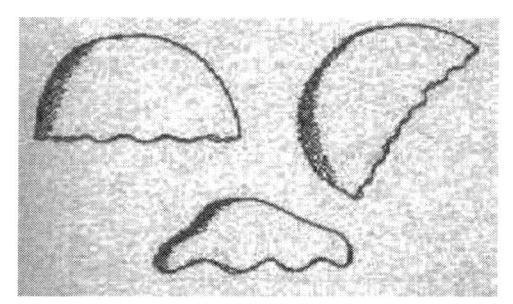

Leonard's Sketch, p.61
Some munched on pie chunks left by aliens.

On pages 61 and 124 (and plate 9) Mr Leonard publishes the Lunar Orbiter-V medium frame number 125 (NASA 69-H-1206) demonstrating what he calls

Stellar Sterility

domes, constructions, stitches, crater coverings and a large screw on the floor of Tycho.

The following are comparisons between his Lunar Orbiter enlargements and the new LROC higher resolution pictures of the exact same locations. You will notice immediately how easy it is to take high contrast lower resolution pictures and make ludicrousness out of lava flows. Believing domes are homes and crevasses are constructions is just too screwy for thumbs down people too privy to such down right dumb theories.

LROC M190672344RC_pyr

In (A) one can see the Lunar Orbiter picture appears to present objects, but with closer inspection the LROC reveals there are no constructions nor is there any large screws in parameter (B). The supposed screw is actually a series of rough mounds that when extremely highlighted and shadowed (in the Lunar Orbiter frames) looks like a series of screw threads.

Stellar Sterility

The domes he speaks of that are located at one end of the lava flow rille (large crack) are actually crater ejecta. George's theory is so dome, that it should rille crack anyone up after close inspection. Here are some rille cracks and crevasses to consider carefully.

Lava flows and ground disturbances. A. Constructions and "stitches." B. Large screws in square boxes.

Stellar Sterility

Leonard's' LO V-125-M and H2, and the LROC Frame M129369888LC_py

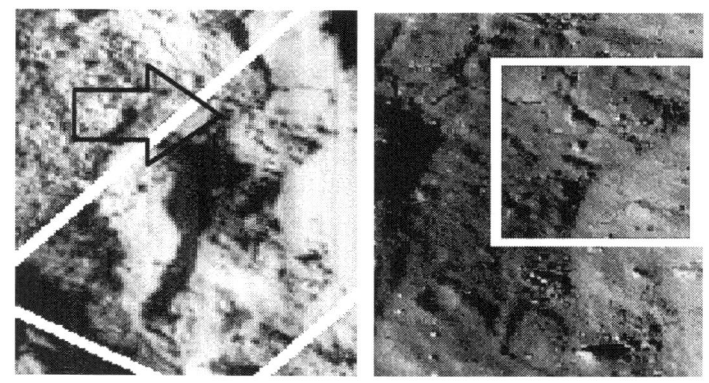

LO V-125-H2 compared with LROC
Supposed "screw" in white box is not there

6
WHITE RAYS OF TYCHO

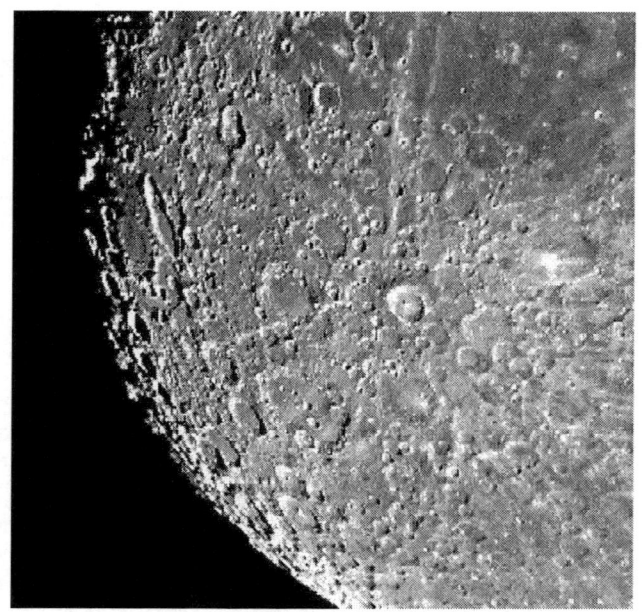

CRATER TYCHO

    The "white rays" of Tycho are a puzzle to George Leonard, maybe because he never bothered to read a lunar geology book. He said they are of alien manufacture, because they do not fit the geological model of crater development, such as a big fat asteroid smashing into the Moon. He gives a long diatribe of 10 counter arguments of why all the rays are artificial.

    He said they are too big, too long, some not radially ejected from the center, many overlap and cross each other, others are tangential to it, and some end with a small crater. Nevertheless, he does admit that the rays have the same albedo as the white rocks and dirt from inside the Moon's crust. Of course, Mr Leonard, what else are they be made of?

George spends several pages trying to prove from NASA controversies over their makeup, that they are not naturally formed, and that *"the old explanations cannot be correct"* (p.108). Therefore, he gives us a new explanation for the crater rays.

The Moon of course, according to George, has been occupied for thousands of years by aliens. To explain the criss-crossing of the rays tangential to the center, he said, alien spacecraft caused them by leaving dust trails. As the UFO's fly back and forth, in all directions from the crater — for God knows what reasons, the dust sticks to the craft until it is loosed upon the surface and left behind. The dust then settles down to form all these weird directional white rays (p.110). The UFOs must have statically carried a heck of a lot of white regolith powder one minute and then somehow dumped it all as they traveled.

He goes on to quote a NASA preliminary scientific report from the Apollo-12 mission that talks about the static nature of the dry lunar dust sticking to objects, but eventually detaching from such places as the bottom of the Apollo landers.

It seems that no logical geological process can account for these anomalies of the rays. The directional nature defies all present theories - so far. They are, he said, still *"seriously debated today"* and *"we are still very much in the dark as to how the Moon's craters were formed"* (p.113). Furthermore, of all the lunar anomalies, the moon rays are the most baffling.

Google searches and photographic studies reveal that the cause of the rays are natural lunar and astrophysical processes and by nothing alien.

Dr Ray Hawke contends, "*The nature and origin of lunar rays have long been the subjects of major controversies.* [Nevertheless, in recent times, since the 1970's] *We have determined the origin of selected lunar ray segments utilizing Earth-based spectral and radar data (from Clementine, etc.)* [And found that the] *lunar rays are bright because of compositional contrast with the surrounding terrain, the presence of immature material, or some combination of the two. This fresh debris was produced by one or more of the following: [1] the emplacement of immature primary ejecta, [2] the deposition of immature local material from secondary craters, [3] the action of debris surges downrange of secondary clusters, and [4] the presence of immature interior walls of secondary impact craters*" [2]. The above definition explains the alien crater rays.

What else is going on in Crater Tycho asks George Leonard. He said NASA is supposedly covering up classified details. What we really have, according to our artifact hunter is a crater filled to the rim and brim with astro-industrial waste. George contests that alien gadgets lie on the ground built by very intelligent brains, ones at least as smart as ours.

Shockingly, we are to believe, with the less brains than a donkey, that there are things that look like letters of the alphabet and huge artificial crater coverings. We are to believe there are space vehicles, constructions of all sorts, an octagonal covering with a glyph on it, and on the inside of the rim at about two o'clock, see a long pole like object sticking out from under the lunar surface material (p. 118-120).

Stellar Sterility

Octagonal "oval" shaped covering with "PAF" glyph lettering (p.120)

We are to see other glyphs that look like an A, an X, and a P, and others, which are everywhere. One such object has the letters P.A.F. inscribed on it (Drawing on page 120).

If one looks harder, they may find little wooden alphabet blocks, too. The aliens supposedly used these letters to mark areas for communication, as they are found all over the moon and do suggest letter tagged location markers (p.162). This is pure bunk! They are all photographic development dust particles, hairs, and micro scratches.

Down the rim from the above "glyph" inscribed rock, we see another design with "*overall straightness,*" with the right edge having some interesting objects suspended from its rim; some of the objects have scalloping on the edges.

To the right is a geometrical cylinder-shaped object (p.122). Further detailed reveals no such objects. Here are some structures, he said, that look like oval shaped solar energy panels. The huge oval hemisphere-shaped coverings could be flat panels soaking up energy as electricity converters for the aliens.

The ovals have "*cilia like protuberances spaced at even distances on the underside,*" are perfect segments of circles or ovals, with evenly spaced 'nodes' on the

Stellar Sterility

rims. Some half-orange shaped blobs with cross-weavings or stitches around the edges lay nearby. George said these are remarkable objects, for they appear in more than one area on the Moon. One is a familiar item that tilts at one side, has two nodes, a cord and a bell-shaped object.

Unfortunately for George, it is also remarkable that this can be whatever the imagination wants it to be. Otherwise, unremarkably, we find some misshapen rocks, lava flows and highly contrasted light and shadow demonstrating no alien devices.

The floor of Crater Tycho is a contortion of ground disturbances as can be seen from these close-up shots. For further domes, groans, ditches and stitches let us turn to George's big screw-job, a giant digging device left behind by the aliens – obviously too big to remove.

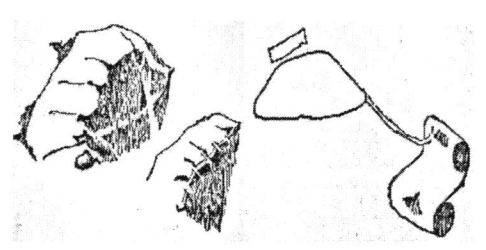

Cross-weaved stitched objects. (p. 123) and an object in the Highlands, north of Tycho. (p. 124)

Stellar Sterility

NASA 67-H-1651

More floor of Tycho Crater Tycho is littered with "alien" looking geometric shapes, all formed out of the geothermal disfiguring of the crater floor.

7
THE BIG SCREW

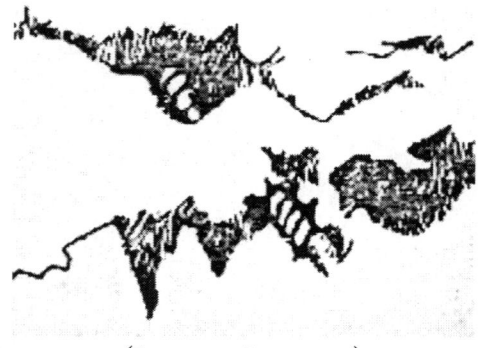

(Leonard p. 124)

There are also abandoned giant alien digging screws. Leonard wants us to believe the aliens had a huge screw job they were dealing with, and that giant twisted stitching screws are stitching together the ripped crust of the lunar surface.

Aliens could be trying to repair the Moon (a giant space ship?) with Cyclopean screws. Why this would be the case on the floor of a giant crater is unanswered. George does not care to show there are plenty of other groovy places to screw with the moon, such as all the rilles, crustal cracks, and splits (long, narrow depressions in the lunar surface that resemble channels) hundreds of miles long. These are much more in need of repair than this small area in Tycho.

Screwing the Moon back together is an enormous weight and labor even for an advanced race of aliens! On the other hand, this alien artifact is just more twisted volcanic geology distorted from twisted observations and logic.

Stellar Sterility

This disturbed area is really formed by lava forced through spiral openings and solidified. Even Mr Leonard agrees that lava can be *"forced through a spiral opening"* to create that kind of shape. However, the location he said casts doubt on that explanation. The location though would be perfect for screwy looking lave formations, since the impact that caused it surly would upset subsurface lava chambers.

The following are the only Lunar Orbiter shots of a screw looking device found so far. Leonard never shows the image area, so these are probably what he was discussing in his book. Their locations seem to fit the descriptions he gives. It is the same geological anomaly from different frames. There are no LROC fames of this spot taken up to this date.

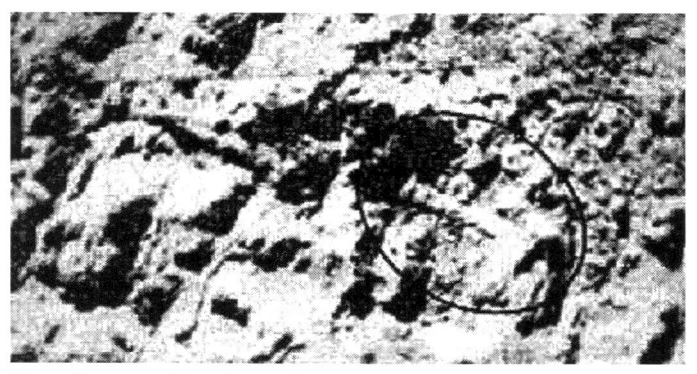

Big Screw Device -1

Stellar Sterility

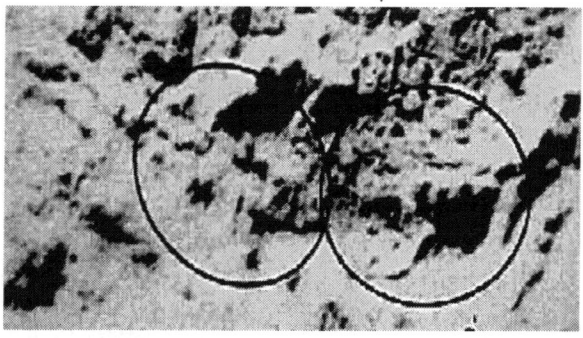

More of Big Screws

LO-V-126-M
The only big "screw" observable in Tycho

Lava can create all kinds of wacky looking contortions such as screws, twists, blobs and geometric patterns, and rectangular shapes, even square columns and blocks. Here is a fantastic lava formation that could represent alien buildings and columns on the Moon if seen from a far.

Fingals Cave, Staffa, Scotland illustrates organ-pipe type basalt columns topped by a mass of rock, the result of different rates of cooling of basalt lava when it erupted about 60 million years ago. Now, increase the size 100 times and place them in a crater and take a blurry picture orbiting the Moon, and you might see alien structures.

Stellar Sterility

In the highland areas north of Tycho, Leonard said there is an area *"seven square miles"* in size..."*ALIVE with construction and activity.*" There are *"symbolic glyphs," "artificial objects," "some power source plates,"* some with knobs and a cord, and some spaying gas, etc.

My God, it is a Disneyland of Alien activity! He said, it is a construction site with *"advanced technology"* that stands out vividly and *"can be taken for nothing else"* (p. 127). What else can we understand it to be? Can it be piles of odd shaped dirt and rocks?

It is interesting that the earth also has examples of such "screw" shaped lava tubes in nature. Sleeping Pele on Big Island, Hawaii is just one example of "twisted" lava flows.

(Left) Fingals_Cave_Staffa_Scotland. Organ-pipe type basalt lava.

(Right) Sleeping Pele is an example of "twisted lava" flow on Big Island, Hawaii.

8
CROSSES OF KEPLER

LO-III-162-M (NASA 67-H-201)
An oblique view of crater Kepler: Location 8 degrees N, 38 degrees W. See LROC photo:
NASA gov/images/content/513108main_012511b.jpg]

Crater Kepler is also another interesting location for galactic garbage gatherers. It is located northwest and next to the Sea of Insularum. It is west of Crater Copernicus and Reinhold. See NASA Apollo-12 photo, AS12-52-7747. The blocky rim, radial ridges, and field

Stellar Sterility

THE CRATER KEPLER (LROC)
[nasa.gov/images/content/513108main_012511b.jpg]

of satellite craters are typical of young impact craters. Kepler and its related features are superposed partly on mare and partly on rugged terra that may be part of the outer rim of the Imbrium Basin.

Insularum consists of many bright RAYS or ray systems. This is perhaps the best location of Leonard's lunar domes, situated north of Crater Hortensius. Other domes lay west about the Crater Milichius that are seen with a telescope. Leonard said that these "spraying" craters are always associated with crosses. *"The crosses are perfect, (and) intersect in the exact middle, and most of the time are tipped with one end on the ground and the opposite end rose up so that they cast a shadow"* (p.65).

The angle is very oblique and raises questions as to whether the shape is actually a CROSS or is a physical anomaly caused by surface shape and angle

Stellar Sterility

distortion, a regular shadow and light play in the lunar mission photos. These examples of crosses, he said are not Latin or Celtic, but intersect exactly in the middle, and are usually tipped on one end with one end sticking up into the air at an angle, suggesting an alien and artificial design and placement. No natural lunar chunk of ejecta dirt would place itself in such a position, at least according to Leonard. He also said there are other crosses of different shapes on the moon: *"IT (the Moon) ABOUNDS WITH THEM."*

There is a cross near Crater Kepler, four miles in length that sticks up at an angle of 1/2 mile. It is in a triangle or square area and looks like a Roman shaped cross. The following are some of his drawings of lunar "crosses."

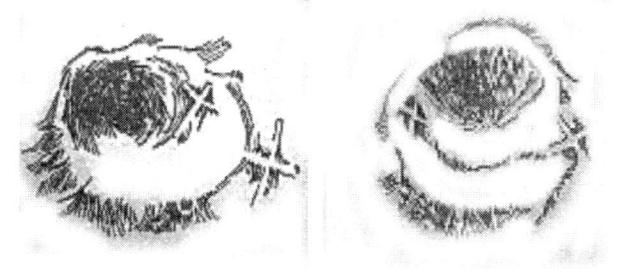

Other "crosses: scattered across the area

He said, *"Outside almost every crater being sprayed out, a gleaming cross can be seen abutting the rim. Possibly every spray crater has a cross"* (Leonard. p. 132, photo 67-H-201). Notice what Leonard said is a cross-shape sitting next the Crater Kepler is actually another natural geological formation. Nevertheless, his drawing helps to mold the mind to see what is supposedly in the photo. Yet, what is actually visible in the enlargement is an irregular shaped rock "boulder" with a semi-flat surface lying at an angle.

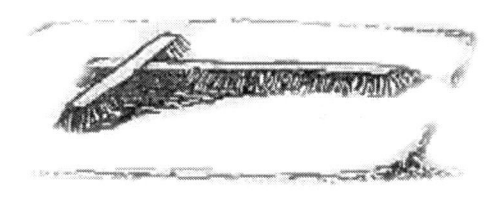

LO3-162-M
Leonard's Latin shaped cross lying at angle. (p.62)

What exactly are these cross shapes? Should we claim they are artificial religious structures before we investigate to see whether there is some natural explanation? What do NASA's lunar geologists say they are?

If we enlarge the 67-H-201 photo, what we really see is an illusion that resembles a cross, derived from a low angle sun highlight that casts a funny shadow, yet not that odd that a reasonable eye cannot detect true geology.

Look closely at the shape of the DIRT MOUND or odd shaped blob. It appears that it is just ejecta mass most assuredly ejected out by one of the surrounding impact craters. After all, the area is known for its crater rays and ejecta debris. Therefore, it is most probable that this is some sort of ejected mass rather than some artificially designed monument. The Gazetteer's reproduction of the Kepler area actually shows the extent of ejecta material and from an aerial view.

Crater Kepler has spewed out ejecta in ALL directions (like Crater Tycho) and any form or odd design can take an alien shape when it comes dropping down. Some masses fall upon others causing odd

Stellar Sterility

clusters, which when seen from an angle may LOOK artificial.

Other artificial structures are visible around the rim of Kepler. Alien lunar landscapers see all sorts of geometric shapes, buildings, walls, platforms, tilled fields, strip mining areas and other anomalies. Notice what we can manufacture as alien activities if we use our imagination.

From this angle, Kepler has all kinds of geometric looking shapes

Large flat platform
1. Shelf
2. Tilled field
3. Geometric blocks
4. Mine pits

5. More pits
6. Tilled field

9
THE BLAIR CUSPIDS
EXTRA-TERRESTRIAL TOWERS

Lunar obelisks are another oddity to arm chair geologists. What is an obelisk? An obelisk or "pointed pillar," is a tall, four-sided, narrow tapering monument , typically having a square or rectangular cross section and a pyramidal (some geometric shaped) top, set up as a monument or landmark. They can range in size and height from a few feet to 555.5 feet.

There are at least two sites with what appears as obelisks on the Moon. One site was found by the Russians and the other by Americans.

On Nov. 20, 1966, Lunar Orbiter II captured an area with six statuesque and mysterious shadows. It is a clustered area of rocks the selenographers call obelisks. They are name the "Blair Cuspids," after the researcher William Blair, who found the mysterious spires. They are in the UPLAND basin area, about the western edge of the Sea of Tranquility, near Ariadaeus Rille. The area is a wondrous place of weird highlighted and shadowed shapes and other anomalies. A meandering monument hunter can go wild in this area.

Well, now we know 'what' an obelisk is, but what does an obelisk mean? What did the ancients understand an obelisk to mean and what is so important about them?

Obelisk making evolved over a long period of time. They were first built around the time the sun cult of Re became central in Heliopolis in the 4th Dynasty.

Stellar Sterility

Therefore, the ancient Egyptians were erecting obelisks around the time when they were building the pyramids at Giza. The first really tall obelisk was built by the Middle Kingdom pharaoh Senusret I or Sesostris I (1971-1926BCE, 12th Dynasty at Heliopolis. It was 65 feet tall and was cut as a single block of Aswan red granite. History shows that the ancient Egyptians perfected obelisk making over a period of more than 1,000 years.

A   **B**

(A) In the center of Saint Peter's Square, there is the 4,0year-old Egyptian obelisk, erected at the current site in 1568. The Egyptian Obelisk was brought to Rome by Emperor Caligula in 37 AD/CE. (B) The obelisk was originally erected in Egypt by a black African pharaoh of the 5th dynasty of Egypt between 2494–2345 BC/BCE during the Old Kingdom. The obelisk in the photo on the right is in Washington D.C.

What did an obelisk mean to the Egyptians? The ancient Egyptians didn't write a full explanation down, so we can only surmise. Stephen Quirke, in "The Cult of Ra," writes that obelisks were mainly associated with the sun. They emerged in Heliopolis at the time when the sun cult became central. Nevertheless, he feels that they might have had phallic meanings too, and that Egyptians associated both ideas, because both were considered full of creative power.

Stellar Sterility

The ancient Greek historian, Diodorus, reports that Queen Semiramis erected a 130-foot obelisk in Babylon and it was associated with sun worship and represented the phallus of the sun god Baal or Nimrod. Some Masonic researchers say that the word 'obelisk' literally means 'Baal's shaft' or 'Baal's organ of reproduction'.

Pagan religions love to worship the creature rather than the Creator (Romans 1:25); and much of the onus is on the sex act, hence most temple priests were "*temple prostitutes.*" Both Satanism and Freemasonry absolutely love the Obelisk. The Egyptians created the obelisk, believing that the spirit of the Sun god, Ra, dwelt in there. (3a)

Satanists and Invisible Masons depict the obelisk as the erect male organ. According to one Satanic author, "*The lingam [male phallus] was an upright pillar*" (3b) Masonry admits these pillars of the obelisk were used to represent sex. (3c)

In the Bible, we see the 'Baal's-shaft' in use in Baal worship. Listen to the account in which God destroyed the perverted pagan phallic obelisk: "*Jehu said to the runners, and to the captains, 'Go in, smite the Baal worshipers, let none come out;' and they smite them by the sword, and the runners and the captains cast them out; and they go unto the city, to the house of Baal... and break down the obelisk [standing-pillar] of Baal, and break down the house of Baal, and appoint it for a drought-house unto this day*" (2 Kings 10:25-27).

Needless to say, if these lunar "phallic-shafts" are real then we have actual extra-terrestrial phallic worshiping, male-organ pillar building perverts on the Moon at least 2000 years before Christ.

The Blair Cuspids are located above Crater Cayley and a little east of Aeriadaeus B (at 15.5 degrees), and

Stellar Sterility

north about 1 degree above a secondary small crater (unnamed), between two tiny craters (see red square above). Using Lunar Orbiter and Clementine imagery, and especially LROC co-ordinates, they are actually located near 5.1 deg. N by 15.5 deg. E, which is within the footprint area [3].

The region of the moon where the shadows turned up is just to the western edge of the moon's Sea of Tranquility. It is an area just north of the moon's equator, slightly to the east or right, of center.

This is near the area of the Apollo 11 landing site. The Cuspids were accidentally found while doing a lunar site selection survey by Lunar Orbiter II.

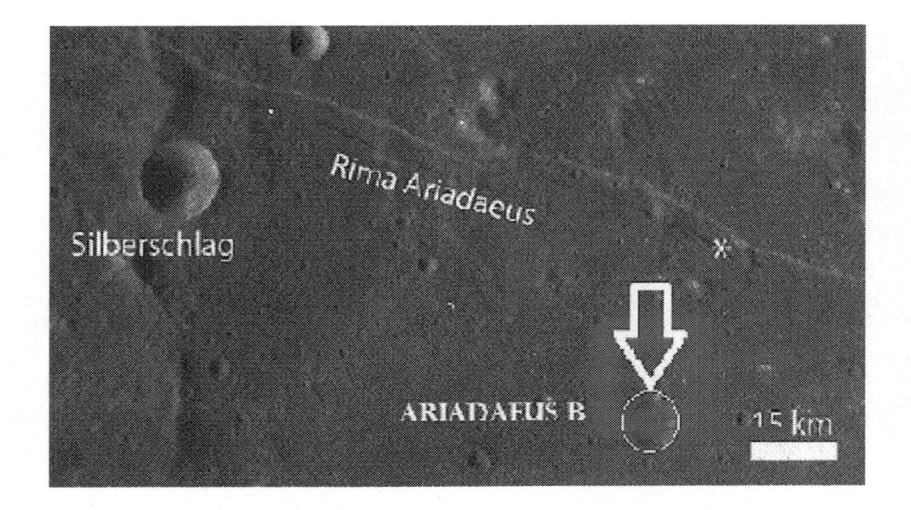

Stellar Sterility

The principal photographs are LO-II-62-M, LO-II-062-H3 & LO-II-61-H3. Never mind the two "white swimming pools." The shadows cast by the Cuspids in Aeriadaeus B

Notice the group of objects with TWO bumps or raised towers. They are casting shadows like the Washington Monument on a sunny late evening! Mr Leonard noticed that the shadow of one obelisk casts over a crater obstructing the sun light on the rim.

The Cuspids consist of eight pointed spires or obelisks. They are not directly visible in the photos, but the long shadows are very visible. The largest of these according to George is 50 feet (15m) wide at the base

Stellar Sterility

and about 70 feet (21m) high, while the largest is about the height of a 15-story building.

NASA changed the ID numbers of all Lunar Orbiter photos. LO-II-61-H3 became the new number replacing 67-H-758. There was another image on the Orbiter's' footprint. LO-II-62-H3 included an overlap, which also showed the Cuspids! The Lunar Orbiter-II photo ("M" or medium shot) shows the area north of Cayley and the crater Aeriadaeus B, and the Cuspids are in the bottom right half of the frame. The features stand out as bright points of light against a dark sloping lunar surface

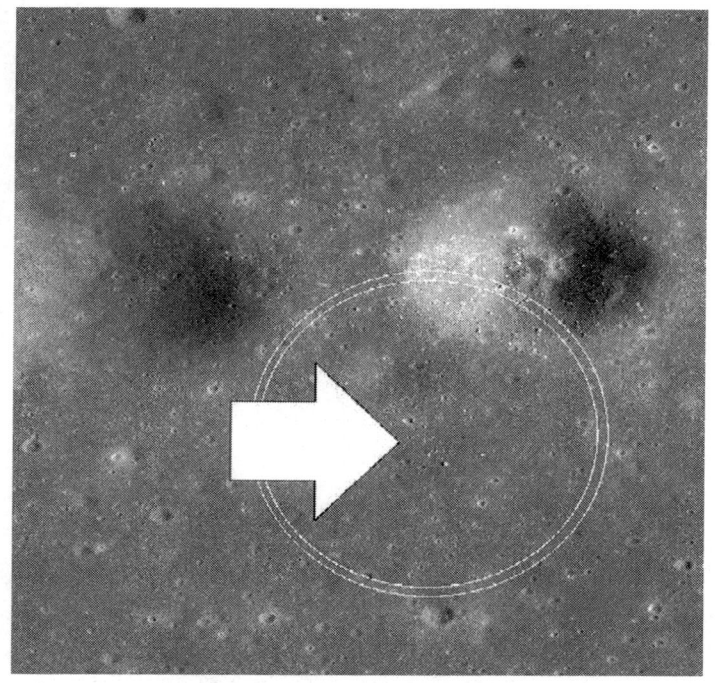

LROC M159847595RC_pyr
(The two small craters and the location of the obelisks.)
Notice in this high sun angle that there is very little "long" shadows. The obelisks are not as tall as was thought to be.

Stellar Sterility

LO-II-61-H3, Leonard's 86-H-758

Enlargement of medium shot This Lunar Orbiter image shows a part of the moon surface within a crater upland basin where there are unusual large protuberances.

Mr Leonard said the objects represent an edifice of intelligent construction for living and working," or a vehicle of some kind, maybe even a form of life (p.96). Sorry Mr Leonard, they are not either, nor are they obelisks, stone phallic monuments or some form of life. They are simply large boulders as we shall see.

Nevertheless, one scientist described the needle-like shadows as the moon's *"Christmas tree effect."* Another described it as the *"Fairy Castle."* One scientist called the region the "Valley of Monuments" [4]. We suggest that it is just a "fairy tale."

The area presents what appears as a complicated geometry of some sort. Leonard said; "It is not enough, nor is it truthful, to simply say that the picture is 'interesting'. It is beyond the realm of possibility for the objects in it - to be of natural origin." His explanations "tower" above all else that have written upon these supposed alien monoliths.

He observes that the bigger objects *"all have two symmetrical bumps"* or *"double points"* and that the

Stellar Sterility

towers *"are roughly equidistant from the middle tower."* Furthermore, the outside objects are *"all single-bumped and are spaced like sentinels."* He said there are markings, tracks and conduits for maintaining atmosphere or communications (p.97).

George suggests that some of the bumps resemble "eyes", while others look like sensors. We could even take his proposition further and say that these positioned rocks look like the Cheshire Cats smile. Now, after viewing the drawing, go back up and see if you can "see" the smile without the white lines. The brain is a funny animal and it see just what it wants to see.

Are they Christmas trees? Yes, people have seen the newspaper photos and referred to them as such. George figured them as more than just trees. They were vehicles of some kind as indicated by the markings on the ground. They are devices using *"connecting cables"* and making *"tracks."*

Be careful with the imagination, for Mr Leonard warns us *"This is strictly opinion, of course - we have no right to inflict our values of design on the occupants*

Stellar Sterility

of the Moon" (p.98). Nevertheless, he goes on to inflict his values of design on the objects and say that "*these bumps have rounded appendages* (and) *may even be having sexual contact!*" (p. 100). It is also a good strict opinion that we should be careful not to inflict our values of design on the "geology" of the Moon – not to mention playing "gametes" and inserting sexual organelle into regolith where they do not belong.

Furthermore, other clues point to these objects as being a form of life or some mechanical substitute, because the appendages are of varying sizes, groupings and have eyes. However highly unlikely an answer this is, he said, it is more an unlikely answer that they are boulders. Well, as we shall see, it is more an unlikely answer that they are anything other than boulders!

No, they are not boulders, Christmas trees, eyes, mouths, towers or croquet balls. They are pyramids! This is what Argosy Magazine said. This rag-mag's argument goes that numbers 3, 4 and 5 are in exactly the same arrangement as the pyramids of Cheops, Chephren and Menkaura at Gizeh, Egypt. "*There are quite a number of strange material things that strongly suggest some extra-terrestrial origin or influence.*" However, they are not pyramids either. Could numbers 2, 4, and 5 rather be three giant stones the size of the pyramids?

The Russians have already figured this out, George! Soviet Space Engineer Alexander Abramov came up with the startling geometrical analysis of the arrangement of the objects by calculating the angles. He asserts they constitute an "Egyptian triangle" on the moon - a precise geometric configuration known in ancient Egypt as an abaka. He said, "*The distribution of these lunar objects,*" states Abravov "*is similar to the plan of the Egyptian pyramids. The centers of the spires in this lunar abaka are arranged in precisely the same way as the apices of the three great pyramids."* [5]

Mr William Blair, who found the "stones", said that they have *"geometric patterns"* and possibly represent *"prehistoric archaeological sites like those in the Great Basin of the arid West and the Great Southwest."* He said, *"If the spires were the result of geophysical events or forces,"* then *"one would expect to find them distributed randomly"* [6]. The geometry also included a *"large rectangular shaped depression or pit,"* indicating *"four 90 degree angles"* like the profile (walls) of a pit structure.

Most astrogeologists believe the spires are the result of some geophysical event and discount the ravings of lurid lunar pyramid hunters. Whatever the 'spires" are (and we now do know what they are), they all agree that the towers are *"natural phenomena"* based on some geophysical event, and likely rock outcrops or large ejecta fragments (boulders) deposited by meteorite impacts or volcanic material ejected through faults in the moon's crust.

The structures might also be eroded cones of old volcanoes - an interpretation Mr Leonard calls *"most unlikely."* Well, finally, the man is right at least once in his book! The pyramidal towers are actually not tall, say the USGS scientists, but are much shorter. The "towers" rest on a surface that is tilted away from the sun and therefore, this accounts for the long shadows. An enlargement of the photo shows the boulders are more stunted, are similar to millions of others scattered all over the lunar surface, and have no more of a geometrical mystery to them than a tossed set of dice.

Stellar Sterility

The following extreme close-up enlargements demonstrate that the sun is further above the horizon, and the LROC camera is shooting at a top-downward angle and records imagery where there are no long shadows. Notice also in the following LROC M159847595RC_pyr photograph that these are only large boulders and not alien towers, skyscrapers, lunar pyramids or any other strange artificial phallic creations. They are simply MOON BOULDERS! The long shadows in the Lunar Orbiter photo images suggested these were huge towering structures, because the sun was closer to the horizon.

As we can see from the following shots the pyramids, towers, ziggurats, temples, eyes, sexual appendages or giant triangle markers are only huge boulders. See the following extreme close-ups.

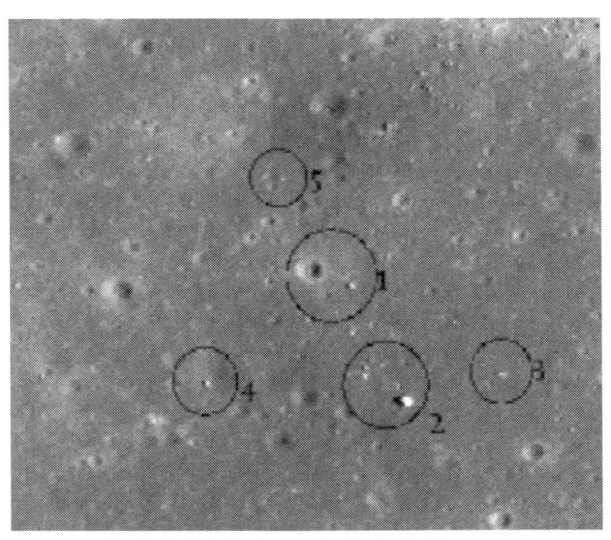

LROC M159847595RC_pyr
BLAIR CUSPIDS (Closeup-1)

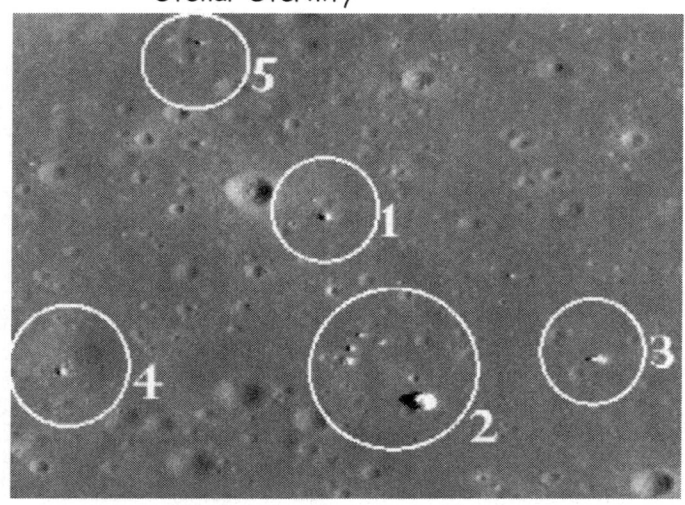

Stellar Sterility

LROC M159847595RC_pyr
Extreme Closeup-2

No. 2 Rock

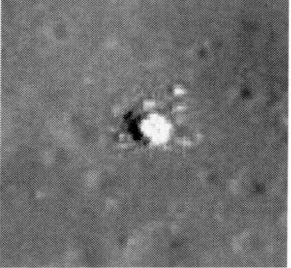

No. 4 Rock

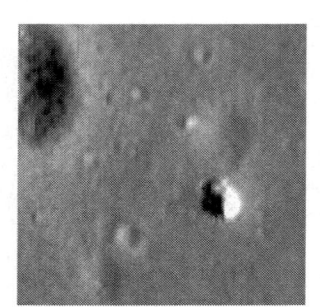

No. 1 Rock (Extreme close up)

10
THE MOSCOW MONUMENTS

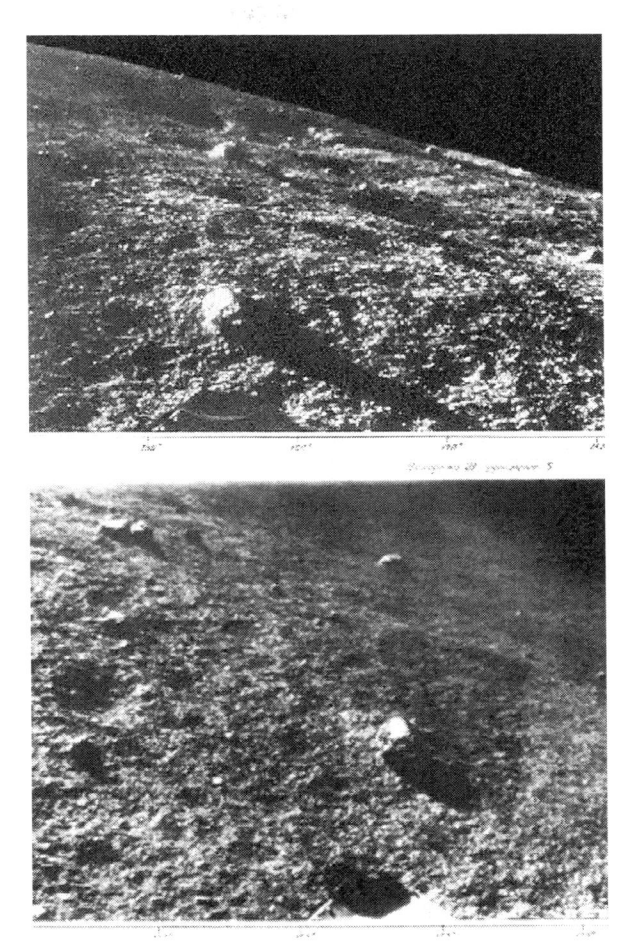

Notice the sun is at a very low angle, causing the object to cast an unusually long shadow. Notice all other long shadows in the photograph! Note also, as LUNA 9 continues to take Photographs, the sun rises changes angle and the shadows decrease in length. As the sun rises further the shadow decreases more exposing the objects true size. The boulder is not as "tall" as some claim.

The Russians say they found a site full of obelisks built by aliens before we did! It seems that no matter what Americans come up with, the Russians need to have their version and they usually claim they are first.

The Russians acquired a set of photos showing obelisks in the Sea of Storms a few months before NASA captured the image of the Blair Cuspids. They are about two thousand miles away from the American site.

These are the Russian Cuspids. Of course, we see that this began a great storm on the subject of what these objects were.

The Soviet probe Luna-9, after its soft lunar landing in February 1966 photographed a series of three panoramas depicting an odd set of stones that appeared to be all the same in size, shape and geometric formation.

The obelisks found by the Russians also presented a strange geometry similar to the American obelisks. The "towers" cast very long shadows suggesting the objects are very tall and thus unusual for lunar geology – even though the!

The three photos presented here show this shadow cast by a small solid object, one of a number of unnatural looking "markers" staked out across the lunar surface by ET's. The photos show an unusual arrangement of stones, many of which are the same size, shape, and set at identical distances apart.

After careful study of the LUNA-9 photos, the Russians concluded that the distance between stones 1, 2, 3 and 4 is equal. The stones are identical in nature and in measurement, they said. They further claim

Stellar Sterility

there does not seem to be any height or elevation nearby from which the stones could have rolled and scattered into these geometric positions.

The objects appear to be arranged with definite geometric laws and mathematical patterns. Some alien intelligence must have arranged them according to definite geometric patterns for a particular purpose.

Leonard claims to prove that the cluster of bumps in the American photo, display a real geometry, just as the Russians claimed for their obelisks.

The monuments are arranged in such a manner that the relationship between the outer objects and the inside cluster are equidistant from the inside mid-point. Leonard mentions that at the time of release, the Russians believed they were artificial and even published an article relating the associated geometry as unusual for lunar ejecta.

Both the American Lunar Orbiter II photographs and the Luna-9 pictures were widely published in the Soviet Union, whereas very few people in this country even consider them. Russian scientists have always been extremely interested in the pursuit of any evidence of extra-terrestrial life. They are now asking whether intelligent beings could have visited our moon long ago and erected these permanent monuments. The LROC photos answer this question.

Are these "obelisks" of alien origin? Who built them? Maybe the Chinese Emperor Yao and his scientists built these obelisks. Maybe ancient Hindu scientists got there first and erected them. Perhaps the Druids build them as a lunar version of their Stonehenge. Maybe the Egyptian gods Ra and Amen, in their heavenly sun boats went to the moon and built some big types of wankers like the ones in Egypt. Maybe it was someone's grandmother. How about

Stellar Sterility

maybe one of those Arabians on a flying carpet? The other scientific possibility is they are natural lunar geophysical results of crater ejecta - "boulders" tossed out of impact sites after meteoric bombardment.

LUNA -9 IS STILL LOST

LO-IRP Restored version of LO-III-214-M
Restored image of Planitia Descendus, possible landing site of Luna 9.

The following image was taken by Lunar Orbiter III in February 1967. This oblique photo shows the region around the crater Galilaei and Planitia Descensus in Oceanus Procellarum (the Sea of Storms). In the upper center of the image you can see the Great Wall of Procellarum. (Lunar Orbiter Image Recovery Project. See MoonViews.com)

Stellar Sterility

Lunar Orbiter Image Restoration Project (LO-IRP), captured this shot of Planitia Descendus (the low hills and ancient craters on the peninsula at the near center along the western Oceanus Procellarum) restored from the original Lunar Orbiter III (LO-III-214-M) in Feb. 1967 [7].

Somewhere in this image is the landing site of Luna 9, the first successfully soft-landed vehicle on the Moon, sent by the Soviet Union and arriving the same month a year earlier, on February 3, 1966.

The Luna 9's landing site is disputable, because the Luna 9 images show a flat region with what may be hills some distance away, though official coordinates (296.0° E, 7.0°N) put the egg-shaped 82 kg lander right in the middle of the rough terrain of the mountains of Descendus. There is obviously not much flatland at that location. So where is Luna 9? There have been many excellent, educated speculations over the years.

Just this side of the largest distinct crater on the flat plain of the Sea of Storms, at the top of this image (west of Reiner Gamma, out of view) is the resting place of Luna 8, which may have landed successfully Dec. 7, 1965, though all contact with the earlier Soviet probe was lost at or near landing [8].

The Russian Luna 9 spacecraft came to a soft landing on the moon February 3, 1966 in the area Planitia Desscentus "Plain of Descent," a borderland along the western edge of Oceanus Procellarum. No one is sure the exact location, but there are possible candidates, while others are not. The best, but still inconclusive spot is located at 8.48 degrees N, 64.47 degrees W. in an area that covers approximately 14,300 square kilometers. The possible landing spot is one of only a few locations with one focusing on an "object of interest" at sunrise (Photos 1 and 2) when the sun was less than 5 degrees over the east horizon. Notice at this time of morning the long shadows allowing for depth perception. In the following LROC photo, we see more than one long shadowed object (Photo 3).

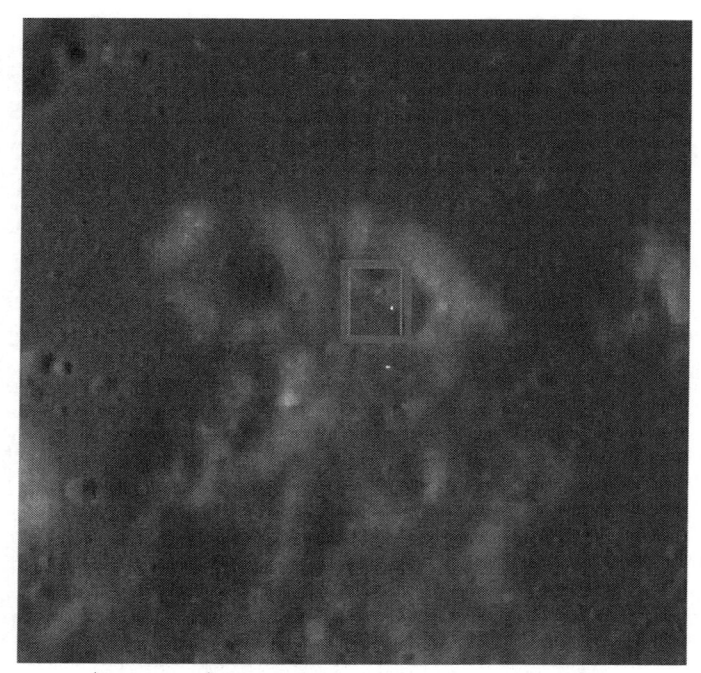

(Photo 1) Supposed Luna-9 Location
M132071202LC_pyr
Planitia Desscentus "Plain of Descent," borderland along the western edge of Oceanus Procellarum.

The following shots are taken from, enlarged and enhanced from LOC NAC M132071202L orbit 4597, taken June 25, 2010 at high resolution. The arrows point out possible locations. The white circle "1" identifies the most probable "object of interest." The object certainly casts a shadow, as anything would

Stellar Sterility

when caught in morning sunshine on the moon. It is also a bit brighter than the other similarly sized objects (2, 3, 4 and 5) caught in this 238 meter-wide field of view.

The object appears to be sitting within the west slope of a 10-meter crater with a darkened interior. Since Luna 9 landed not long after local sunrise, this does not help identify the lander. Another bright object to the southwest (Object 2) is too large to be Luna 9, but its size and location to the circled object in the crater makes a case for it being part of the larger lander bus [9]. Object 5 is also too big, while objects 3 and 4 might hold some future possibilities. These might end up being large crater ejecta boulders making this is a wild goose chase.

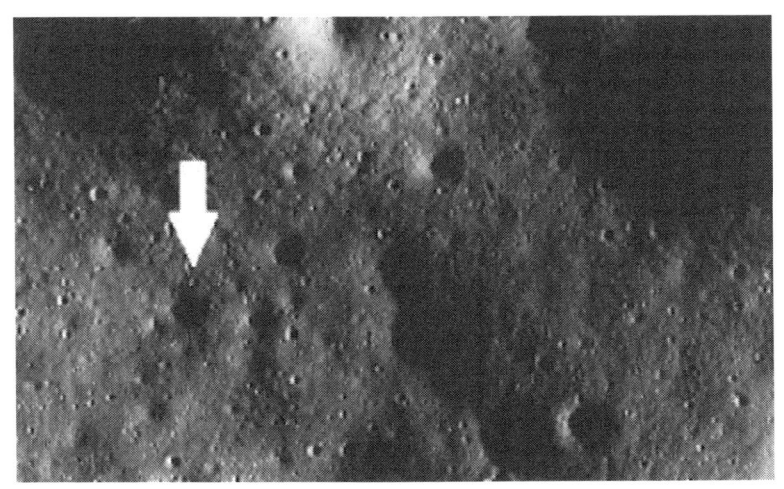

(Photo 2) M132071202LC_pyr

Stellar Sterility

(Photo 3) Possible area of Luna 9 landing site

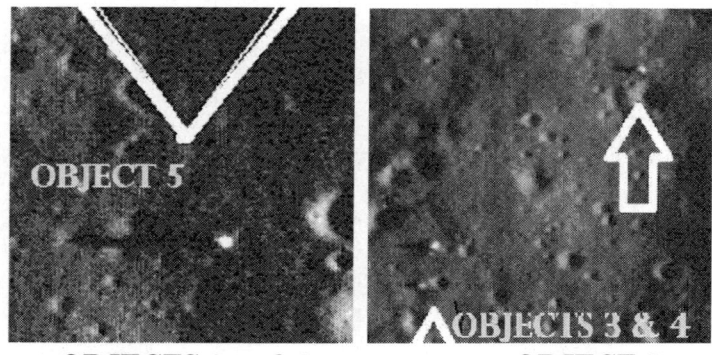

OBJECT-1 Luna 9   OBJECT 2 Lander Bus

OBJECTS 3 and 4   OBJECT 5
M132071202LC_pyr Enlargement

11
THE L-BOW SHAPED OBELISK

67-H-187 (III-012-H1)   Leonard's Drawing p.169 reinterpreted.

If it is not plumbing, mechanical legs, rolling dirt flingers, it is obelisk elbows! The following photograph was taken on Feb.15, 1967 and shows something rising about 100 feet and then for no geologically known reason, turns and juts out a horizontal BAR at a perfect 90 degree angle. Our friend George shows us his drawing of what he believes it looks like from photo blow-ups. Actually, the following photo presents an interesting set of objects that appear to have some symmetry and artificiality.

George calls this object one of the *"sentinels in the wasteland"* and a good example of a single tower *"rising straight up from the ground – not on peaks or highlands."* He continues that in some cases there are towers spaced several miles apart and perfectly aligned. His favorite one is the above LO-III-012-H1 "L-Tower,"

Stellar Sterility

which rises for perhaps a hundred feet or more and then turns suddenly horizontal at a perfect 90-degree angle (p.168).

LO-III-012-H1 High resolution

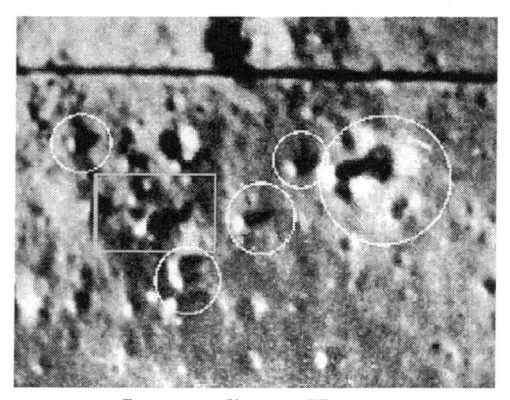

Leonard's 67-H-187
PLATE- 21 L-BAR" is in square. Other "towers" in circles appear to be surrounding pyramidal structures to complement the "L". (Kopal, Pl.308, p.280)

The area (Longitude 35.27 degrees, Latitude 2.71 degrees) shows a cluster of stones, where one tower has a horizontal jut. It shape is like the letter "L". The

Stellar Sterility

object casting it (in the yellow square) looks like a mechanical star shaped 'thing', which is drawn and mapped out. The photo has a directional arrow pointing up, letting us know Leonard published the photo upside down. Nevertheless, the following extrapolations derive from the lunar Stonehenge.

First, we draw a grid with perspective lines. Secondly, we draw the lunar surface and use a 30-degree angle for sunlight. Then, chart out perspective lines and add objects and shadows according to the grid pattern, the shape of the rocks, and the structure and morphology of the lunar surface. The lunar surface in the area has crater ejecta fields and mounds of ejecta running among these boulders. One main ridge runs along the tip of the "L" shaped shadow. This causes a false interpretation of the top of the object.

Looking from another angle, we see the tip of the "L" shadow rises up onto a raised surface area. The topography of the area is not perfectly flat, but irregular. At about 30 degrees, the sun is casting a shadow from the rock that makes it look taller than it is and the tip of the shadow runs "up" and against a raised area, which helps to bend the top. Yet, the top of the rock probably does have an odd shaped "jut" to give it that "L" look. However, it is not as "geometric" and tall as one would think.

Stellar Sterility

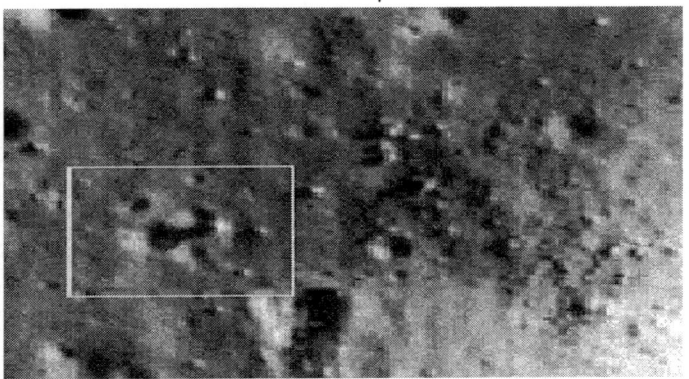

"L" Shaped Tower

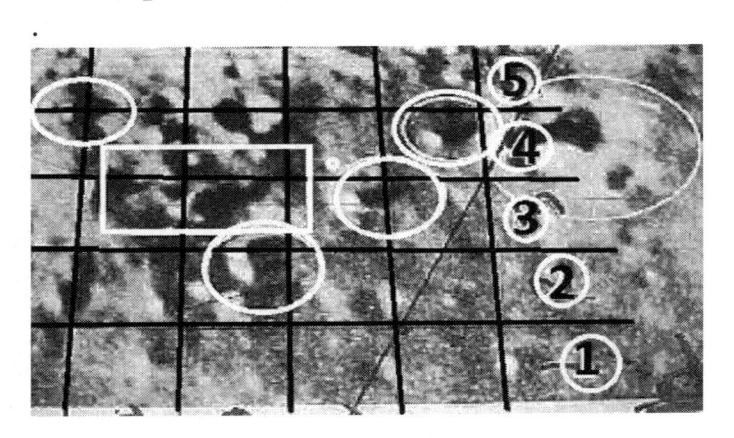

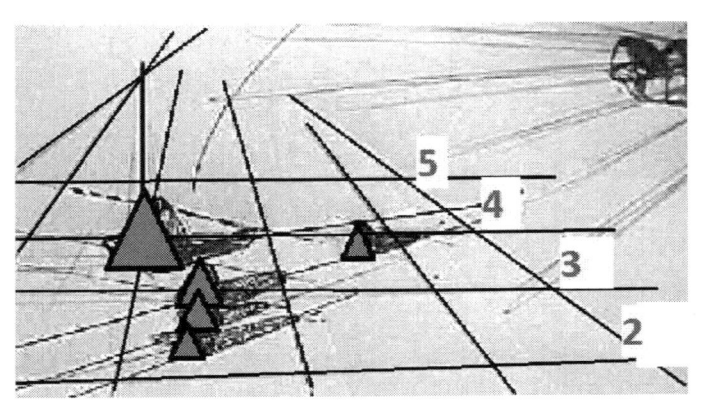

Stellar Sterility

Top View

¾ oblique view

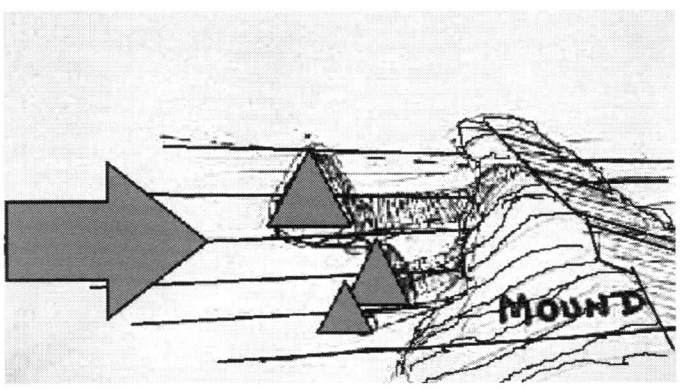

Side view

Stellar Sterility

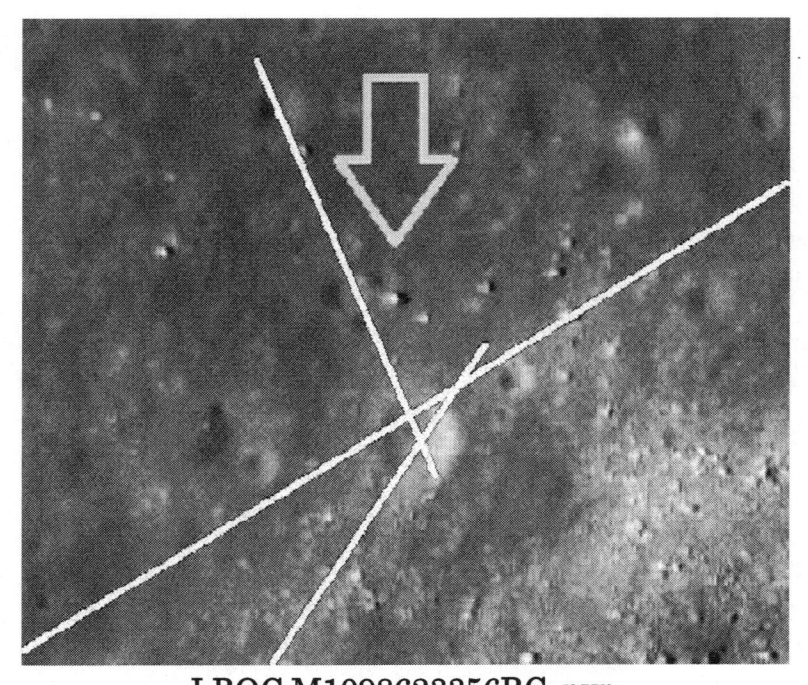

LROC M1098623356RC_pyr
"L" Shaped Tower or "boulder"? See any "L" shaped obelisk?

12
OUT OF "CONTROL" WHEELS

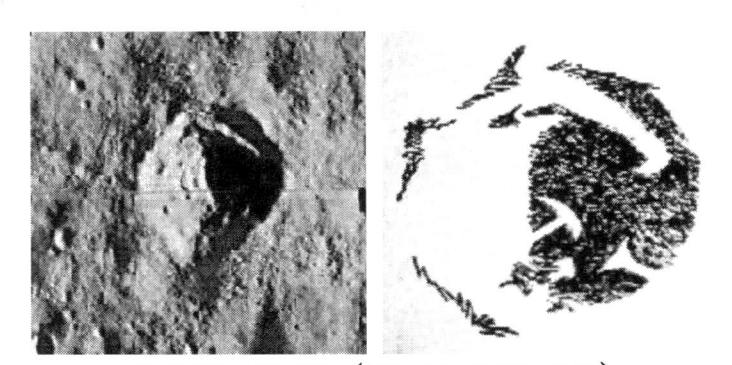

LO-II-182-H1 (NASA 67-H-041)
PLATE 27 Control Wheel in crater. (Drawing p. 179)

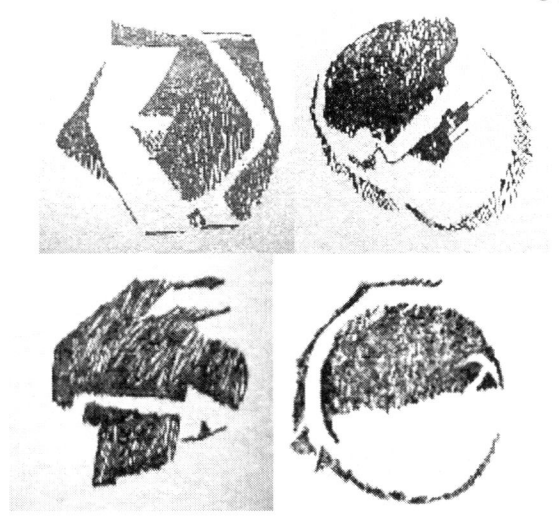

Machine tools.
T-shaped pointed device. Diamond shaped tools.

One of Leonard's' favourite habits is to throw a moon artifact at us, write up a page of imaginative speculation and then refer us to a NASA photo

Stellar Sterility

supposedly supporting the description. Unless you have superhuman eyesight, you have to trust Leonard's out of control interpretations. Such is the case with the following mechanical "control wheel." (p.145)

Southeast of Crater Kepler there is a site with lots of smaller craters and in one of these, George said, has some weird mechanical thing jammed in it. This is the Kepler Crater ejecta site in (67-H-041). The ejecta area around the crater is full of oddities like many craters on the Moon. He said there is one with a "control wheel" in it that seems to have some kind of "shank" or "screw" attached to it. He suggests it is some kind of plumbing.

We have to beg the question further, for only crater rocks, debris, and collapsed crater rims exist. Here are some examples where George jams all kinds of mechanical things into craters for our enjoyment.

After slamming NASA for teasing us with obvious alien artifact photos, George directs us to other photos teasing us with even more details.

This Apollo photo shows more "control wheels" if one can see that close. However, without enlarging the crater, Leonard draws it for us (p. 179). This "spoke wagon-wheel" is near a small crater in the Fra Mauro area. It looks like a wheel lying flat from an airplane's view.

Is it an entrance to an underground community? Is it a doorway to the hollow moon? Maybe this wheel opens some hidden door. On close inspection, it is shadow play along a collapsed crater rim and Mr Leonard is wheeling and dealing a big spoof. A twist here and a twist there, add some fantasy, and POOF! We have an extra-terrestrial steering wheel the size of New Your City.

Stellar Sterility

Humbug, Mr Leonard! If your grandmother had wheels, she would be a wagon too! It looks like one of the thousands of other craters with collapsed rims. Here are some examples of wheels and other oddities in small craters. The material found in the craters is crater debris and nothing else.

13
CRATER RINGS, DOMES AND SPEAKERS

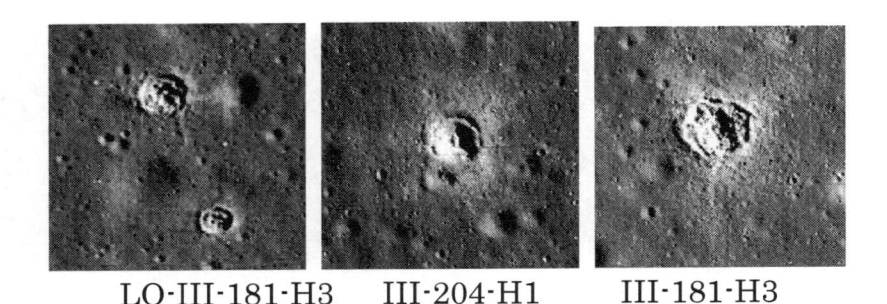

LO-III-181-H3   III-204-H1   III-181-H3

There are not only craters with spoke wheels in them and pieces of plumbing pipes, but there are craters with double rims called "double craters." These are craters with what appears to be a smaller crater on the rim of a bigger one. They can consist of double impacts or collapsed crater rim. The following are some examples:

[Ref. Moon Viewed by Lunar Orbiter (p.69)]

168-H3   III-118-M[2]   III-118-H2

Gruithuisen K   125-H1   AS8-12-2052

Stellar Sterility

We have seen craters with double rings, convex and concave craters, and crater domes. So, how about crater CIRCLES? In previous examples, we saw stones that look like obelisks, stones that look like Stonehenge, stones that roll around, and stones that appear arranged in circles, even stones that roll up and down hills. Some of these sites look like scattered beds of boulders cleared out in the center in certain areas, creating meadows, with a perimeter of stones jumbled around them.

Here we have another geological anomaly that represents craters with rings and round mounds inside. The ringed craters look like giant "speaker cones." Could the aliens be listening to us? They are concentric craters, ones with more than one crater rim, caused by double impacts or other volcanic geological processes.

The double ring or inner "second" ring (as depicted in the above picture of Gruithuisen K crater and the succeeding four photo frames) is more likely the result of successive volcanic eruptions, but could be the result of slumping around the entire wall **[10]**. A combination of both processes may also configure such boulder layouts.

What about domed craters in LO-III photo 118-M? Could they be some alien domed habitat structure? The origin of convex floors or crater mounds are rather thought to have been produced by elastic rebound through stress waves reflected by competent strata or discontinuities beneath (Schultz. P46). As to alien listening devices, it is becoming evident that listening to Leonard's devices is becoming more incredible.

14
BLOBS AND BLACK CROSSES

In photo 67-H-935 (LO-IV-187-H2) we have what appears to be an alien construction site (Leon. p.146 plate 29) abandoned 1000's of years ago, or at least right before the Lunar Orbiter took the picture.

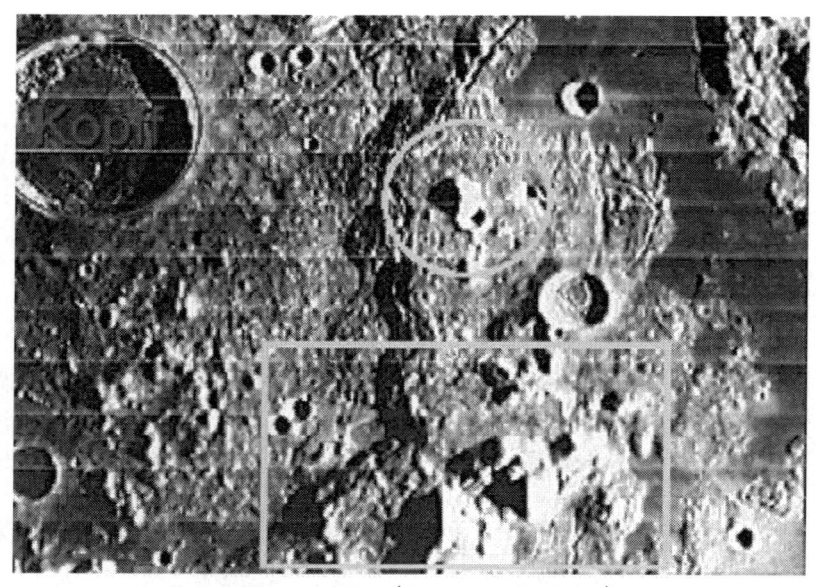

LO-IV-187-H2 (Leonard pl.29)

This site is located in the areas of Mare Orientale, Mare Veris and the Rook Mountains. The Oriental Sea is stuffed full of lunar geological oddities. It presents in extreme close-ups and enlargements every possible geological events that could take place. It is a geologist's dream world of exciting shapes and geophysical processes. There are cracks, crevasses, ridges, mounds, bumps, lumps and lava flows to

Stellar Sterility

mention just a few. In the upper right of Leonard's photo, we cannot help but see a rock with some kind of black cross, stamped on it.

Here at the Orientale site under construction we also find a crater with some blob "mound" covering 1/2 of it. Just strolling about Mare Orientale is exciting enough, even without the alien constructions concept. There are tons of weird (alien) looking constructions scattered everywhere.

There are more oddities here for the geologist to ponder on than Carter has liver pills! For example, LO-IV.187-2 has all sorts of "things" sitting around on crater rims and in valleys between the hills. It is a wonderment why Mr Leonard did not compose an encyclopedia.

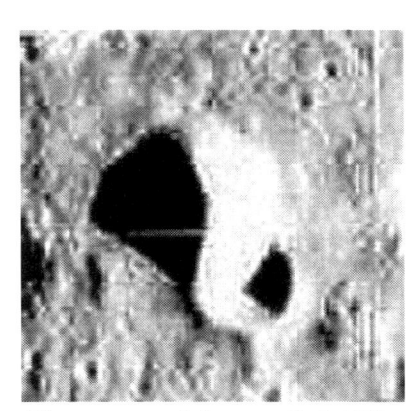

Close-up of frame LO-IV.187-2

Hummocky terrain has numerous mounds, blobs, cylinders and other odd shaped geological features. This one isolated peak (in the above photo) happens to be sitting over part of an older craterlet. They are more likely crater ejecta that have landed on a crater rim or are volcanic in origin. See Steckling section 2

Stellar Sterility

concerning "domes" for a complete analysis of this object.

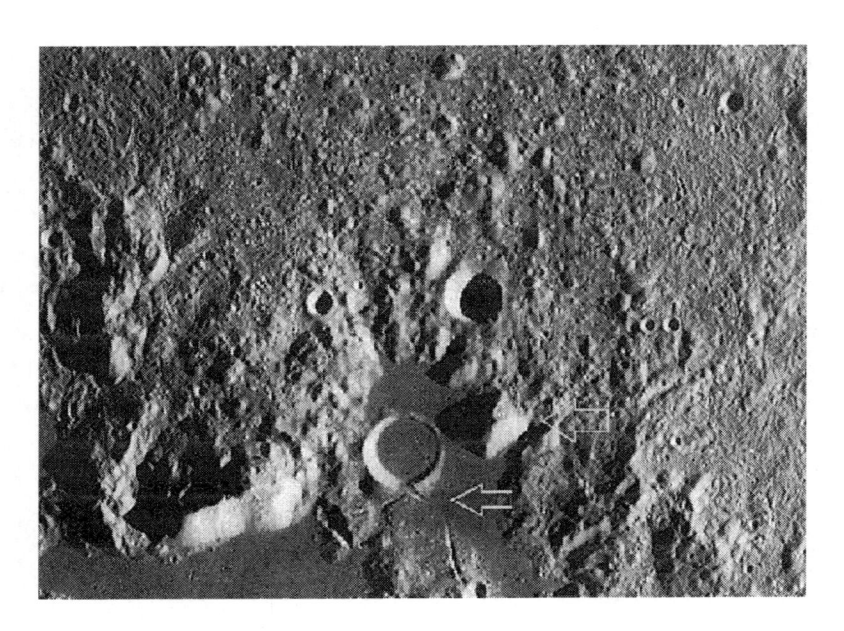

LO-IV-187-H2

These are some of the close-ups of the area. Black crosses, domes, odd shaped mounds, patterns, spheres and shelves, cracks, crevasses and craters.

Are these decent explanations? Leonard gives us his description of the area. See if you can see these things. He said there are parallel walls with an arch between them with sun shining beneath the arch.

There are nodes, raised markings at exact symmetrical spots on one wall, and *"each node (is) on a line with the inside line of the two walls, each an exact distance from the corner, each with the same size shadow."* Could these alien towers actually be crater ejecta boulders? What has the highest probability of

Stellar Sterility

being true: crater ejecta boulders or alien artificial objects? After just a short survey of lunar photography, we have found nothing. If the moon is littered extensively with extra-terrestrial trash as Mr Leonard said, we should find at least one conclusive example of proof. Let us continue, for galactic garbage may not be all that common.

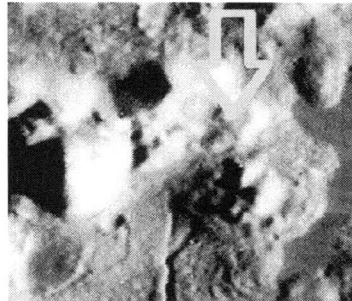

LO-IV-186-H3 (Black Crosses) IV-187-H2

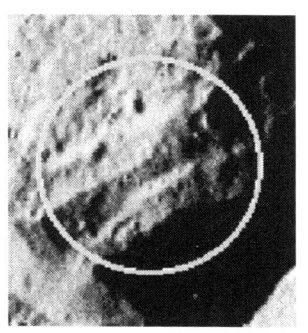

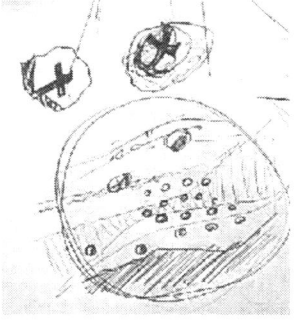

IV-186-H3   Pattern of "spheres"

15
PROCELLARUM PYRAMIDS

AS15-M-2087, 2088, 2089

How about pyramids on the moon? We have seen possible domes, roads, pipes, crafts, machinery and supposed obelisks. So why not pyramids? George Leonard's photo, p.147, plate 30 [from NASA frames 71-H-1300 and 1765, AS15-M-2087 to 2089] is a good example of what could be Egyptian like structures. The three tall and one small pyramid-like towers stand out above all other surrounding objects. They cast long shadows as compared to the other "mounds" and "blobs." These tall piles of planetary pyramids are taller than the previously discussed obelisks. They are located (at the right edge of the photo) of the Herodotus mountain range and Oceanus Procellarum.

Stellar Sterility

Why are these areas interesting to Mr Leonard? He finds in them every kind of construction imaginable. The area, he said, is loaded with *"dozens of mountain masses, a long valley, many constructions and domes."* Obviously, these are alien habitat!

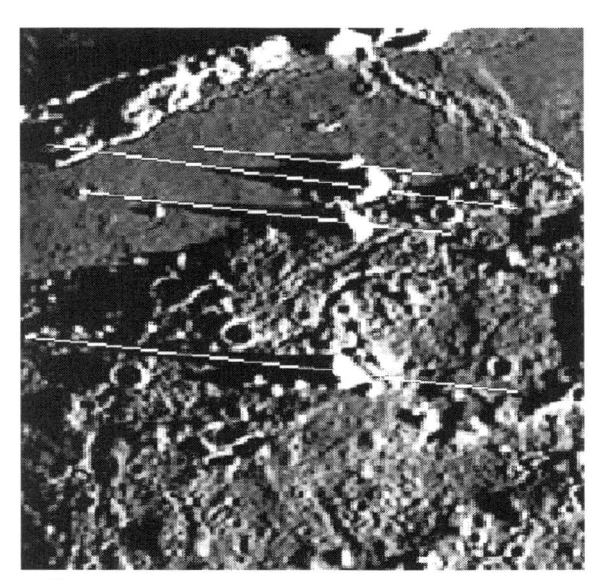

AS15-M-2087, 2088, 2089

Leonard's drawing of the "domes" p.182

He said that the domes are not natural as those created by *"volcanic swelling"* or mantle stresses that are usually *"low and irregular,"* and often adorned with a small crater at the summit. These domes are true mounds and therefore artificially made.

If artificial, then why round and why are they many times built in or next to small craters? NASA has the answer, if Leonard does not. In support of lunar habitation, craters are perfect for possible living quarters.

They are round like a quarter, have a rim like a quarter and are somewhat flat like a quarter. They just as well are "two bits" as domes or pyramids. Nevertheless, they may serve as already made fox-holes to bury the rooms. The builder does not have to spend extra time and alien currency at digging. Mother Moon or the 'Man in the Moon' has done it for us. It is half way already built for us – or the aliens.

This kind of shelter is also good shielding against outer space dangers such as radiation. Radiation on the Moon is very different to that on the Earth because the Moon lacks both a strong magnetic field and a thick atmosphere. Thus, natural radiation protection on the surface of the Moon is non-existent. Radiation hazards on the Moon come from two sources: Galactic cosmic rays (GCRs) and solar energetic particle events that produce extremely high levels of radiation.

These events are serious radiation hazards and the heavy GCRs and the secondary particles require the use of extensive shielding [11]. Moon craters are perfect foundations for this! All that is needed is a little regolith cover-up and George has done a real good job at this – covering up the real facts about ET foxholes.

It is not far-fetched to say that the environment of the habitat influences its design. Regolith (lunar soil) is the perfect shielding for radiation and the "dome" shape is the perfect geometric structure for such a

Stellar Sterility

dwelling; and NASA is in the process of designing these type domed structures for the Moon.

Leonard would have us believe that the moon-monument makers beat us to the punch. Whether they have or not, NASA is engineering dome structures as the perfect shape for lunar habitation and considering putting them in craters. This surly reminds us of those large domes sitting on crater rims and inside craters.

Extreme environments, whether they are the frigid nights of the Polar Regions, the burning heat of the desert or the harsh environment of Selenia, the goal is to provide radiation protection, while also providing an aesthetic living environment.

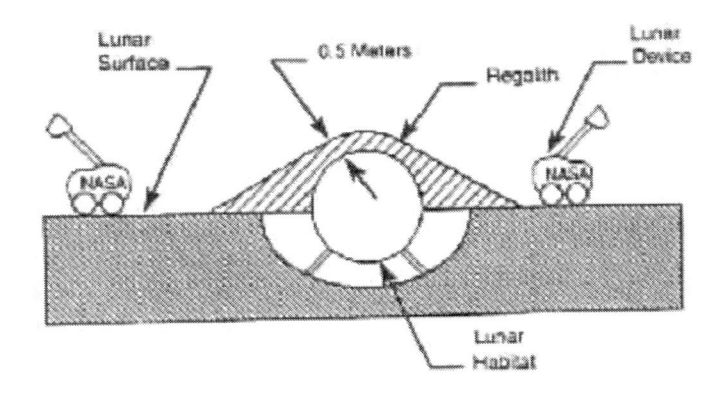

Domed habitat using crater for protection.
("Domed Cities..." p. 4)

NASA said, *"Because of the need to provide for radiation protection, a unique structural system... was created. The system uses cable networks in a tensioned structural system, which supports the lunar regolith used for shielding above the facilities. The system is modular, easily expandable, and simple to construct. Additional innovations include the use of ... various sized craters to provide side shielding. The reflective properties of the fabric... are utilized to provide diffuse illumination. The use of craters along with the suspended shielding allows the dome to be utilized in the Moon's hostile radiation environment... construction techniques for large domes...100's to 1000's of meters - for the design of habitats for long term use in extreme environments*" [12].

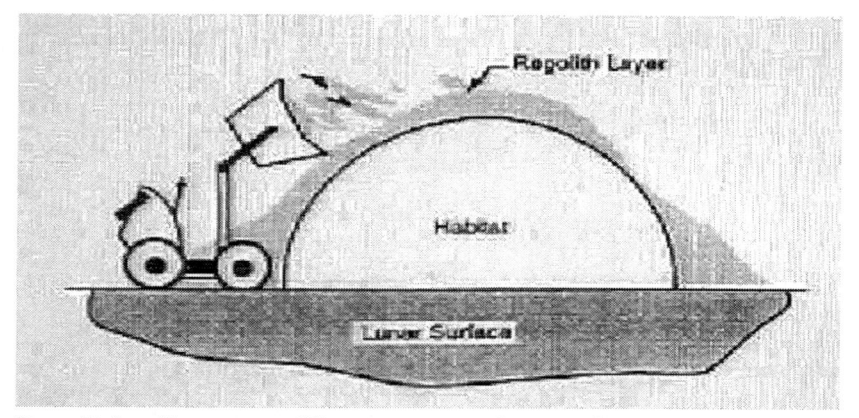

Regolith "lunar soil" covering dome for shielding purposes. ("Domed Cities..." p. 20)

Do we see alien domes in craters in the enlargements of the frames? Closer examination reveals nothing of the sort. There are no Genesis Moon camps, alien airports or crystal domes. When blurred and cheaply published, the lunar photographs shadows and highlights offer an interpretive frenzy to the imagination of dome hunters.

When properly examined in detail, these shadows and highlights reveal a clustered jumble of rock piles, lumps of crater ejecta and mounds of dirt with shadows

Stellar Sterility

and highlights playing "cloud formation" tricks on the eyes of the selenographer. Now properly exposed, Leonardeans should return to their clustered pile of NASA prints and re-examine the lumps, bumps and lunar dumps they call domed houses and reconsider their fancies. They must also answer why the aliens built many of their domes on the "rim" of the craters and not inside. Maybe they missed calculated layout and placement and missed the mark?

Speaking of getting high on the highlands and finding more of these lunar domes, let us now turn to the Alpine Mountains.

Stellar Sterility

16
ALPINE CONSTRUCTION, CO.

LO-IV-115-H3 (See also LO-IV-H-116-H1)

The Alpine Valley is located on the near side of the moon and on the extreme northern edge of Crater Plato. The Alpine Valley looks out upon a vast mare or plain area, while in another direction lie mountains. This is the region of the surface dwellers according to George.

In moon maps and charts, you will notice a big crack close to Plato. It looks like an old scar after a decade of healing. There are supposedly constructions near the Rock Mountains and Schickard Crater in the southeast region. George has come up with the crackpot idea that the above ground builders built domes on platforms around this area and beyond the Ocean of Storms.

Stellar Sterility

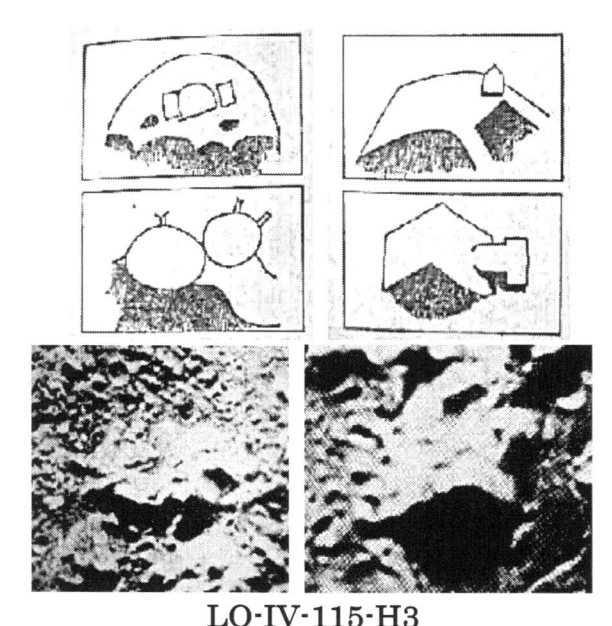

LO-IV-115-H3
Dome, top left; 'sawhorse' dome top right.

Leonard finds many interesting constructions lying around this area, as indicated in the above picture. He calls it the Disneyland of the Moon. In one area he sees an odd shaped "domed structure" (p. 187) sitting on a platform and identifies it as "an abode" rather than a spaceship (p.188). There are many "oval shaped" objects and other "circular" areas scattered about the area suggesting that many of these are dwellings, living quarters and platform style terraces.

Almost all of them said Leonard are not natural but artificial structures - *"Platform after gleaming platform..."* each with its own brand of similar size dome and approximately one half mile high, and some six miles wide (p. 189). They are so big (six to ten miles wide) that you could stuff a small town inside one.

Stellar Sterility

Mr Leonard could have imagined hydroponic gardens and recreational areas as well. Here are some of the structural shapes in the area identified by him as alien dome habitat, platforms and other dwellings.

The Valley formed by volcanic and erosive geological events in the early development of the Moon. Leonard said that the alien occupants keep watch on us with their telescopes from this area. To him, alien artifacts pitched from balconies clutter this whole area. He said, "*A residence of one of the habitats high up on a sculptured platform has a view out on a broad mare, or plain. In another direction there are mountains, and everywhere there are the carved aesthetics so tied to his existence...they live above ground in domes and other sculptured geometrics carved on top of gleaning flat platforms.*"

George then asks whether the following drawing of a weird artificial object "sitting on a platform" intrigues you. He calls it a dome and not a spaceship, because it is funny shaped and is not changed in the photos, taken months apart. In the area, there are ovals and circles rising to beautiful geometric peaks. He goes to conclude that NONE of the mountains and platforms is of natural origin. All the platforms range from 6-10 miles across, while all the domes are about an equal size of 2 miles.

IF you look hard enough, which Mr Leonard neglected to do, you can also find a sculptured "duck" and what looks like a huge animal "skull."

Stellar Sterility

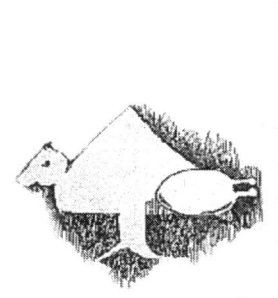

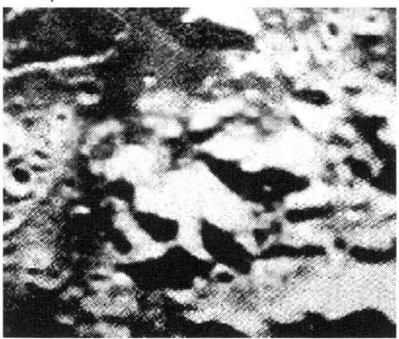

Leonard's drawing (p.188) of alien orb. This is rather, a sculptured "duck" with "egg."

The duck sculpture seems to be intact as opposed to the other, and apparently has some "egg-shape" nested next to its tail, which is Leonard's odd shape dome. The skull's "right" ear has collapsed from decay and erosion, while the nose area has fallen to its left side. After reconstructing and attaching the dilapidated remains to their proper locations with computer graphics software, we obtain the formal original shape of the icon or sculpture of a mouse face.

The only connection found between aliens and rodent-duck iconography, which Mr Leonard also neglected to research, lies in human history. It is obvious that the ancient lunar inhabitants were observing our ancestors, and if the 'duck" and "mouse" icons are contemporaneous; they may still be doing so - the duck icon dates this site as modern. It could be that later (surviving) Selenian generations connected the modern "duck" to the more ancient mouse icon, being that the duck carving is much less eroded and thus appears more recent.

Why mouse and duck imagery on the moon? What could a mouse mean to lunar inhabitants? I shall attempt to give an explanation based upon the

Stellar Sterility

postulation of alien presence. Apparently, if aliens have been on the moon they have been there for a very long time.

The "Skull" or renamed "Mouse Rock."

Rodent iconography found on the moon would indicate the artifacts approximate age relative to human history, placing such relics about the time of ancient Greece and possibly back to at least the last dynasties of Egypt. There were rodent iconography and ancient mouse cults in Greece thousands of years ago, as well as in Egypt, India and other places.

Alternately, there are no such formal mice cults (religions) in the modern age, yet some entertainment images contain the vestiges of zoomorphic veneration. Mr Leonard's suggestion that areas of the Moon are Disneyland-like was prophetic of these newly found discoveries.

Besides the mouse (*vamana* "vehicle") association with the god Ganesha, we have those of the Sminthean cult of Apollo. Historians are unclear as to the exact connection of the mouse with Apollo and are not sure what Apollo's original mouse-smintheus association was. Nevertheless, there are allusions relating to a possible connection to disease and the power to control it. Mice seem to control disease by being carriers.

Stellar Sterility

Smintheus could be a geographic name derived from Sminthia, an ancient city in Asia Minor, which Strabo said is near Hamaxitus. Homer places other early cult centers of Apollo Smintheus in the cities of Chrysa and Tenedos. Depictions of Smithean Apollo show the god standing with a mouse under foot. He is also shown with a mouse in hand.

The only relationship between rodents and the god Apollo is that both caused and averted plagues. It seems, according to the ancient Greeks, that any god who could control mice could also control disease, pestilence and starvation, as well as defeat in battles. Apollo Smintheus protected farmers from field-mice depredations, he therefore staved off famine, pestilence and thus plagues.

Here's what Strabo has to say about the connection of the sminthean mouse (Strabo XIII). "*The temple of Apollo Smintheus is in this Chrysa, and the symbol, a mouse, which shows the etymology of the epithet Smintheus, lying under the foot of the statue... They reconcile the history, and the fable about the mice, in this following manner. The Teucri, who came from Crete,... were directed by an oracle to settle wherever the earth-born inhabitants should attack them, which, it is said, occurred to them near Hamaxitus, for in the night-time great multitudes of field-mice came out and devoured all arms or utensils, which were made of leather; the colony therefore settled there... Heracleides of Pontus said, that the mice, which swarmed near the temple, were considered as sacred, and the statue is represented as standing upon a mouse*" [13].

So far, no evidence of "living" aliens is found on the moon, though tons of suggestive astro-archaeological evidence is supposedly surfacing to the contrary through photographic analysis. Nevertheless, the possibility that aliens visited earth, contracted diseases and viral infections, and even brought them back to the moon is highly suggested in this lunar iconography. Could it be that the aliens, just as Apollo had done, tried to harness the power of Apollo through colossus size mouse iconography (talismans or Amulets) to battle diseases? This icon and the fact that no living aliens have been found, suggests that the Selenians were wiped out by pan-lunar plagues.

The plot thickens: Whence the duck? Could the duck represent survival, celebration and Vanity Fair? It is tempting to think that the plagues were as water off a ducks back and that they have survived to this day, maybe hiding this time underground.

The photographed area under close inspection actually reveals that the above hypothesis is a regolithic load of alien Poo, based on the same hyper-dimensional over the hill synthetic thinking of the Leonard type mind set. With a little wine bibbing and some crackers and cheese, it is not hard to "dream" up ideas out shadows cast from moon light, like a small child spooked in his bed after the lights are turned out - especially after watching a good horror movie.

The area amounts to the same contrasting, varied and congested lunar geological formations. The platforms are just huge areas with bright spots; the domes, either ejecta or swelling uplifts, while all the other "constructions" of supposed ducks, mice and other such zoomorphic fictions are only varied and assorted lunar morphologies, radiating shadows and highlights in a Disneyland of geological deformations.

17
CAUSEWAYS IN CRISIS

AS16-121-19438
Apollo shot of northern rim of the Sea of Crisis

Dr Farouk El Baz, another alien artifact supporter, who taught the astronauts geology, said there are spires on the moon much higher than any constructions on Earth. He also mentions funny unexplainable and unnatural flashes of lights: *"There's no question about it, they are very tremendous things: no comets, not natural."* They were also seen by Ken Mattingly on Apollo 16 and by Ron Evans and Jack Schmitt on Apollo 17. (Leon. p. 150)

Ivan Sanderson, another early alien theorist of Agrosy Magazine, stated flatly that the Moon is littered with constructions everywhere. Many Russian

Stellar Sterility

Scientists, at the time the photos hit the market, saw all kinds of wild things in the photos. Joseph Goodavage (a prolific writer, journalist, remarkable astrologer and humorist) believed he saw alien artifacts, artificial constructions and ancient ruins in lunar photographs! Most of the larger and more prominent artifacts are located specifically around MARE CRISIUM - the Sea of Crisis [14].

Mr John O'Neil, a science editor for the New York herald Tribune reported in 1953 that he saw (through his telescope) *"a gigantic natural bridge having the amazing span of about 12 miles from pediment to pediment"* [15].

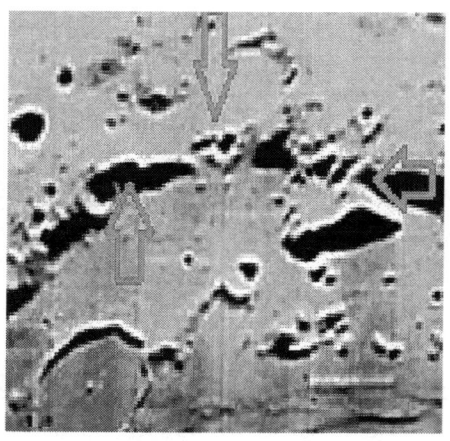

LO-IV-1191-H3
THE CRISIUM CAUSEWAY

O'Neill observed his bridge at the thin neck along the western shore of Mare Crisium just where two promontories come almost together. The rim of a large ruined crater to the left, a smaller bright crater, and two ridges in the mare probably appeared as arcs of shadow and brightness, giving rise to the interpretation of a bridge.

He said the main bridge between the two promontories was *"straight as a die"* and cast a shadow beneath. O'Neill's Bridge was popularized at a time when amateurs thought anything was possible, including UFOs. Percy Wilkins, in his 1954 book, "Our Moon", confirmed the existence of the bridge and hinted that it could be artificial. This, and other embarrassments, was too much for the other members of the association and Wilkins was forced to resign as Director.

Nevertheless, bridges apparently do span huge fracture lines in what geologists refer to as scarps, such as can be seen near Crater Lalande, 50 miles west of Sinus Medii (AS16-M-0849).

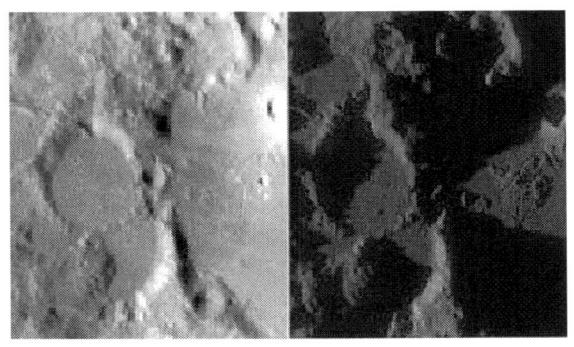

Mare Crisium Promontories at different times.
(Compliments of http://www.the-moon.wikispaces.com/)

Stellar Sterility

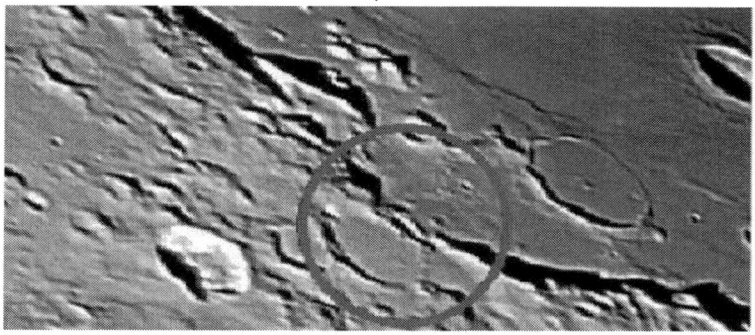

Mare Crisium "bridge"
(Compliments of Alex E. Norman, 'Anacortes Astronomy Club')

The "bridges" appear as two narrow filaments crossing the black expanse of the scarp's trough. Areas of the plains on both sides of this scarp are covered with patterns of parallel and linear ridges, some making sharp 90-degree turns. The general appearance is similar to a high-altitude aerial photograph of a city, connected by bridges across a canyon to the outlying suburbs [16].

Twenty-four years later, Mr George Leonard stumbles upon these same "bridges" in his telescopic observations, while Moon cruising for alien activity. He relocates the bridges that O'Neil saw decades earlier, running along the northern mountainous ridge area of the Sea of Crisis. The bridges he saw were huge.

Mare Crisium is located in the Western Hemisphere of the near side of the Moon. It sits above the equator and looks like a large SEA area, or plains.

It is part of what we see making the face of the "Man in the Moon." To the North and along the lower rim edges of the old crater rim of the Northern Highlands one is supposed to see what appears to be bridges spanning over from one area to another. It is a

Stellar Sterility

jumble of assorted geological mishaps, erosion and weird ridges.

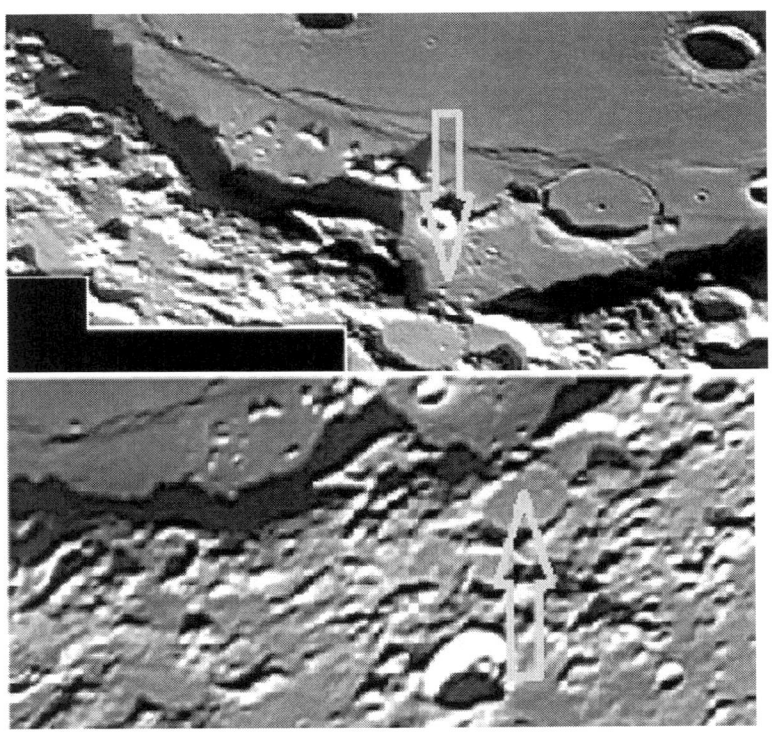

Mare Crisium "bridge."
-- Anacortes Astronomy Club

It truly is "the Sea of Crisis" as far as the Highlands are concerned. George said, "*The entire area is filled with constructions of various shapes which rise into the sky. Some are bent over, not touching the ground. Others touch the ground and become 'bridges,'* (plate 1)." To clarify his observations, he presents what he thinks he sees. Here are some interpretations and the close-ups to match.

275

Stellar Sterility

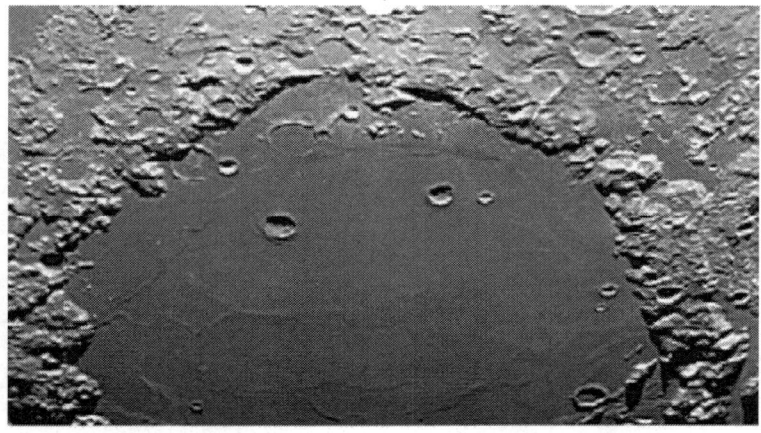

The Sea of Crisis

Nevertheless, it is also a major crisis for the reader to see what he claims. After showing my local astronomy friend what I was looking for he managed to capture some close-ups of the bridge. Yet, from the above NASA Apollo photos just about any ridge, raised area or highlighted odd shaped crater rim can be made into an alien manufactured artifact or bridge.

If one looks long enough through a telescope, he will eventually see what appear to be O'Neil's bridges. It is hard to catch and capture, but it can as the above pictures prove. Nevertheless, what we see are not bridges, skyscrapers, high-rises or decaying crystal domes, but sun light playing off crater rims. We see a beautiful view of the point where Prom Lavinium and Prom Olivium almost touch.

Stellar Sterility

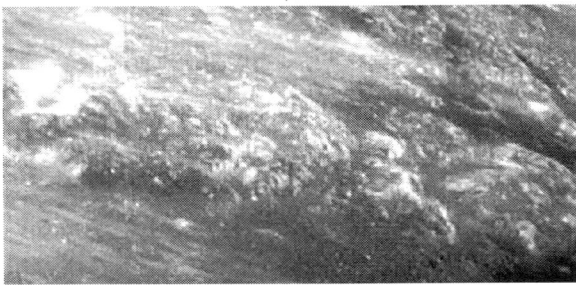

AS15-94-12753
Close-up of Mare Crisium, (Apollo 15)

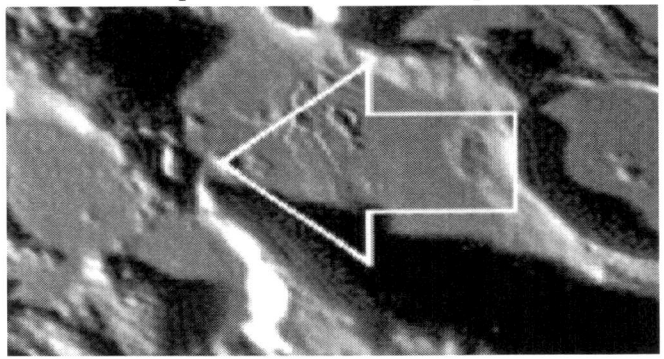

THE BRIDGE   Anacortes Astronomy Club

This is where Wilkins supposedly confirmed his 2 km wide suspension (or portable) bridge. What we see is the so-called bridge between the Promontories ruined, collapsed, and aged from constant micrometeorite bombardment. With a good smaller telescope, the bridge is often seen by the way the sun hits the mountains and craters in the area.

Of course, no one knows exactly what O'Neill saw because the "bridge" does not exist. However, what do exist are shadows and some highlights. A crater on Promontorium Olivium close to the gap can resemble a natural arch under the right lighting conditions, when seen with south up. Remember, significant changes in the shape and appearance of lunar surface features can

Stellar Sterility

occur in as little as a half-hour of watching, so be patient and have fun. Some other views of Mare Crisium can be viewed in Lunar Orbiter photos, IV-061-H2 and IV-191-H3.

18
BRIDGE ON THE FAR SIDE

NAC M113168034R "The Bridge"
Another amazing bit of lunar geology revealed by LROC!

Just when you think you have seen everything, LROC reveals a natural bridge on the Moon! NASA has discovered an amazing lunar bridge on the far side of moon, something Mr George missed. This one just happens to be a real bridge and is visible in the following photos no matter what time of day.

The discovery was made by the Lunar Reconnaissance Orbiter Camera, which sent pictures of the King Crater on the moon's far side to NASA. The bridge is seven metres wide and 20m long and spans a canyon between two and four stories deep. NASA speculates that the formation may have come about following the collapse of a lava tube. Of course, George would have identified it as some kind of alien mining operation.

Stellar Sterility

The LROC confirmed a theory that the moon's surface may hide a vast network of lava tubes — alien tunnels! The discovery of the bridge seems to have further bolstered the theory. Scientists believe that the astonishing formation is the result of ancient lava flows, which left hollow tunnels, which in this case have fallen away and created the bridge like structure, NASA said.

This above photo is one of two natural bridges found within the melt pond region north of King Crater that probably formed by the collapse of the upper cooled surface layer of a subsurface lava tube or chamber [17]. [See also images: M130863593L/R [a, d]; M113168034R [b, c]; NASA, GSFC, and Arizona State University]. The bridge is approximately 7 meters wide on top and perhaps 9 meters on the bottom side, and is a 20-meter walk for an astronaut to cross from one side to the other.

Is this feature truly a bridge? Look closely and you can see a little crescent of light on its floor. That patch of light came from the east, under the bridge. In other lower resolution images, you can see light passing under the bridge from the west. Therefore, there must be a passage.

How did this oddity "bridge" form? Did aliens dig it out and then abandon it? The impact melt deposit on the north rim of Crater King, which is over 15 km across, was emplaced in a matter of minutes as the crater grew to its final configuration. The most likely answer is the surface collapsed into a lava tube leaving a small bridge connecting the sides, while the other sides collapsed into holes. This reminds us of Florida sink-holes that have gobbled up houses and even people.

Stellar Sterility

From the Apollo era and the new LROC images, we know now that lava tubes formed in the ancient past. These images have raised the tantalizing prospects that lava tubes remain intact to this day. In fact, there are possibly many more. The same NAC (Near Angle Camera) image revealed two natural bridges – not just one! Wow! What a great thing to find. We can now design tunnel villages for future lunar colonists and not have to transport digging machines to the Moon, like the aliens foolish did.

To understand lave flow tunnel tubes we have to understand the geological properties of crater development, crater ejecta and lava flow. The impact melt that was thrown out of the crater, pooled on the newly deposited ejecta, allowing its interior to stay molten for a long time. As the local terrain readjusted after the shock of the impact, the substrate of this massive pool of melt jostled to some degree, thus collapsing sections of the upper part of the lava tubes.

Local pressures built up and the melt moved around under a deforming crust. Then the melt was locally pushed up forming the rise and the magma found a path to flow away, leaving a void, which the crusted roof partially collapsed. This is common, for the whole area has pot holes, sink-holes, collapsed lave tube pathways and many other odd sunken spots, ditches and pits [18]. But, strain as we may, we find no alien canals, holes, ditches or transit systems.

There are actually six NAC images in which you can find the bridge under varying lighting: See LROC photos M103725084L, M103732241L, M106088433L, M113168034R, M123785162L, and M123791947L.

19
ALIEN ARTIFACTS ARE THE PITS

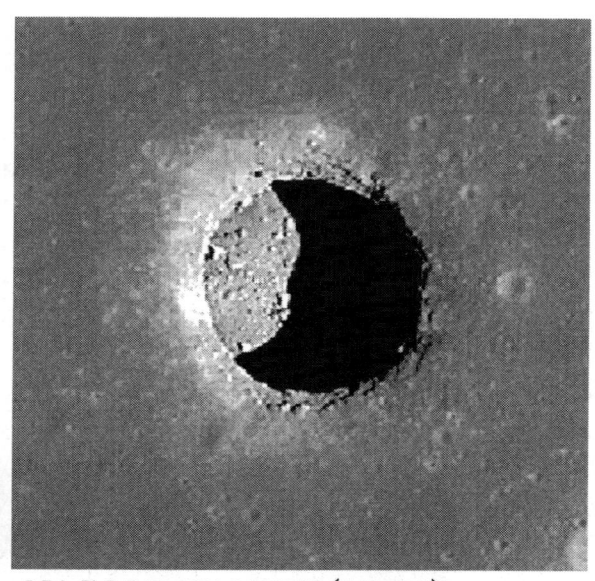

NAC M126710873R (LROC)
"Bullet Hole" or the "Pit" in Mare Tranquillitatis.
[NASA/GSFC/Arizona State University] (LROC)

There are many other related types of lava tube collapses. There are potholes, portholes, sinkholes, collapsed lave tube pathways, and many other odd shaped sunken spots, ditches and pits. There are even what appear to be "bullet holes." The LROC has now collected the most detailed images yet of at least two lunar pits, quite literally giant holes in the moon. Scientists believe these holes are actually skylights that form when the ceiling of a subterranean lava tube collapses, possibly due to a meteorite impact punching its way through. We can be sure that the floor of these

Stellar Sterility

"holes" are the tops of the ceiling that collapsed, leaving a large underground mound. Thus, the floor we see is not the floor of the lava tube, but the collapsed remains of the lunar surface.

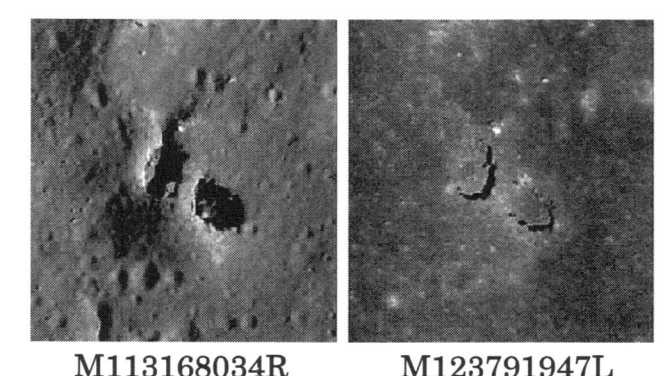

M113168034R   M123791947L

The left close-up shows the bridge when the Sun is 42° above the horizon and the right is the same area when the Sun is 80° above the horizon (near noon). M113168034R on the left, and M123791947L on the right, both are measured across at 128 meters.

THE MARIUS HILLS PIT
M133207316LE, M122584310LE, and M114328462RE
Image Credit: NASA/Goddard/Arizona State University.
(LROC)

The above bullet hole shows a spectacular high Sun view of the Mare Tranquillitatis pit crater revealing boulders on an otherwise smooth floor. Image is 400 meters wide. North is up [19]. See the little alien golf balls littering the floor of the pit?

The Japanese SELENE / Kaguya research team observed one of these skylights, the Marius Hills pit, multiple times. With a diameter of about 213 feet (65 meters) and an estimated depth of 260 to 290 feet (80 to 88 meters), it is a pit big enough to fit the White House completely inside. The image featured here (middle, above) is the Mare Ingenii pit. This hole is almost twice the size of the one in the Marius Hills and most surprisingly is in an area with relatively few volcanic features [20].

20
LUBINICKY JUNKYARD

AS16-M-2493_med. Bullialdus-Lubinicky Area

The Bullialdus-Lubinicky area is another left-behind alien graveyard site of busted, decayed, and disintegrated Moon moving machines! For Mr Leonard this site offers the finest in used E.T. technology!

The area is located between Mare Cognitum and Mare Nubium, a little South and East of the exact center of the Moon's Near Side. It is exactly South of Cognitum and West of Nubium. It is a medium size crater and somewhat of an old one submerged in the mare surface, and locating it through a telescope can sometimes be an eyesore. Looking for it in NASA

Stellar Sterility

photographs can also "crater" the eyesight and after long hours of searching for artifacts can cause a false sense of dyslexia as well as fantastic hallucinations as is demonstrated in "Somebody Else is on the Moon."

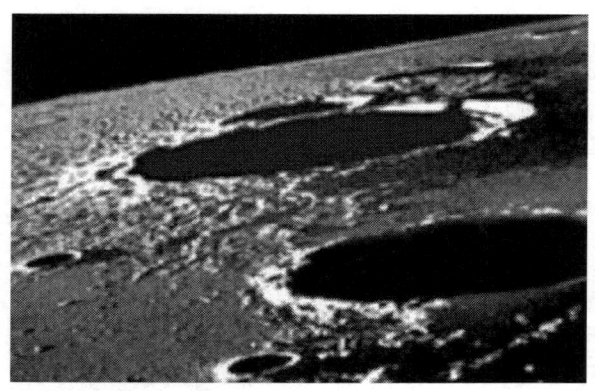

AS16-124-19901

This shot of AS16-124-19901 is one of the most beautiful landscape shots NASA has taken of this area and it is a favourite public relations photograph of the Moon. The area is full of hills, mounds, dirt piles, crater ejects, pits, gashes, crater rays and all sorts of other morphological shapes that are exciting to the eyes, especially to Mr Leonard and other extra-terrestrial imaginers.

Mr Leonard describes the area as screaming of underground inhabitants - from all the seismic rumbling and general ground disturbances there - and contends that the area is more than (as his associate criticizes) just "Mountainous rubble." He boasts that just two square inches of the photo *"can keep one busy for weeks."*

He said the thirty-seven-mile-wide crater Bullialdus sits in the middle of the southeast quadrant

Stellar Sterility

of the Moon and claims that between Bullialdus and Lubinicky E is *"the most fantastic area on the Moon"* riddled with all sorts of *"macroscopic engineered objects"* (p. 40-41).

He continues that the crater with the sun-struck rim (Lubinicky A) has what looks like a shaft of a gear sticking out. We are told to look just below it at the remains of another larger "gear" that apparently has been ripped away, exposing its inner teeth. He estimates that it measures about five miles in diameter and would be a real blast if dropped on Manhattan: *"It would obliterate everything from midtown to the Bowery"* (p.42). Next, we are to note the *"perfect symmetry"* of the underside arc, the *"absolute perfection of the teeth"* in the small gear and the *"four perfectly spaced rows of teeth"* in the larger one. He points out that the shaft *"sticks straight out for at least two miles."*

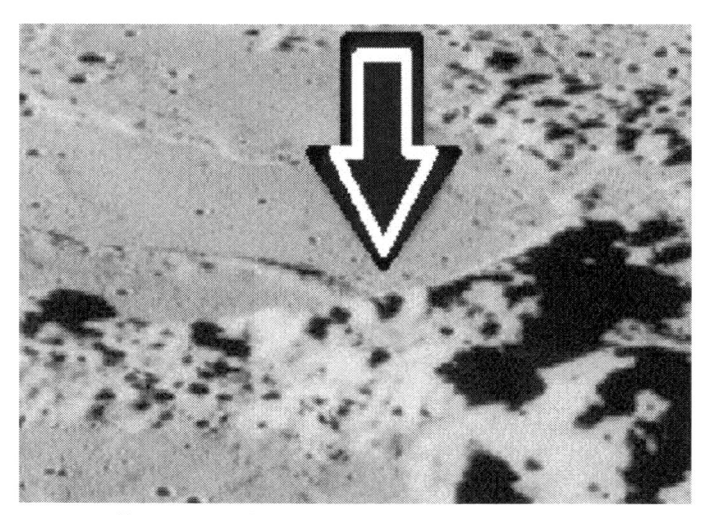

Gear teeth sticking out of rubble?
Notice the crater rim with the smaller meteorite impact holes.

287

Stellar Sterility

Without showing the reader any photographic enlargements as evidence, he sketches for us some of the engineered objects to illustrate for us what he sees. After enlarging the photo, rather than finding "stitches" the observer will contract stitches or at least a few chuckles. Nowhere are there any artificial machines. There are no parts of equipment or other alien devices. All that can be seen is dirt, rocks, hills, mountains, craters and crater ejecta. Sorry, the alien mystery is unfolding as an alienating scam to distract people away from truth, while making boring space science more fun and exciting!

More gear teeth.   (Unknown)

Moon artifact hunters consider this site to contain leftover parts of alien machines. Mr Leonard, (especially) sees HUGE gears and shafts complete with their housing encasements. He said they are artificial constructions and not natural ones, because the sun shadow angle outlines the symmetry of the underside arcs and the perfection of the gear teeth as they cast shadow lines on the mounting plate for the gears, in regular, spaced rows. One is said to be able to see two shafts and their shadows, sticking out over the lunar landscape. The gear housing is at least 3-MILES

Stellar Sterility

LONG. One shaft is at least 2 Miles long. Others may be longer or shorter, depending on the design and usage. As far as other machine parts, there are metal T-Scoops, pipes, busted hoses and - well, the list is endless if you have an endless imagination.

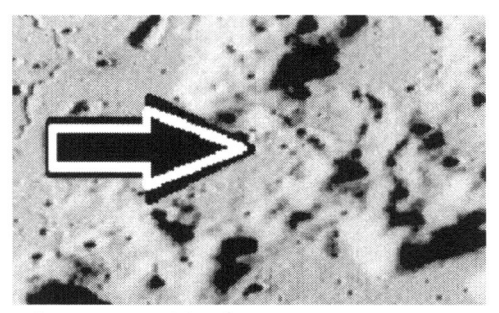

The 3-mile Shaft

AS16-M-2496_med

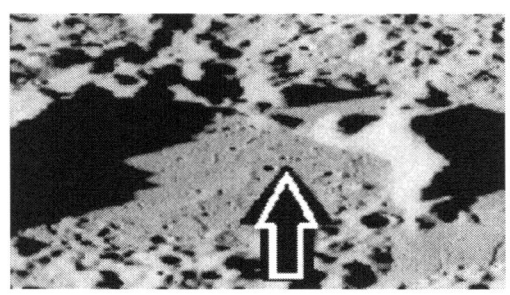

Unknown object

Stellar Sterility

Our alien imaginer said the Lubinicky area is comparable to Crater Tycho in abounding with piles of junk. However, this is not all the junk seen there. There are what appear to be "stitching" or big bans that hold two large chunks of the lunar crust together, as if to hold it from splitting apart. Craters Tycho and Lubinicky are unique spots for these stitching's.

He said, *"These (photos) are some of the best examples of 'stitching' the skin of the Moon in the Bullialdus -Lubinicky area. You cannot fail to notice the precise regularity of the stitches. They are identical in length, identical in distance apart, and identical in width."* Now, how an alien can stitch together loose, dry regolith dirt and hold it together is way beyond science and reason – it is fantastic!

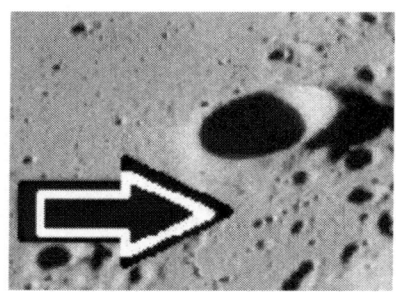

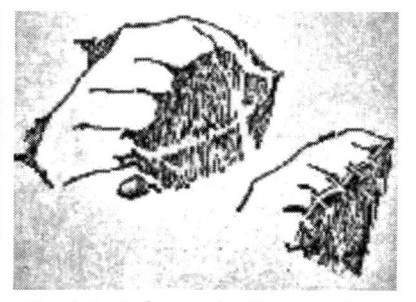

Tycho "stitching's"   Lubinicky stitch marks
(Un-located)   A crater chain?

Leonard draws up some sketches to illustrate what he sees, rather than enlarge the photo and show what is really there. After enlarging the photo, rather than finding a "stitch," the observer finds a switch. Nowhere can any artificial machines, parts of equipment or other alien devices be found, so we see why George switched the photo of the stitch with a drawing. All

Stellar Sterility

that can be seen in the photo is dirt, rocks, hills, mountains, craters and crater ejecta. All that can be seen in his drawing is everything but dirt, rocks, hills, mountains, craters and crater ejecta.

Examples of "stitching" in the Lubinicky Area
(Cannot locates these)

After establishing the reality of the stitches, ditches, screws, shafts and gears, George sows together another tapestry of industriousness, explaining why aliens were stitching up the lunar surface. He finalizes his polyester yarn with the moon being an interstellar spaceship that had received major damages for some reason and was (is?) in the process of repair. This is at best a flying carpet story if you think twice about it or can think at all.

21
KING OF THE FARCE SIDE

M119062083MC
King Crater on the far side of the Moon - LROC

George Leonard seems to be the KING of the dark side of the moon studies. King Crater is another of his "fantastic" locations. If two inches of Lubinicky regolith would keep you busy for weeks, this King of the farce side will have you buried in it for years to come.

King is a prominent lunar impact crater that is located on the far side of the moon, and is not viewable directly from Earth. It forms a pair with Ibn Firnas, which is only slightly larger and is attached to the northeast rim of King. To the northwest is the crater Lobachevski, and Guyot is located an equal distance to the north and northwest.

Stellar Sterility

King Crater is 77 km in diameter and has what appears as an unusual claw shaped central peak or what looks like a fat tuning fork.

This is not an alien amphitheatre, but is the rebound material going up and collapsing to the North. There are landslides in the walls and impact melt ponds on the floor with some volcano like domes with vent-like craters in their tops running in a line along the northeast floor. The south wall of King has collapsed inward onto the floor and external to the rim on the ejecta blanket to the east (right) it has a flow like appearance. There is just so much to see and describe in this King of the far side of the moon crater.

Leonard also presents some interesting alien machines digging about King Crater and the surrounding areas, and something coming out of a small crater in the unnamed highlands. He said, this mysterious object was *"coming up out of it."* This was the largest "spraying" ejecta stream he had ever seen other than Yellowstone's Old Faithful. He finds in the King's area what he calls digging machines or "X-Drones" and other industrial items. Now, we will have to take his word for most of what follows, for all the looking in the world is not going to find one alien iota.

There are other reasons why we will never find what George sees. The first reason is that most people are not brain surgeons and cannot tell what is happening in his imagination. Secondly, many of the "artifacts" cannot be relocated because he used the public affairs office press release renumbered photos rather than the NASA photo code numbers. The number system was there for reference and it is obvious why he did not use the code numbers: He was too lazy and cheap to make a trip to the NASA archives,

Stellar Sterility

buy the negatives and run detailed copies off from the original masters. This is the reason for the blurriness of the pictures, which by the way lends support for the blurry obscured alien evidence. One would have to search for weeks to relocate most of the little craters. Some of these craters are located, but most are impossible to find unless a cross-reference is located.

M103739394LE
Deceleration lobes in King crater ejecta. LROC Narrow Angle Camera (NAC) observation, LRO orbit 453, August 1, 2009. Lighting incidence angle 61°, field of view approximately 1,200 meters across.)

As far as alien artifacts are concerned, just looking at the quality of his photo plates begs the question of stretching the truth or rather "stitching" truth together out of mental phantasms. The objects are mostly blurry and the ones that are clearly defined look

Stellar Sterility

not much different from millions of other so-called "alien artifact" objects scattered over the moon.

A Constellation Program Region of Interest near the northeast edge of the unusually large melt pond adjacent to the lunar far side crater King. The boundary between the dark, coherent impact melt rock at the lower left of the image and the bright, pulverized ejecta blanket to the upper right is clearly visible in the floor of a smaller crater that formed at the boundary between these two units. Image width is 1.3 km, pixel width is 1.29 m

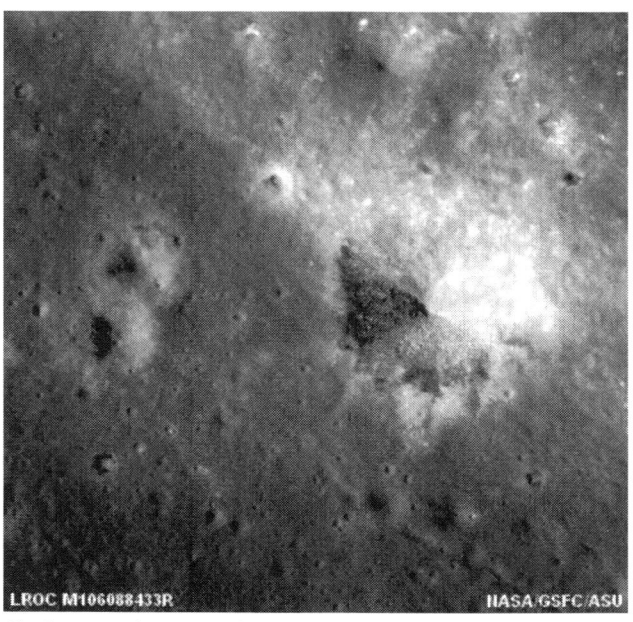

Subset of NAC frame M106088433R

Stellar Sterility

M115529715RE

A fault scarp separates two zones of impact melt within the King Crater (5.0°N, 120.5°E) north interior wall. LROC Narrow Angle Camera (NAC) observation, LRO orbit 2159, December 15, 2009; solar illumination incidence 75° from the east, field of view approximately 900 meters (north is up). [NASA/GSFC/Arizona State University].

22
REGOLITHIC ROVERS

NASA is working on what they call a Spray-puff, smoke blowing EXO-robotic recreational vehicle or colossal size roving regolith mineral sampler. Leonard anticipated the invention decades before by speculating that aliens had invented them long before man ever even had the wagon, not to mention the steam engine to pull it. In the following photo of Crater King, he locates such devices to prove his point.

AS16-120-19228

PLATE-12. Circles indicate several small craters in the process of being worked with marking crosses on their lips and spraying drones inside. Also X-drones and puff- like orbs circled in this shot of area near King Crater.

Stellar Sterility

In the above Apollo 16 photos, Leonard mentions the *"technological glory of the Moon occupants"* left behind. This junk apparently is in use and lies between King Crater and an unnamed crater with a "pond effect" bottom. *"There is a long ridge separating King Crater from the ... crater with a smooth pond in it."* (Many craters do have flat bottoms according to lunar morphological process).

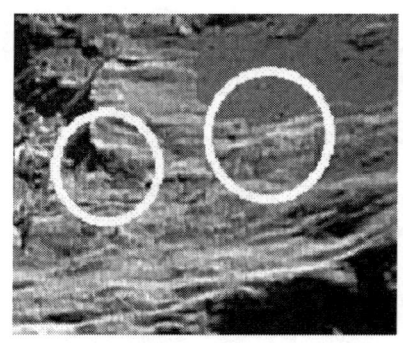

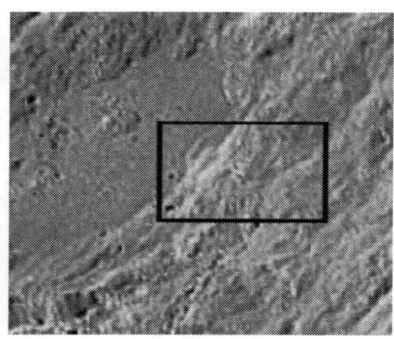

72-H-834 (AS16-120-19265)   AS16-120-19226
X-drone with no spraying.   Another X in King Crater

He said, in one of the areas of King *"Something was coming up out of it"* outside of the rim area. This was the most dramatic *"spraying effect"* of all the spay-jobs he would ever find. Mysteriously, this active "spray" in photo Frame No. 834 does not exist in the later photo No. 839 taken two days before. He said this was his amazing proof that an artificial spraying was taking place. Now you see it and now you do not. What is more amazing is that No. 839 (AS16-120-19228) is a more medium shot, further away, and if you look at it closely, you cannot make hide-nor-hair out of it, let alone make out any alien "hides" or E.T. "hairs."

Nevertheless, Leonard has found more astro-goodies to support his quackery! With

Stellar Sterility

super-human eyesight, he locates an X-drone in the earlier 839 picture and said it must have been *"working in exactly the same spot from which the spray would emanate in the 834 picture taken two days before."* With even more super-human vocal cords, he squawks further of X-drones digging for minerals and such, but speculates other reasons for excavation activity. The aliens need locations for "landing berths," "archaeology," "recreation" and places for "competition" games.

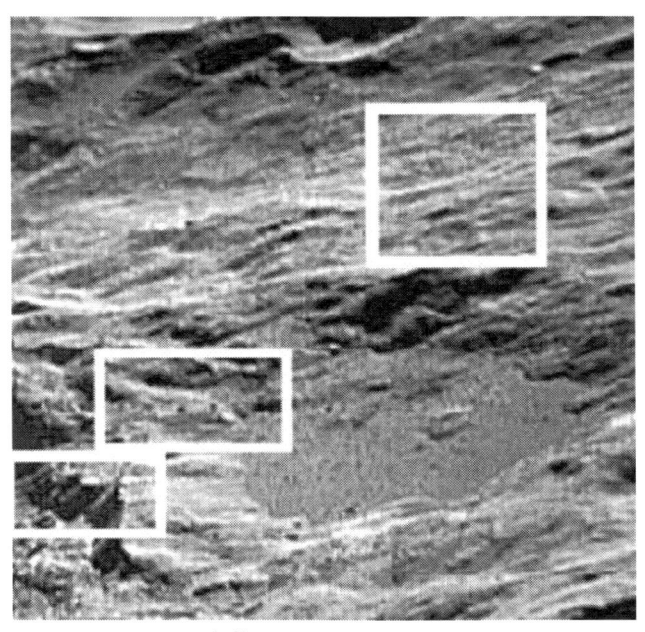

AS16-120-19226
Top box shows the "sprayer" unit.

At the base of the ridge (in the above picture), he said are *"some old friends and some new oddities"* that *"began to jib together, and to click."* Yet, after some detailed observations of the photograph, we find rather

Stellar Sterility

George has conned his new friends by fantastic interpretations of old oddities. One little jibbing object looks like a large *"oversized cannon"* that could be used as some kind of recreational game device, a mechanical digger or *"some functional rig in a complex series of steps undertaken by a complex culture."*

No matter, after showing the reader a pathetic drawing of a blob with a suction hose for a nose, he begs his readers to come up with their own ideas about what it is. Since he cannot tell a moon mountain from a molehill, he suggests that we supply a label, since he does not have enough evidence for an idea of what to call it or what its function may be.

What a pleasure it is to suggest a few things for Mr Leonard. It is an astro-biogenetic dinosaur size Anteater! Yes, take a new look. It looks like a huge colossal size balloon bag. Could it be a 747 size alien colonic bag? Could it be a gigantic trampoline, part of a bagpipe or a giant exo-kazoo? What if it is only a brightly sun lit funny shaped rock? Maybe we will never know, because the photo is too grainy to tell.

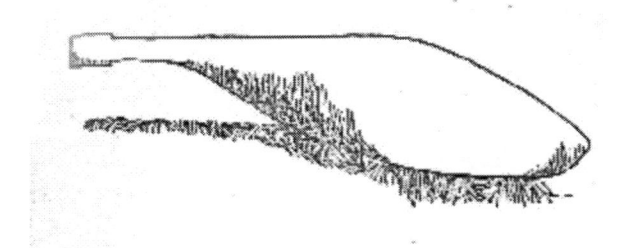

Leonard's attempt to draw an alien device in Crater King. Looks like a huge "cannon." (p.80)

Stellar Sterility

23
X-DRONES OF CRATER KING

Here are the drones drawn by Leonard. The arrows point to their locations according to his specifications.

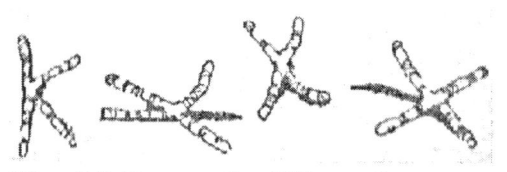

The "X-Drones" of King Crater

Leonard did not publish very good versions of the NASA Apollo 16 King Crater photos. There is more than one King Crater photograph. I picked the best available from all the frames. Since he also did not enlarge the areas I thought it necessary to show my readers the big "X" drones, where the X stands for an "X" out"! Whatever he saw in the photographs are not there now. Here are some extreme close-up enlargements of the three locations.

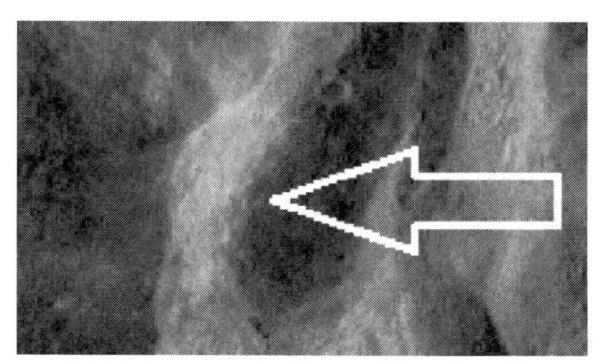

AS16-M-0890_MED -- CLOSEUP -1

Stellar Sterility

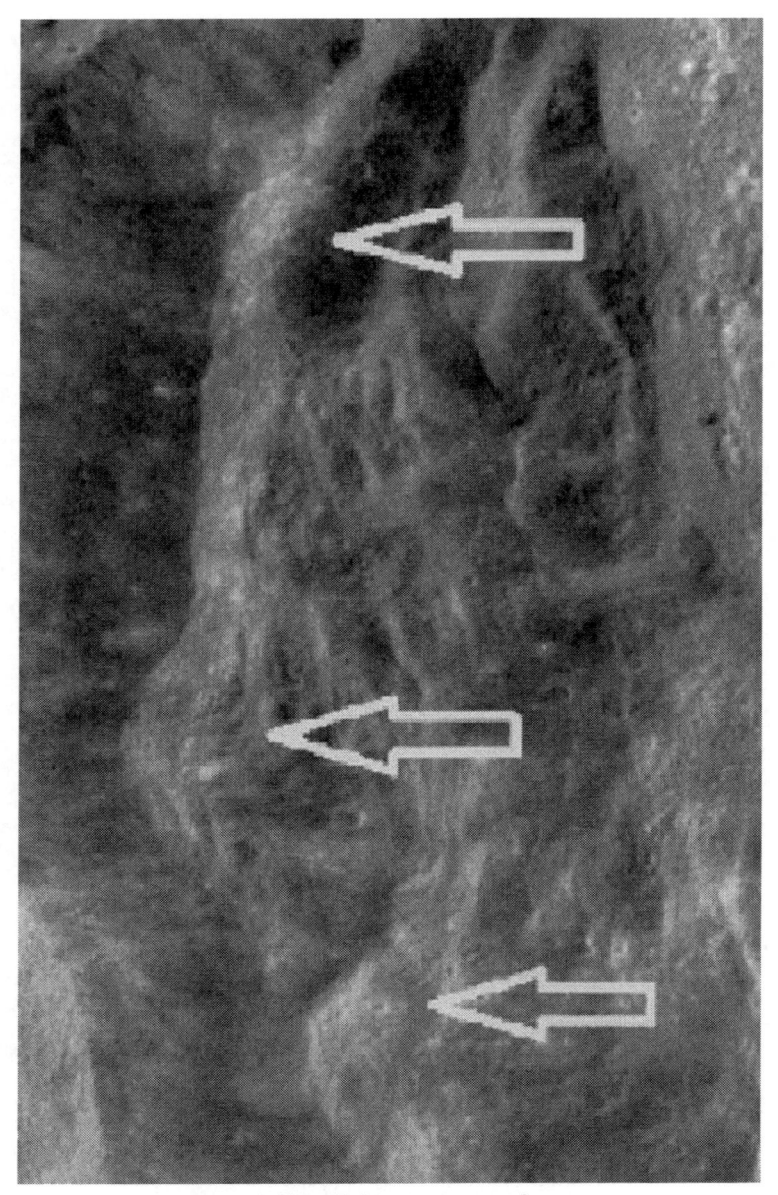

AS16-M-0359_med
Location of the X-Drones

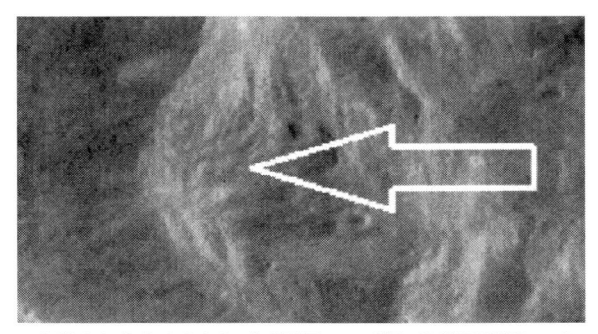

AS16-M-0890_MED -- CLOSEUP - 2

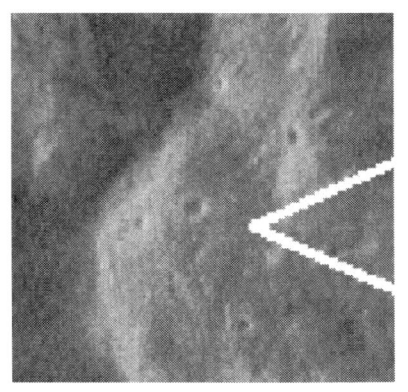

AS16-M-0890_MED -- CLOSEUP – 3
Nothing!

So, we have evidence of crater sprays, ridges being knocked down, huge X-Drones slaving away for their masters, grey lava sludge smoothing over lunar surface , pounding effects stirring up dust, little puffy orbs and nipple like extrusions - what more could be found? The possibilities are endless depending on the excessiveness of your imagination.

Stellar Sterility

24
DINOSAUR CRATER

LO-I-136-H3
Unnamed Crater on far side of Moon. Nickname:
"Icon or Dinosaur Crater"

Stellar Sterility

The DRONES are radical in "*GOUGING OUT A MOUNTAIN*" in search for minerals, said George! These "X-Drones" are gobbling up and spewing out lunar dust. Leonard suggested (in plate 8) that drones are "*raising dust on the rim of King Crater*" and in other locations, such as the above. Furthermore, he mentions there are many types of regolith rovers and other machines, such as Big-rigs, T-scoops and super rigs in various places on the Moon.

George directs us to the far side of the Moon for further proof to a location north of Crater Tsiolkovsky. After consulting the surrounding area other oddities appeared (See Leonard, plate 4). Leonard suggests that super rigs are working an area at the base of the walls of an octagonal double crater (pp. 49, 50 and 51).

The photo in Leonard's plate-4 (NASA Photo 66-H-1293) is of an unnamed double crater. It lies directly north of Tsiolkovsky close to the Craters Love and Prager, at -5.40 degrees Latitude, 129.33 degrees Longitude. This fascinating double crater has a gently domed floor, crisscrossed with linear depressions [about 1 kilometer wide] that give it the appearance of a turtle's back and crudely polygonal fracture patterns occur, formed by upward movement of plastic or liquid material. The doming of the floor of this crater could be the product of local magmatic activity or of slow isostatic rebound. Both alternatives point to probable activity of internal origin in the lunar crust [21].

At the base of the rim in the larger crater at about two o'clock, according to Leonard, is an object, "*which is too wispy to show up well when reproduced in this (his) book*" (p.51). He said the wispiness is due to its structure, whereas it is actually due to the bad quality of the photo enlargement. He illustrates this object in the following sketches (p. 52).

Stellar Sterility

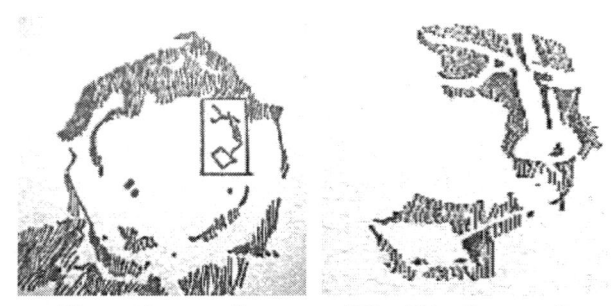

Drawings of crater Big Rig "scoop"

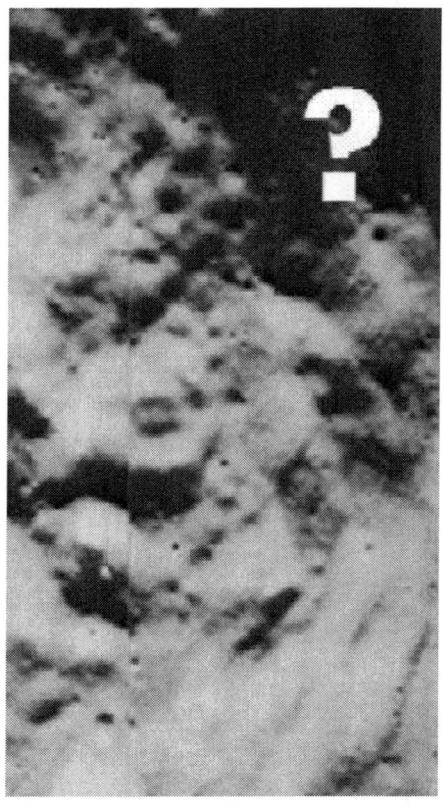

Enlargement of LO-I-136-H3

Stellar Sterility

He continues: *"The object pivots at the junction — just where you would expect it to pivot. Its two main struts rising from the ground (the wispy elements) are very straight and parallel. There appears to be a filament of some sort, which raises and lowers the horizontal piece leading to the scoop. There is a long thin device that runs from the base of the object down the hill toward the center of the crater, ending with an oblong plate, which is found elsewhere on the Moon and is perhaps connecting to a power source"* (p. 52). He goes on to explain the large equipment, the need for resources, minerals, and alien habitat.

Is this really in Crater Pasteur D? Let us look closer at this area and see what we can uncover. Notice in this enlargement that the eye can strain until it bursts before it sees any super rig. Super-size your eye balls, because no normal eyesight is adequate. The following cropped picture is the exact location George is sketching. We will not waist our time looking at the higher resolution LROC pictures, since this one is sufficient to prove the nonsense.

A SUPER-SIZE DINOSAUR

What did George overlook that might be interesting? With little imagination one can carve out of the distorted crater floor and debris geometrical black dotted rectilinear bars, some connected dots (red triangle), a "big-rig" (red square), and three rows of black dots (small red square). There is also a statue of an "eagle" (red oval) and what looks like an artificially "sculpted" super-size dinosaur (red arrow)!

Leonard's speculation of aliens needing recreation suggests that they busied themselves on their off time studying social science, art, and the history of man. It looks like this might be the case for dinosaurs, eagles,

Stellar Sterility

and other animals do not live on the moon and never have. The "occupants" must have manufactured iconographic versions of earth creatures for study and pleasure. Maybe they were toys. It is possible that these so-called icons represent leisure prevailing over labor and that the moon occupants were free to pursue games, recreation, and scholastics.

Locations of objects in red.   Dinosaur icon mound

Carved monument of an Eagle

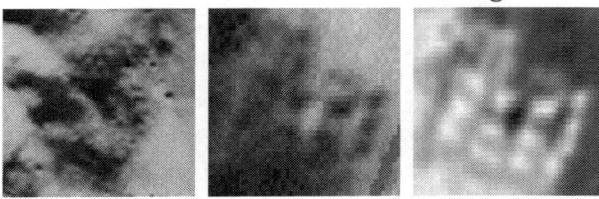

Geometric constructions with bulbs, dots, and "round appendages" along a set of (geometric) ridges. (Mid) 12 Dots pattern. (Right) Negative image of 12 dots.

308

25
CLOUDS OVER LOBACHOVSKY

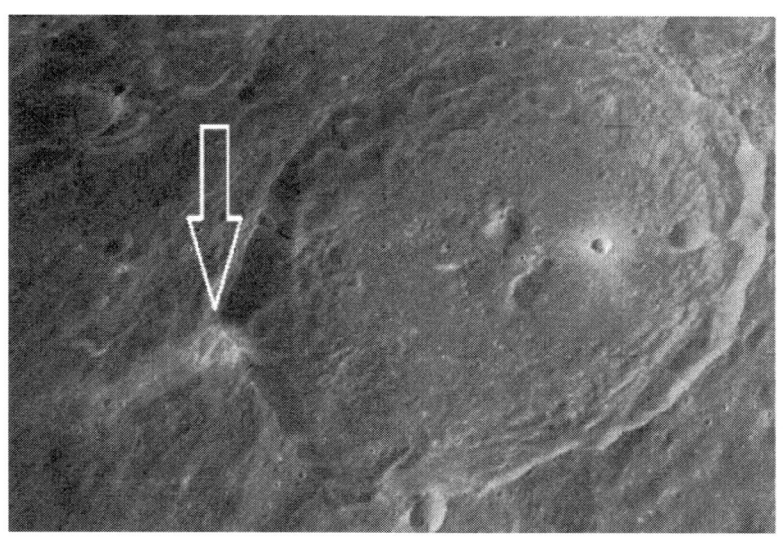

AS16-M-1322

Leonard's 72-H-1113 photograph on page 141 is of Lobachevski Crater; a crater NW of King Crater on the far side of the Moon a 10 degrees North by 112 degrees West.

Mysterious white ectoplasm energy clouds are now the topic with cratered scientists such as Mr Leonard. He said Lobachevski is mysteriously spewing out some "intelligent" energy cloud over its rim, as can be seen in the photograph above. As to whether it is a smart cloud, a gas spray, an ejecta RAY or some low level nimbostratus or stratocumulus cloud, is hard for him to tell. From his reproduction of the photo, it could be some luminescent gas spraying out that brightens by the sun's rays hitting it or a photographic processing glitch. It might even be the smoke spewing out from an

Stellar Sterility

alien chimney on a cold lunar night. It could be whatever you want it to be if you are writing a science fiction book rather than a science fact paper. It could also be exactly what NASA said it is a crater ray made from a highly reflective regolith material.

Leonard insists it is not a "ray" like those of Crater Tycho and Kepler, but must be some form of *"pure energy moving over the crater rim towards the center of the crater."* He further said, *"The band of light is not a ray, it is not a patch of reflected light on the ground. This band of light is like no other light one usually sees on the Moon. It maintains its integrity as a light even inside the rim of the crater, which is in shadow, (because) the topography beneath the band 'cloud' shows through."*

Our crater expert then complains that it is still a complete mystery to him as to what it really is. This is no wonder to those familiar with NASA photos and the different resolution qualities. Poor quality photo enlargements cause distorted images and mistaken interpretations. If Mr Leonard had used higher quality images, he would have clearly observed that the rays were the result of fragmented ejecta material.

Crater rays appear when ejecta are made of material with different reflectivity or thermal properties, usually having a higher albedo than the surrounding surface. More rarely an impact will excavate low albedo material, for example basaltic-lava deposits on the lunar maria. Craters on the near side with pronounced ray systems are Aristarchus, Copernicus, Kepler, Proclus, and Tycho. Similar ray systems also occur on the far side of the Moon, such as the rays radiating from the craters Giordano Bruno and Ohm [22].

Stellar Sterility

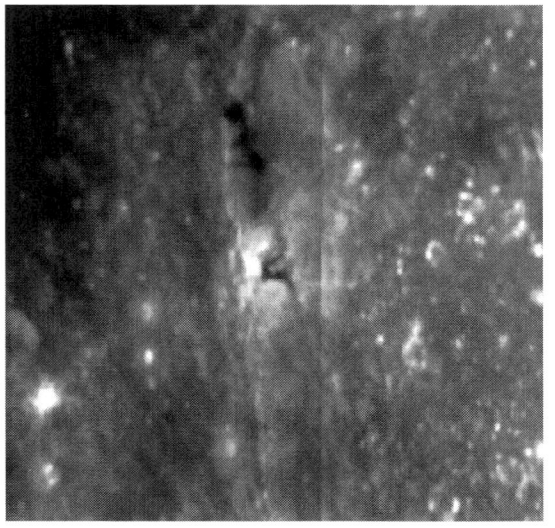

Lobachevski Crater photographed by Clementin

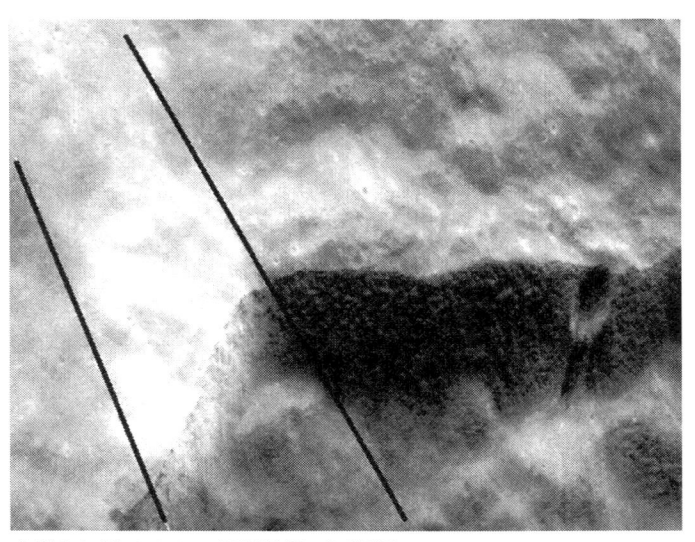

AS16-P-5029_FULL_MED

To cover up the true nature of Lobachevski's ejecta "ray", George took the photo (72-H-1113) and published

Stellar Sterility

it without offering the higher resolution panorama images (AS16-P-4849. See also 4851 to 5029).

The following Apollo 16 photo is an example of higher quality resolution pictures. Here we can clearly see the ray as part of the lunar surface and crater rim, and not some cloud floating above.

These enhancements are sufficient in demonstrating that there is no mysterious exoplasmic intelligent energy bleeding over the crater rim. When observed in high resolution, it reveals exactly what NASA said it is. It is simply a crater ray just like those of Tycho and Kepler.

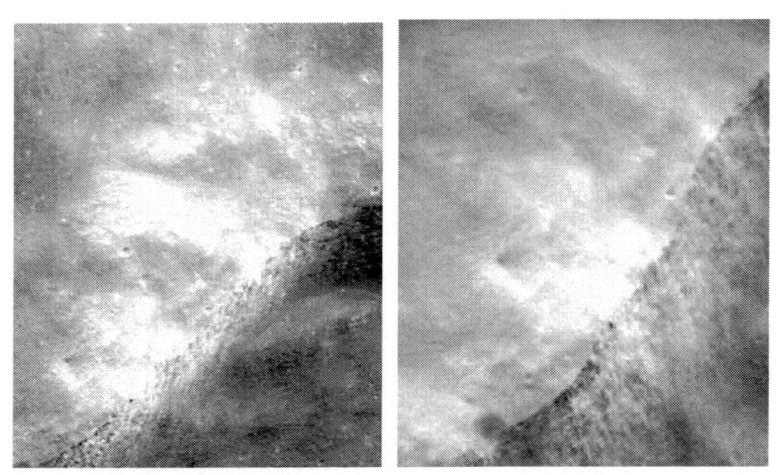

AS16-P-5022_full_med   AS16-P-5029_full_med

Stellar Sterility

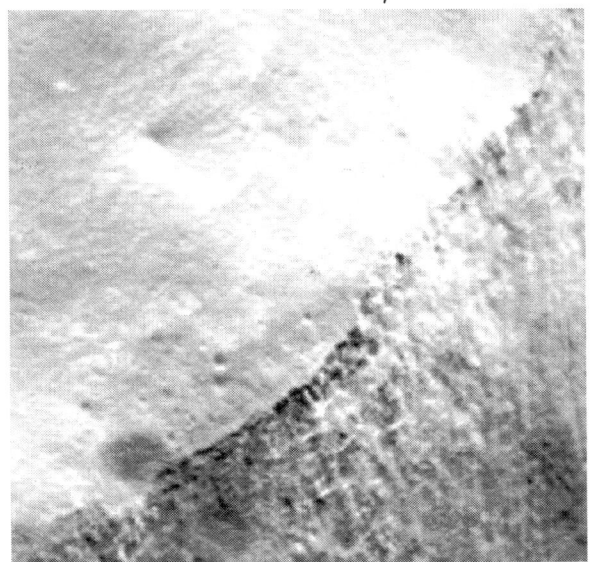

Photo clearly shows higher albedo material as the composition of the crater anomaly.

26
SUPER-SIZE-ME RIGS OF PASTEUR -D

Giant super rig machines are on the moon, said Leonard. In plate 6 of his book, George shows us a photo of a crater with a super rig machine taken by Apollo 14 (See 71-H-781 = AS14-70-9686), which he said is sitting inside. He does not name the crater, but the lunar maps name it Pasteur D. It is located at 9.0 degrees S, 109 degrees E on the lunar far side. It sits on the northeast corner of Crater Pasteur and is viewable in the following frames: LO-II-196-M, AS12-56B-8308, AS12-56C-8308, and AS12-56 D-8308.

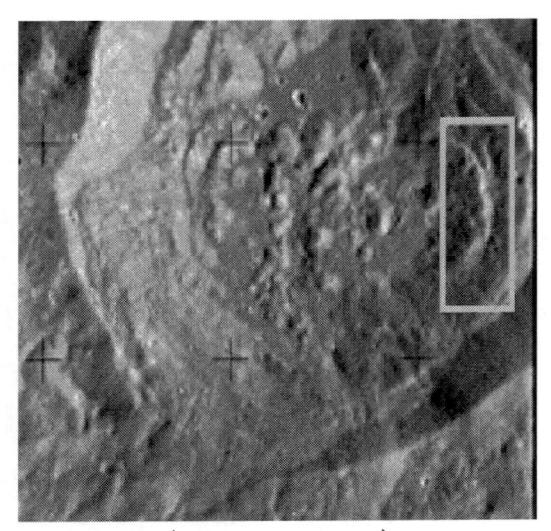

71-H-781 (AS14-70-9686)

George said this is the most remarkable photo ever taken by the astronauts during the Apollo 14 mission. It is the clearest picture of a super mechanical rig on

Stellar Sterility

the moon available. He calls this "super-rig 1971" an almost duplicate of his 1966 rig.

It supposedly sits on a terraced, inside rim (See Drawing A) of an unnamed crater (now called Pasteur D) on the far side of the Moon. It stands up straight and is made of filigreed metal for strength and lightness, thus casting no observable shadow. A "cord" runs from its base (B) down the side of the crater (C). On the right of the terrace two other rigs are working. They are of the same design as the first one mentioned and have two pieces working from a fulcrum. Cords run from their bases. They have made an even cut (D) straight down into the terrace. The cut notch is straight as a die. Something also stretches across the gap. As to the size of these objects, George arrives at the rough estimate of one-half miles for the rig, and at least three miles high for the chunk of ground from crater floor to where the rigs are perched (p. 54-55).

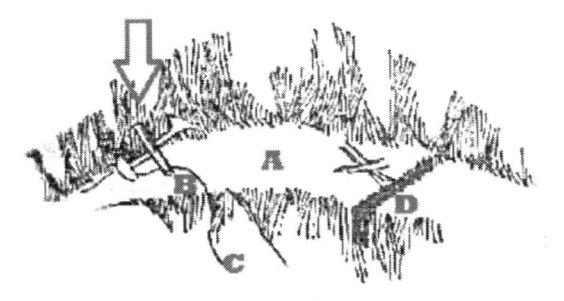

Leonard's Drawing (A) of "super-rigs." (p. 55)

Is this the single most remarkable picture ever taken? There are hundreds of other remarkable Apollo 14 photographs and thousands of other Apollo mission photos. This one photo is only remarkable to George for a specific reason and holds no value to the conservative selenologist. It is no more "remarkable" than the

Stellar Sterility

thousands of other Apollo photos. In fact, it is one of the least interesting of the Apollo collection.

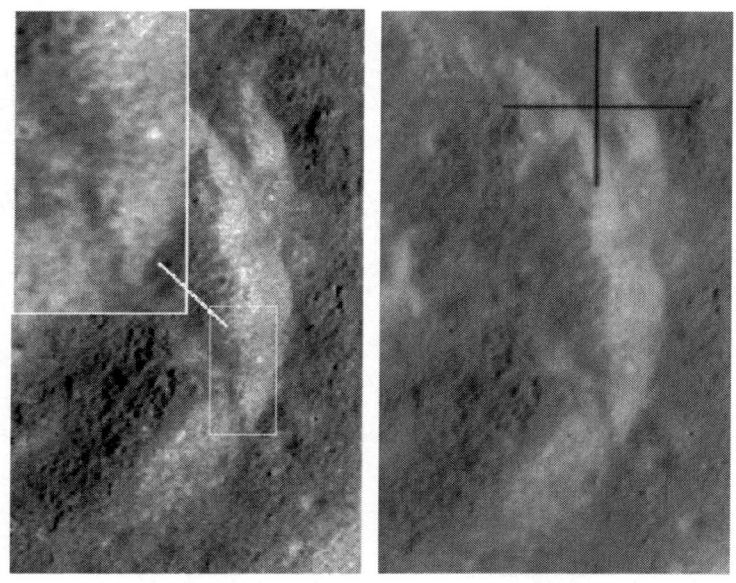

AS14-70-9688   AS14-70-9689

Crater Pasteur D PLATE 6. Leonard's "super-rig." Notice no "super-rig" or machines in frames 9688 and 9689 enlargements.

Is a super-size-me rig squatting on the inner side of the crater? We cannot tell from these frames no matter what enhancements we make. It just looks like a collapsed crater rim with heavy sun highlight glistening off the top edge. Nevertheless, to make sure we are not passing up a grand opportunity in finding a priceless antique, alien spacecraft or mechanical digger, we must look closer. None of the Lunar Orbiter or Apollo photos will help. To see what is there, we must turn to the new LRO photos to obtain higher resolution pictures. Notice all that is visible are rocks, boulders, dirt, regolith and crater debris.

Stellar Sterility

The Lunar Reconnaissance Orbiter camera reveals even less about super-rigs and more details about natural lunar surface processes. In no location do we find any machines, diggers, dirt flingers, terrace cutters or dump trucks.

Stellar Sterility

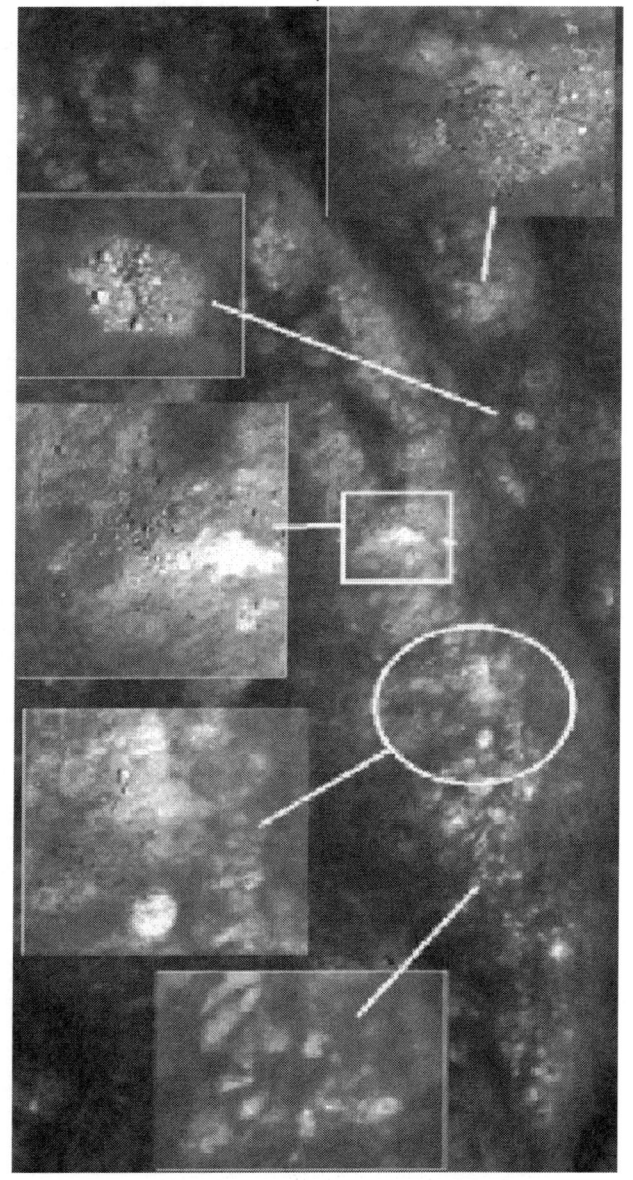

LROC frame M1100495581L
Enlargement of Pasteur D crater rim and "Big-Rig" area.

Stellar Sterility

27
SUPER TREADS AND LUNAR LADDERS

Next we are told to see some "tread marks" in the following Apollo 8 picture of Crater Doppler, located 12 degrees S, 162 degrees W. on the far side of the Moon. If it is not a super rig, it is "super-treads."

69-H-8 (AS8-13-2244)

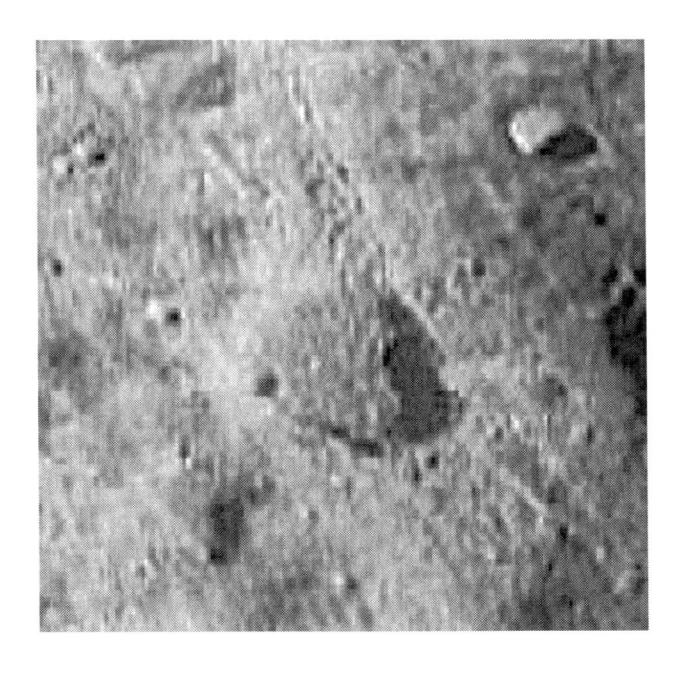

Stellar Sterility

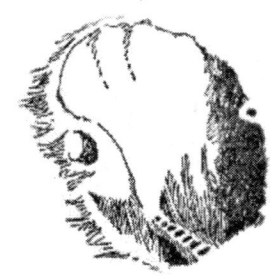

Close-up   Page 175
Drawing of crater "rope-ladder"

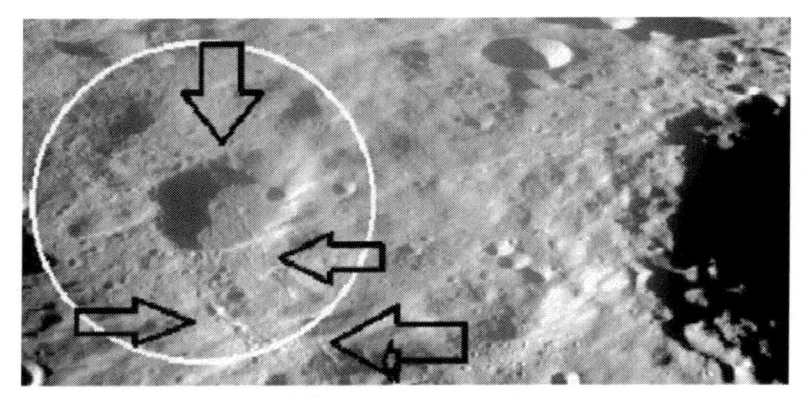

AS17-151-23115
Crater Doppler area

LO-I-036-H3

Stellar Sterility

Leonard goes through a lengthy diatribe explaining some kind of construction was going on here. He mentions, there are *"tread marks on (the) rim and (a) rope ladder extending from (the) rim to (the) floor... visible on the far side crater"* photographed by Apollo 8."

He said, *"In it [this far side crater] is an almost obliterated crater with many parallel markings that run through it with one marking running in the air from the rim and into its bottom."* It appears to be "an enormous rope ladder" or "tread from a very large vehicle." He concludes from a rough measurement that it is four miles long. He draws what he believes he sees and would have us believe and see what he wants us to see, because the picture resolution is too weak to detail it. Do you see a "ladder" in any of these photos?

Stellar Sterility

28
WHAT LEONARD MISSED

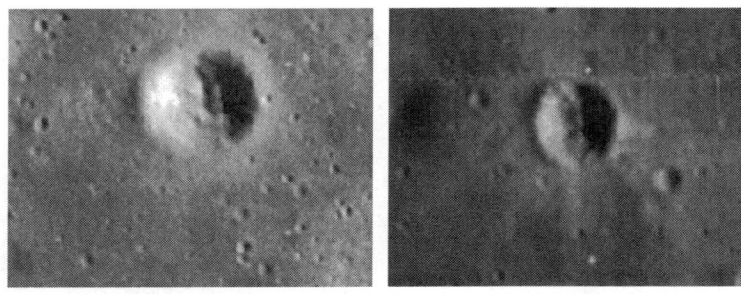

NUMERALS LO-II-037-H1 LETTERS
The number "4" "Exclamation Point"?

Giant "C" Clamp LO-II-108-H1 "Rope, Cable"?

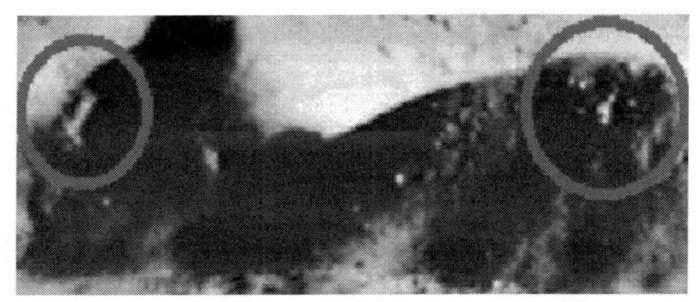

Sledge Hammer? LO-II-162-H3
Weird device, "Clamp"?

Stellar Sterility

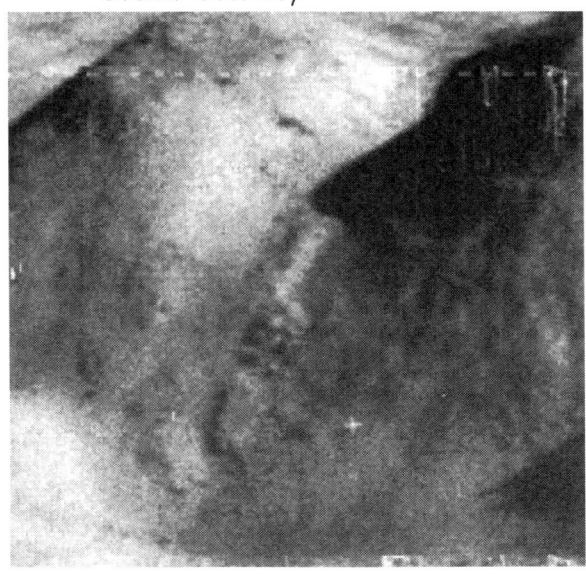

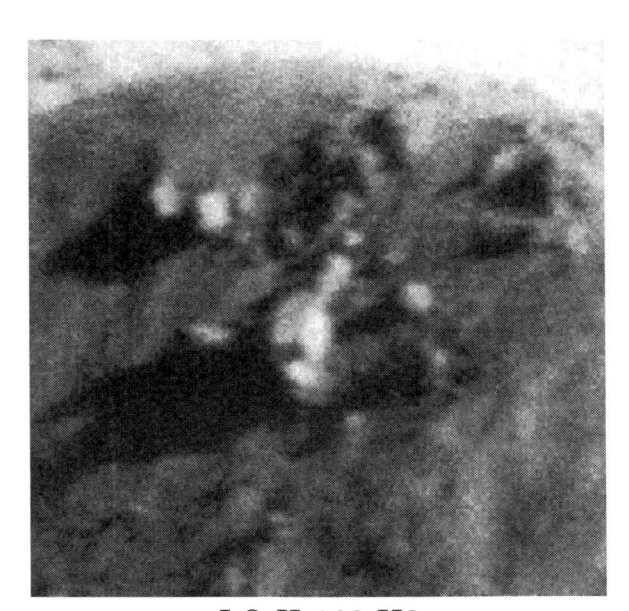

LO-II-162-H3
Just dirt and rocks! (LO-IRP enhancement: "MoonView.com")

Stellar Sterility

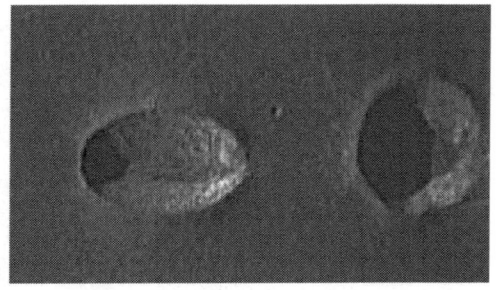

"Coffee Bean" Crater
AS10-33-4906

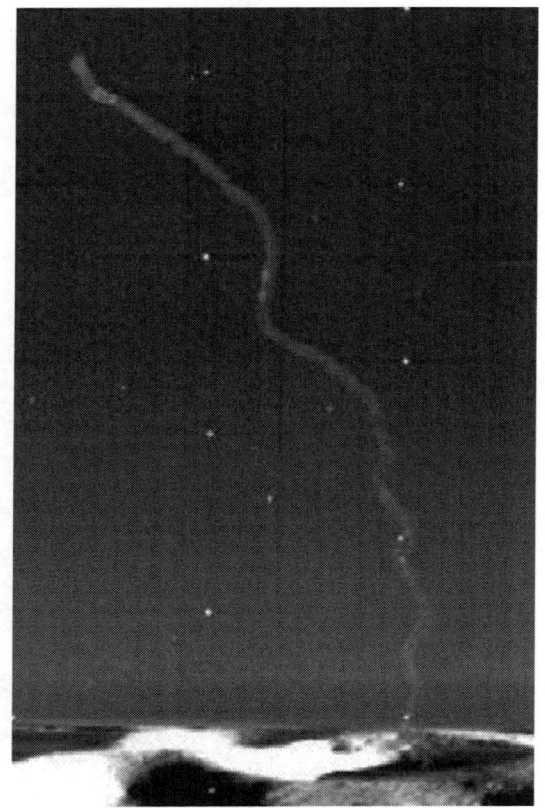

Smoke Spiral Crater
LO-III-213-H1-[2]

Stellar Sterility

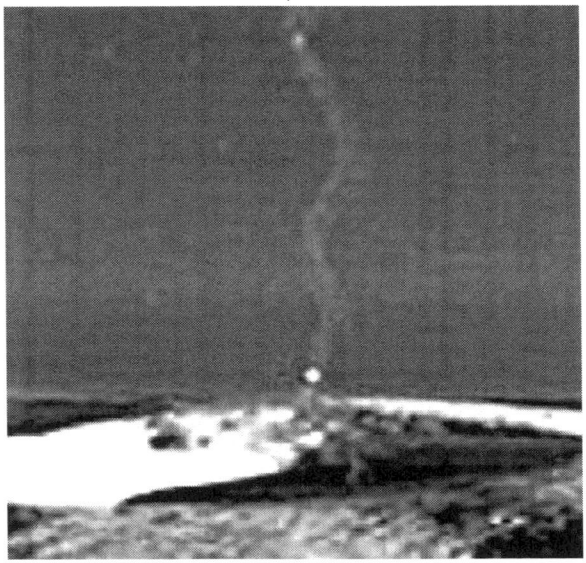

(Close-up)

While enlarged, one can see that the smoke spiral is continuously the same "width". This tells us that it is a "traveling emulsion" blob processing error.

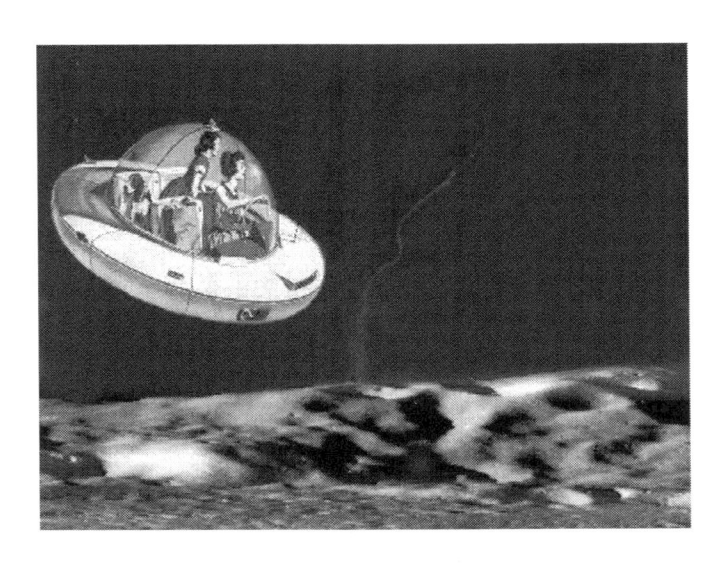

Stellar Sterility

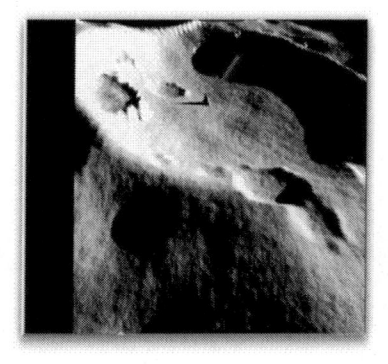

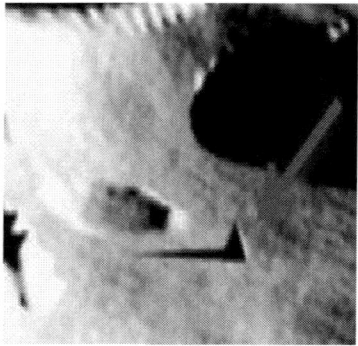

Snap_2011.07.13_11h25m39s_005
PYRAMID IN CRATER (Hubble telescope, 2008)

NASA uses Hubble to look for resources on the Moon too. Therefore, it is possible that it could have taken this picture. Every mention of it on the internet said this picture was "leaked" out from the Hubble classified archives. Of course, no one has traced it to the Hubble. Some say it does not exist in the Hubble archives. It is suspect as a fake because of the perfect symmetry of the pyramid shape [23].

Here is a nice geometric and hieroglyphic shaped object sitting in a plains area. It looks artificial from a distance, but up close shows the same extreme highlight and shadow contrasts as all other similar types.

AS15-M-1554_MED

Stellar Sterility

AS15-M-1554_MED
(Enlargement)

29
MOON MONUMENT HUNTERS NEED BETTER GRAPHICS EXPERTS!

Further studies on Lobachevski Crater found other "things" or rather created things by Moon artifact hunters. Leonard missed one of the best alien "monuments" for consideration by not referring to the higher resolution images. If he has done so, he might have gotten to this "tower" first.

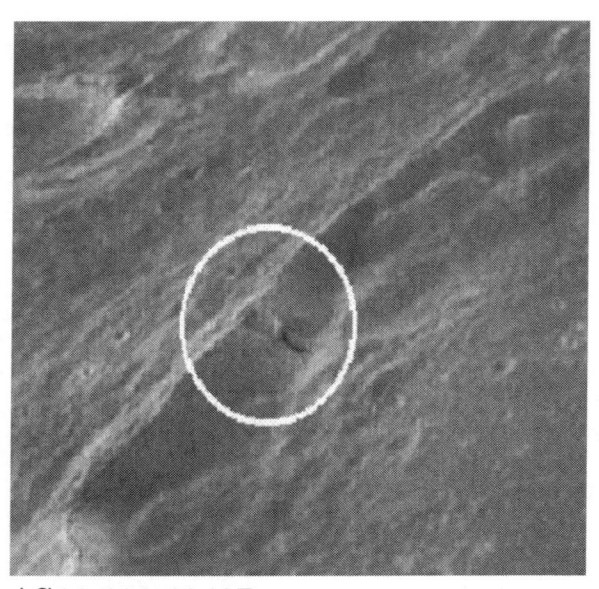

AS16-121-19407
Mysterious crater "tower"

Miss X's TOWER
After lying dormant for decades, this picture of Lobachevski Crater was presented to a graphics

Stellar Sterility

analyst, Miss X (we shall call her, "Miss X") at 'I Wonder Productions' for enhancement and further analysis. Her work quickly underscored ("distorted") the facts and therefore, incorrectly showed some bizarre anomalies in the photograph. Furthermore, because of her special graphics enhancement ability, she was able to show more than one. Nevertheless, for brevity, we will deal with the most prominent so-called object - an object that is actually not there!

Miss X thinks she exposed what appears to be a tiny tower in this Apollo-16 photograph (AS16-121-19407). As the "story" goes, she made it clear that her "multi-stage enhancement process" takes the image to its most critical viewing size "without distorting" the object in question.

Below is her first enhancement of the NASA Apollo 16 photograph. It is a shame as well as a sham that such a so-called graphics expert does not multi-stage "enhance" the available higher resolution pictures.

Her excessive computer enhancement process of a lower resolution image actually creates an optical illusion of a distinct object through pixel distortion of a small highlighted rim area. In the process of excessive enhancement this highlight becomes pixilated into an over blown rectilinear "tower".

The following are a series of improperly (excessively) enhanced frames of 19407. The following picture demonstrates that any lower resolution photo can be distorted to shows unusual images, illusions and fake objects. The excessive enhancement of any photo can easily create objects that are not there.

In stage 1, frame 19407 is enlarged. Stage 2 adds enhancement and sharpening. Picture No.2 above shows the actual impact crater with sun highlight on

Stellar Sterility

the right side of the rim. No. 3 is stage-2 of enhancement process.

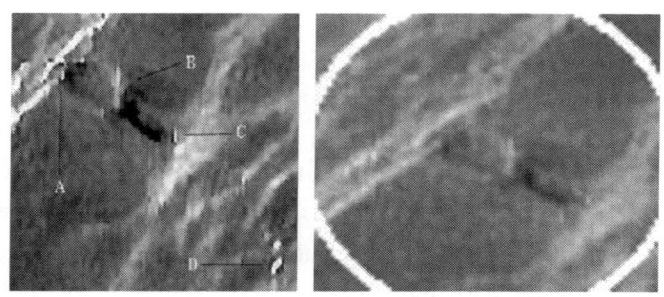

AS16-121-19407
LOBACHOVSHY TOWER
(2.) AS16-P-5024 Enlarged
Notice the rim is highlighted [1] and there is no mysterious object "tower" [2].

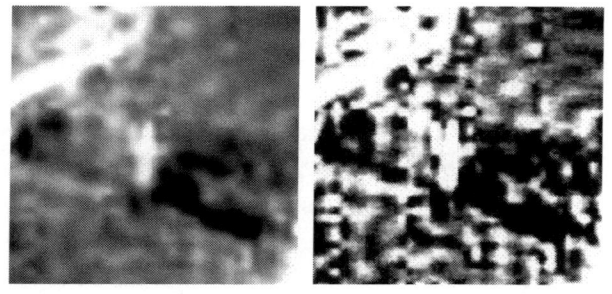

(3.) Ms. X's technique replicated - stage-1 of 19407
(4.) Stage -4 Enhancements
(5.) Enhancement doubled = extreme distortion!

The sharpening is doubled to 210%; radius is 220% and zero thresholds. No.4 is an exaggeration, all enhancements again doubled. No. 5 is all the above doubled again. What we now have is a bright white tower or rocket shaped object, all created out of false enhancement processes. The picture below is the

Stellar Sterility

actual craterlet impact site enlarged to maximum size before distortion begins.

Her false analysis is beyond remarkable, in that it consequently creates other strange and unusual 'objects' or anomalies on the rim of Lobachevski Crater. She describes object B as a spectacular 'spire' soaring perhaps hundreds of meters straight up from the lunar surface and standing next to what appears to be a rectangular shadowed hole or depression running from its base out to the right.

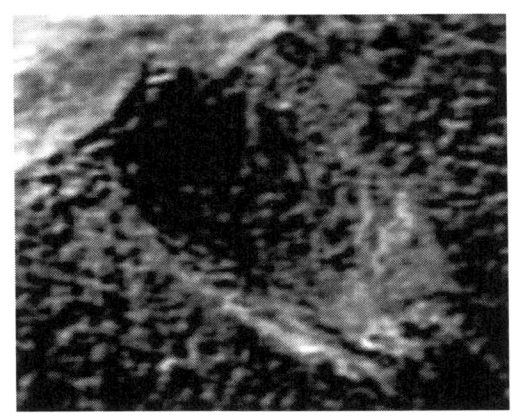

Apollo-16 AS16-P-5024 Enlarged

Lobachevski impact craterlet enlarged. Shows sun reflecting off bottom and right side of craterlet rim. There is no "tower" object. What is visible are boulders and ejecta debris.

This is impossible. The original photograph (AS16-121-19407) shows the lunar surface sloping down to the left and the enhancements are likewise orientated. If the lunar horizon is properly orientated to an observer standing on the surface, it becomes evident that the illusive 'spire' is actually leaning at a considerable angle, possibly as much as 45°, to the observer's right. It is possible that some structure can exist at such an extreme angle, but the probability is almost zero in light of all the facts presented here.

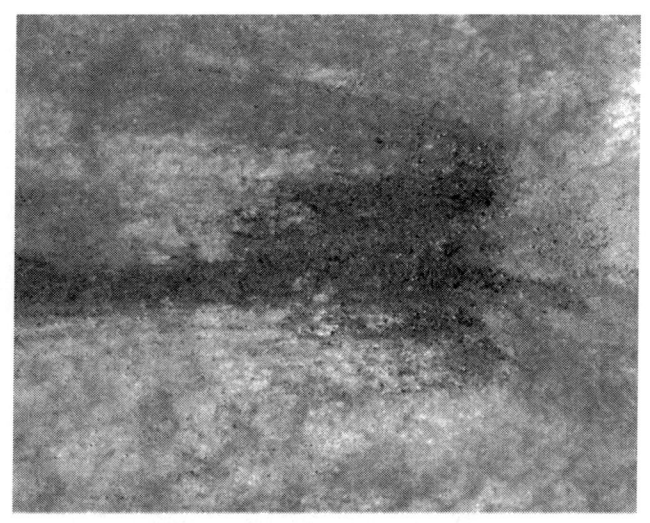

LROC Photo M176899195LC
Crater Lobachevski from a direct 'top view'. No object seen.

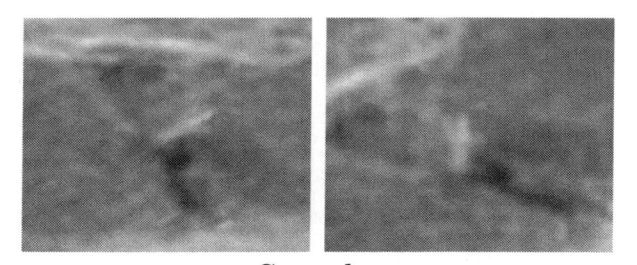

Craterlet
LEFT: Enlargement as done by graphics expert.
RIGHT: Enlargement of true angle – shows highlight at 45%

As for photographic enhancements, pixel streaking is visible nearby, aligned with the vertical axis of the picture, and parallel to the axis of the 'spire'. This is particularly noticeable and suggests that the 'spire' is

Stellar Sterility

an artifact of excessive imaging and enhancement processes.

There is no excuse for persons such as Mss X. Everyone has access to computers, the Internet, imaging software's and NASA photos, and with very little effort and download time good quality pictures of these type sites can be obtained.

As for Mr Leonard's analysis, we might allow him some grace in his speculations, for lack of computer technology and recent higher resolution photos. Nevertheless, he did have access to higher resolution photos of Crater Lobachevski and development laboratories! Pray, maybe we should put the blame on his lack of funding rather than on his lack of credibility. Not really. A good positive print from a sharp negative was more than affordable in those days just as it is today.

30
EXAMPLES OF CRATER RIM LANDSLIDES

The above Lobachevski crater rim collapse is most probably a result of a large micrometeoroid impact. Crater Kant P is an example of crater disturbance rim-slide; caused by a number of reasons, primarily large micrometeoroid impact, where part of the rim collapses downward toward the center of the crater. Kant P lies in the central highlands on the Moon's far side.

The younger, small pear-shaped crater on Kant P's north wall is an excellent example of the controlling effect that topographic relief plays on the shape of an impact crater. Because the small crater was formed on a steeply sloping surface, the ejecta deposited chiefly down-slope and formed as broad rim. The original rim and wall on the up-slope side is obliterated by slumping. The slumping has left a landslide scar and has caused the deposition of talus and scree in the lower part of the crater [24].

At the right angle, the same sun highlight anomaly would develop and present an artificial "tower" looking object. The small impact crater rim of brighter ejecta material would similarly give off a super bright highlight and give the appearance that there was an object sitting there.

When the La Pérouse-A impact excavated material from the side of a hill of highland material (1.5 to 2 km taller than the plains to the west), the slope became unstable and collapsed into the crater La Pérouse A. The result is a crater with one rim 920 meters higher. The Featured Image shows the top of this landslide,

Stellar Sterility

now the new rim of La Pérouse A. The landslide material inside the rim is high reflectance, while the undisturbed section of the highland material is relatively darker.

Despite being the same composition, the landslide material is higher reflectance since it is fresh (recently uncovered). Faint lines along the outer edge of the rim are evidence of highland material slumping towards the crater. Collapse features on the Moon often display slump lines, which show where material has fractured and moved downwards, but not completely collapsed. Subsequent impacts or seismic shaking could cause further landslides inside the crater! [25].

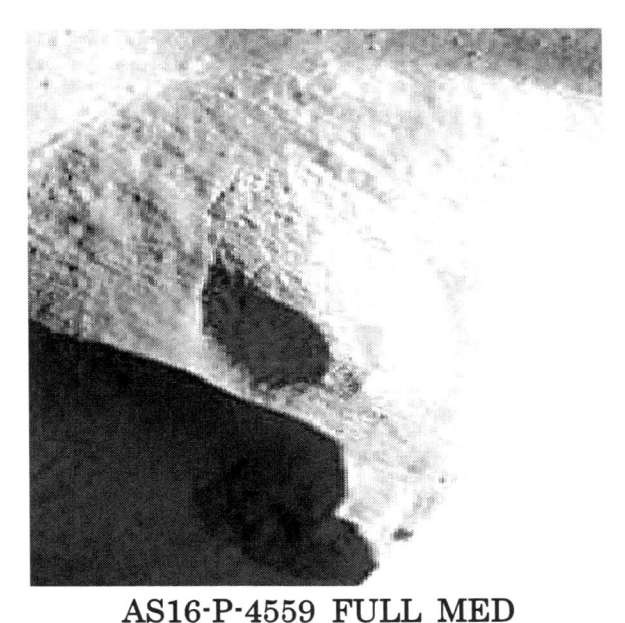

AS16-P-4559_FULL_MED
Crater Kant -P impact crater landslide, rim collapse slide. La Pérouse A Crater IMPACT LANDSLIDE

Stellar Sterility

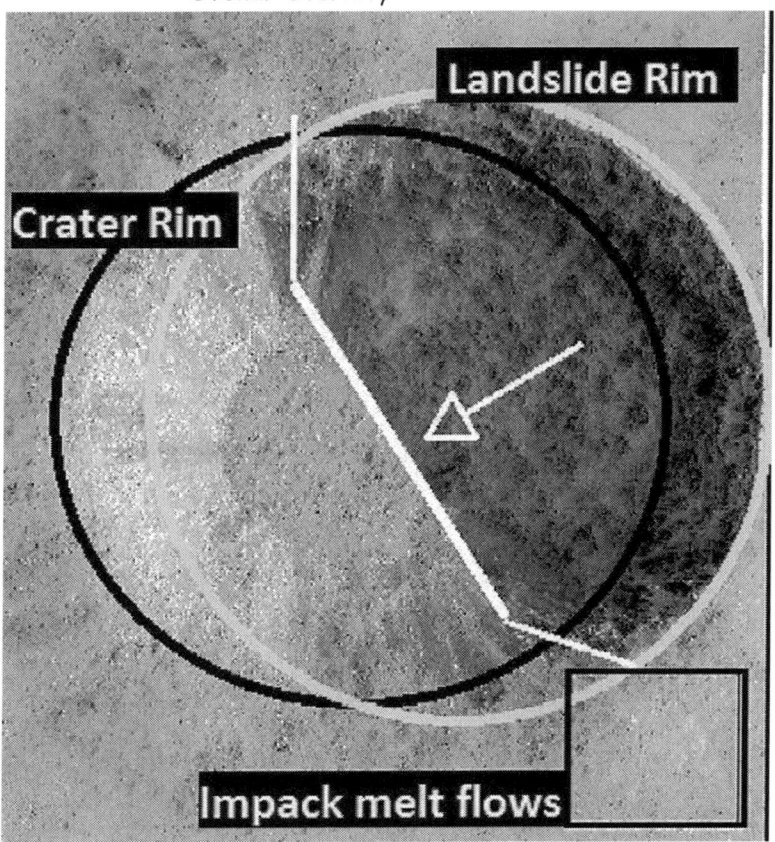

The yellow (white) line shows the crater rim and landslide, while the black dashed line shows the new rim due to the landslide. Reduced scale NAC mosaic (M152390311LR), image is ~4.5 km across [NASA/GSFC/Arizona State University].

31
MR VITO SACCHERI VITOED: TRYING TO BRIDGE THE GAP! CREDIBILITY IS MARE'D

A supposed close friend and supporter of George Leonard and the aliens on the moon theory is Vito Saccheri. Like George, who inspired his alien presence delusion, he also claims to have seen classified lunar photographs at NASA in the 1970's showing all sorts of ET constructions.

He said he observed *"A boulder that seemed to have been rolled uphill, leaving its tracks in the side of the hill; obvious machinery on the surface, showing bolted sections; three dilapidated "bridges" crossing a chasm that reminded me of the Grand Canyon; pipe fittings that looked like four-way T's (or X's) that could be seen in every photo, some with their ends turned up or down as they hung over the edge of a crater; three surprising pyramids... apparent pipelines cris-crossing the surface, running to and from craters; a UFO rising from the surface and photographed directly above a crater; and perhaps the most memorable, the unmistakable figure of a rectangular structure placed squarely in the biggest crater pictured - the structure looked either very old or under construction, but the crater had to be miles wide, and the camera angle gave a perfect three-dimensional view"* [26].

Unlike the Lunar Orbiter Cameras, let us zoom in on the credibility of Vito's claims. Apparently, his claim as to what he saw in the NASA photo collection is no more credible than his story on how he came to see them.

Saccheri recounts a 1979 experience with a Lester Howes in Houston. After they met, he said, Mr Howes, an amateur astronomer and a ufologist, wanted help in obtaining access to secret (classified?) photos held somewhere inside NASA. Saccheri claims: *"Les showed me a small paperback book entitled "Somebody Else Is on the Moon," written by a former NASA scientist, George H. Leonard. Leonard had been working in the photo intelligence division of NASA."*

Is this pure journalism or what? The information is not true at all, for Leonard never worked for NASA, especially the 'photo Intel' division. Leonard never CLAIMED he worked for NASA.

Mr Saccheri claims that after Leonard's' arguing futilely with NASA authorities about releasing the photos, only then was he able to publish them in his book. Again, this is a lie, for Leonard GOT the photographs from published sources – sources everyone has access to. He did NOT ever possess "unpublished" photos.

The Houston Sky continues to point out: *"Reinforcements were called in, and we soon found ourselves having the same conversation with the big boys from administration. None had seen the book, but significantly, one had taken the time to confirm that Leonard was in fact a former NASA scientist-at the Jet Propulsion Lab, from what I could gather."*

Good God! Besides no NASA employee of any quality and integrity would want to be seen with such a comic book, Saccheri blows the fictional account cover when he claims George worked for NASA. But, George Leonard never worked ANYWHERE for NASA and never claimed that he worked for NASA.

As to the mysterious hidden, hard to get to, offbeat, out of the way NASA photo archive, Saccheri continues to lie: *"We were directed to a Building 30, which had not been on the tour and which didn't even exist. Building 30-A turned out to be empty so we walked into Building 30-B and found ourselves in the middle of a high-security area where an existing mission was being monitored. Finally, some serious discussions transpired. The photo library, we were told, had been relocated off site to the 'Lunar Landing Observatory' directly adjacent to the east NASA property on NASA Road 1."*

This is also pure space spam. Building 30 existed, and was on the public tours (it is better known as 'Mission Control'), and was easy to visit, because in 1979 there WERE no NASA manned space missions to monitor. The lunar photographs were (and always have been) in the "Lunar Science Institute", the former Jim ("Silver Dollar") West hunting lodge, east of JSC on NASA Road 1. The name and the 'relocation' are erroneous.

Saccheri continues to fictionalize that *"Two days later, we drove east on NASA 1 past the main entrance of the facility, found a chain-link fence that marked the eastern limit of the property line, expecting to see a building or sign. Nothing but a heavily wooded area! Driving back and forth along the road trying to decide whether they had done it to us again, we noticed a narrow dirt road running back into the woods directly along NASA's fence line. Hung on the chain between two small posts was a sign that read simply: 'No Trespassing.' Instinct told us this had to be the place. We lowered the chain and drove about three-quarters of a mile down the dirt road, which U-turned back toward the highway. Directly behind the trees and camouflaged by the woods was our building."*

Stellar Sterility

This is ever so much bull and chicken droppings, for the building is 50 yards from the road, with an open lawn in front of it. The driveway is paved. Anyone can easily visit the place. There was no chain across any dirt road – not until years later when the institute moved to a new, much larger building about two miles to the north.

And this fairy tale goes on with, *"When we told the receptionist we wanted the library, she pointed toward the broom closet, which as it turned out, opened onto a winding stairway leading down into a dimly lit under-ground tunnel. I'm certain it took us back toward the NASA property line."* Sounds like a scene from some Indiana Jones adventure. The truth is the LSI had no basement, because it was all crawl space. The photo archives were in the library just off the large entry hall – on the ground level.

When it comes to technical details, the lie gets worse: *"He [the fictional librarian at NASA] explained that for security reasons, NASA had split the country into five regions, each with a duplicate set of records and a different code number system. Leonard's numbers weren't applicable in this facility. I asked where the master list was kept, and Roger [the fictional librarian] replied, 'at Langley, Virginia.' Les and I looked at each other."*

This final dose of sleeping potion is the ultimate in misinformation: The NASA Apollo photographs have exactly one code number, but when reissued as part of press releases, they received a public affairs office reference number too. These were the numbers Leonard used, which shows that the pictures had been published. The press office has a conversion sheet for the numbers it uses.

Of course, the Houston Sky article also mentions that NASA would not let him take notes, pictures or make copies of the negatives – a convenient way of Saccheri hiding the fictional BS! [27]

NSSDC Lunar Photo Archives

32
CONCLUSIONS:

Is there really someone else on the moon? Did we find artificial lunar (boulder) crop circles or over blown boulder balderdash? Were there tools in Tranquility or was it all tranquilizing tall tales of Tom Foolery? Were there found any alien crater condos for sale or should we crater the alien condo con job?

Lunar highways and byways or by the way, imagine these highways? We found no technology in Tycho, but only Leonard's fantasies and journalist typos. The alien made rays are not real. They are only George's personal speel. There are no wheels, no axles, no metal or steal, no cuts or crevasses, no ditches or stitches; no eggs, ovals, hedges or wedges. No nuts or bolts and no hammers or screws; no crescents, no crosses, mine pits or food fields. We find not water, not a shower, not cuspids or tower, no Egyptians or Mayans, no pyramids or any geometric angle, no artificial things that mysteriously dangle; no cars and no tires, no gears or spires. We cannot find rings, domes or any giant speakers, let alone skulls, mice, ducks or people. Bridges and causeways really do not exist, no more than junkyards. X-drones and dirt rovers are but pixel glitches, while there are no dinosaurs, eagles, dots, dashes, and super-size rigs and hitches.

Sorry, there are no giant treads, tractor trails or ladders, no numbers or letters, no missiles or rocket towers. When people try to make mountains out of a molehill and tell lies to bridge the gap, time and reason soon expose it as just pure flap. No, aliens have not visited our solar system.

Stellar Sterility

SECTION-3
Did We Discover
Alien Bases on the Moon? NO!
FRED STECKLING Stuck It to Us!
A Refutation of Fred Steckling's
"We Discovered Alien Bases on the Moon"

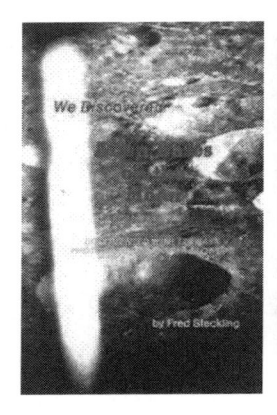

FRED STECKLING
On Nationwide Television (1977)

The second most important study in the chain of alien artifact hunting lunacy, demonstrating alien moon monuments, is Fred Steckling's *"We Discovered Alien Bases on the Moon."*

In this small tome of extra-terrestrial Tommy-Rot, Mr. Steckling attempts to squeeze the atmospheric blood out of the proverbial orbiting turnip - the moon, to prove that we should swallow the moon's habitability, at least along the termination line between the light and dark sides of the Moon. Why? He said, because *"the Moon's atmosphere, nevertheless, is*

Stellar Sterility

dense enough to support clouds and vegetation" (Steckling p.5).

Of course, anyone with any sense of investigation will find this book to be another example of outrageous synthetic selenography and lunatic geology. It supposedly proves alien settlement of the moon thousands of years ago. Steckling tries to follow up and prove the George Leonard theory, which he follows religiously, that there is plenty of evidence that the aliens never left, but are actually STILL digging for minerals, building space domes, and constructing underground cities! Fred Steckling is another Moon Monument Mad Hatter mythomaniac!

According to Fred, we can no longer believe the text book propaganda that *"the Moon is incapable of supporting life, is airless and is, plainly speaking, a globe of dead rock."* For, he suggests further, that with a good magnifier glass and a thorough study of (blurry) NASA Lunar photographs, the diligent student (stooge, dunce, imbecile) will see the marvelous wonders, such as cities that grow, hundreds of artificial lunar domes, constructions and artifacts that cannot possibly be explained any other way than of alien origin.

Needless to say, grab your seats and prepare for an enlightening implosion into the imaginative cratered blunder-world of Fred Steckling's aliens on the Moon. We are going to see that the Moon is as alive as the earth – alive and full of pipe-dreams, phantasms, hobgoblins and witches on brooms! There is no doubt about it, if we take his word for it, that it is crawling and bustling with as much activity as Manhattan, New York, looking not too close with unbiased eyes, we will see that "It's Alive," and screaming of liveliness.

Fred would have us to believe, after pointing out references to centuries of lunar observations, that the

Stellar Sterility

Moon is filled with anomalous and mysterious "crisscross lines" (highways?) within the Crater Schroeter, "tube system" type constructions in Crater Cassendi, and twenty mile long "glass-lined" tunnels running between the Craters Gassendi and those of Messier.

The evidence does not stop here, but continues indefinitely with such examples of space alien manufactures as a black wall in the Crater Aristarchus (supposedly made of glass) that was not there before, and is not there now, and we may as well say was never there to begin with.

Without slamming on the breaks of our un-reasoning, we may Moon-cruise ahead and easily find "artificial mounds," "alien domes" about the Crater Archimedes, and other junk surrounding the local areas, along with various other mysterious geometrical shapes (Steckling p.12).

Furthermore, with no delay, we must orbit forward and keep our hats on tight as Fred blows more hot air artifacts up our exhaust pipes. For, besides these highways and byways, tunnels and funnels, there are more evidences to be obtained. Look closely and you will see UFO shaped objects, elongated cigar shaped crafts, digging machines, mining equipment, pleasure vehicles and lodge dwellings, weird carved symbols, signals, markings and best of all, old rivers and some few lakes, clouds, rivers and even vegetation!

In the Craters Plato, Aristarchus, Messier, Eudoxus and Mare Crisium (not to mention, a million other craters), it can be seen (with not only a powerful telescope, but an strong imagination as well) mysterious moving lights, blazing triangles, flashing signal lights, long cigar-shaped alien craft, industrial platforms and other utility buildings (p.15-16).

Stellar Sterility

Since the moon is alive and well (p.13), let us grab a shovel and straddle our brooms and take off to the moon to see for ourselves. But mind you, it will all be in old black & white, whereas, we shall enjoy the Technicolor of the imaginative interpretations.

Some of the first, and I mean the first of materials to ponder over in lunar selenography, are the locations about the area Oceanus Procellarum, an area between the Craters Seleucus and Hevelius. The Lunar Orbiter photographic equipment captured this area filled with alien debris and what looks like "ponds" as can be seen in (Steckling's choice of) NASA image IV-157-H3. The photo spans 30 degrees, from 30 degrees N X 55 degrees West, to O degrees X 55 degrees West, and exposes what Fred claims are UFO hangers and swimming pools, if not agricultural ponds.

Of course, these ponds could also be hydrogen fuel supply tanks, since UFO's run off of hydrogen power! And, because that equals "water," they could be alternately used to swim in on hot days, after heavy work schedules. They were obviously piping the water underground from such places as these lakes. Anyway, let us see what Mr Fred has to say, and then do some studies of our own for verification to determine if there are such structures on the moon.

1
ALIEN UFO HANGERS PONDS, CLOUDS AND OTHER MYSTERIOUS MOON BLOBS

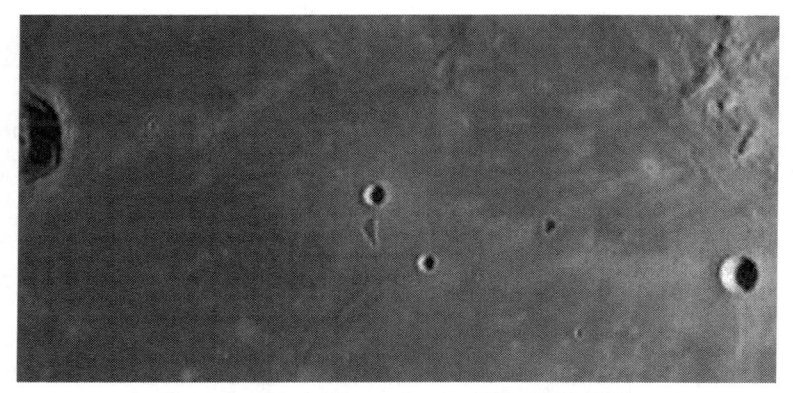

Lunar Orbiter photo IV-157-H3

Fred imagines he sees an artificial crater and a twenty five mile long artificial "UFO Hanger," a large structure supposedly used for hiding alien space craft. Fred said that "without a doubt, nature does not produce objects positioned like the well-formed crater and this perfectly straight mound or hanger, which is some 40km (25m) long." (p.85)

Fred's photo reference (157-H-1) on page 88 is NASA's Lunar Orbiter photo L.O. IV.157-H3 and 157M (Medium shot). The above photo shows Fred's unnamed crater and an alien "UFO hanger" nearby. Below are my additions of a "pond" and a crater. My selected pond is taken from one of those emulsion splatters, which we shall talk about soon.

Stellar Sterility

Fred imagines he sees an artificial Crater and a twenty five mile long artificial "UFO Hanger", a large structure supposedly used for hiding Alien Space craft. Fred said that, *"without a doubt, nature does not produce objects positioned like the well-formed crater and this perfectly straight mound or hanger, which is some 40km (25m) long."* (p.85)

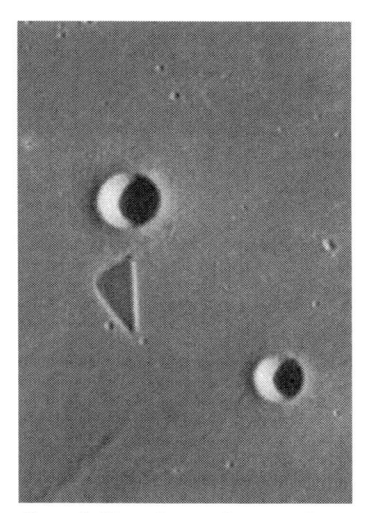

Steckling's enlarged and cropped image

Composite Photo

Stellar Sterility

It is a weird photo with an odd shaped structure and is very impressive to the untrained eye. For those who trust Fred's reproductions in the book, the evidence is undeniable! With a little effort and some money to purchase NASA Lunar Orbiter Photo negatives, the student of alien artifacts can easily see that it is a farce!

The book uses horribly blurred and under developed photos. The reproductions are as bad as bad can get. With such photo copying anyone can take a NASA image and demonstrate anything he desires. And this is exactly what Fred Steckling did. For instance, here are some other examples of "blurry" clouds, giant finger-print like space bacteria, alien craft and other such nonsense.

Mr. Steckling knew exactly what he was doing and also knew exactly of his lying and forging "evidences" when it comes to supporting alien activity on the Moon. He was writing a book to sell and he had access to the best negatives NASA had, just as anyone of us today can have. All a person has to do is order the same materials from the NSSDC. (The National Space Science Data Center), which is the repository of most everything NASA has available on the subject of planetary photography and see for themselves that Fred is full of Lunar hot air! Never mind the men behind the curtains of NASA, who will argue the real truth, Fred has it all figured out for us! NASA is the one lying and covering up secret data and knowledge from the public.

In reading Fred's research on page 85, we are to believe that what is depicted here are alien objects, located at 55 degrees West by 25 degrees North. This is an area that covers some 319 by 81 miles of lunar surface! This location lies west of craters Aristarchus and Herodotus. This coverage is located in Oceanus Procellarum, east of the craters Seleucus, south to crater Hevelius.

On the NASA print one can easily see emulsion 'splatter' marks and development "blobs," and running completely off the image area into the non-image edges between the frames! Surly, without doubt, Fred was not blind nor was he ignorant of this, as to completely miss such a stark raving white emulsion splatter flying across his negative! It is unheard of short of forgery to misrepresent an obvious photo development flaw. This is perpetrated hoax.

Stellar Sterility

2
MOON MICROBES

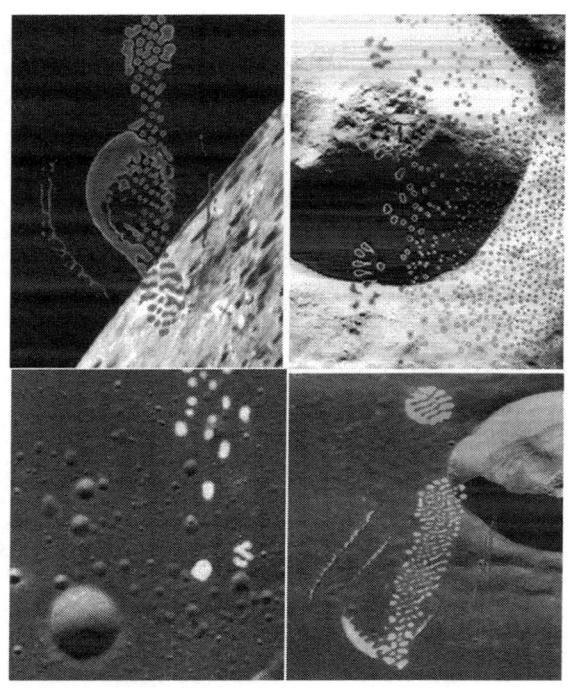

Alien UFO's, Lunar Clouds, Moon Microbes?

Mr Steckling's photo on page 88 and plate 61 is presented to prove without doubt that there is a 25-mile long "Hanger" structure adjacent to the Crater. Actually it shows no such artificial structure at all, but only a development flaw! So what is it? Is this alien evidence in the negatives or evidence of alien emulsion splatters on the negatives? Even with a two digit IQ it is not hard to see that these are development imperfections.

Stellar Sterility

If we compare NASA's IV-157-H3 with other Atlases of the Moon photographically covering the same area, we will notice the missing emulsion splatters in the Crater Plato. Comparing the atlases with the NASA negative reveals just what we suspected and were looking for: The truth is the alien construction is not there, was never there and is not there now. The only place it can be found is literally "on" the NASA negative!

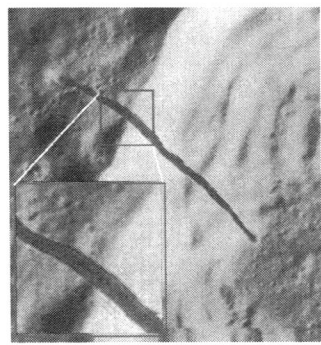

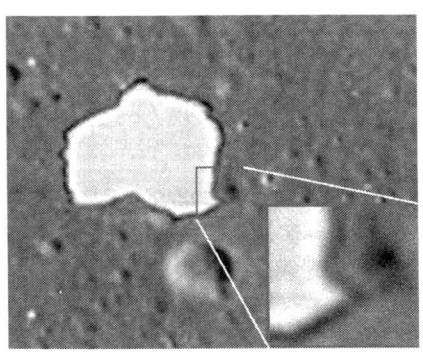

IV-128-H3   Water slide   III-194-m Pond or blemish?

Researching the atlases produced no substantial results in support of Steckling findings. For example, the so-called "alien hanger" or "pond" that appears to be alien in structure, after careful analysis was not alien at all. It turned out to be an emulsion artifact and is only alien to the NASA image.

Steckling presents one of the worse examples of the area. It is so blurred and printed with the dot matrix pattern in such a spoor fashion that anyone ignorantly trusting Steckling will obviously believe he sees some alien activity site. Even I was somewhat surprised of this before I saw the same image in full resolution clarity directly from the NASA print!

Stellar Sterility

Further studies proved that this alien structure was no alien artifact after all. When the 157-H-3 Neg. from NASA was scanned and enlarged, the so-called "hanger" was revealed as an emulsion Blob! Just another water or moisture emulsion 'blotch'! Here are some more processing artifacts:

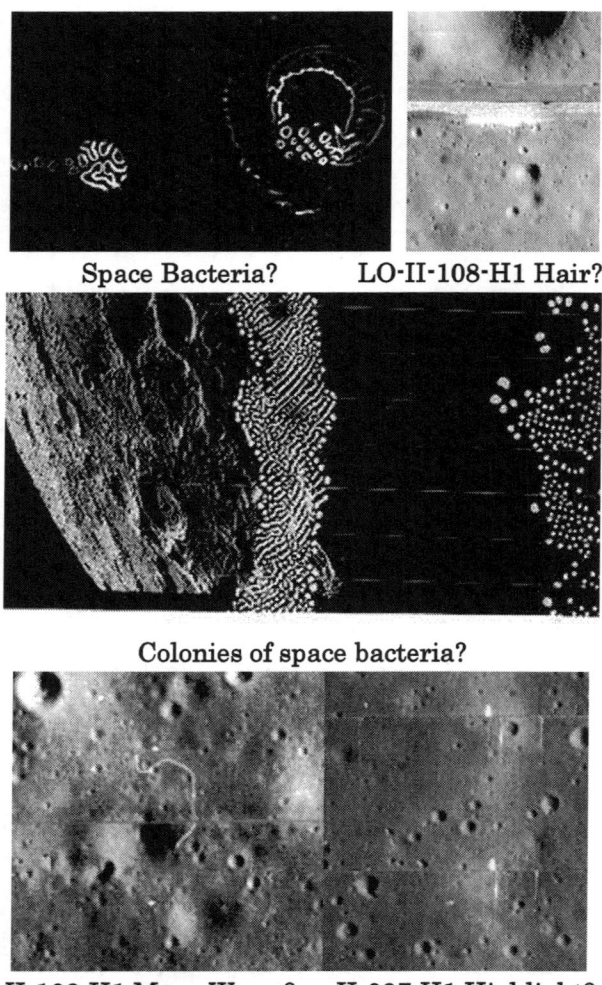

Space Bacteria?   LO-II-108-H1 Hair?

Colonies of space bacteria?

II-108-H1 Moon Worm?   II-037-H1 Highlight?

3
ALIEN IRRIGATION PONDS

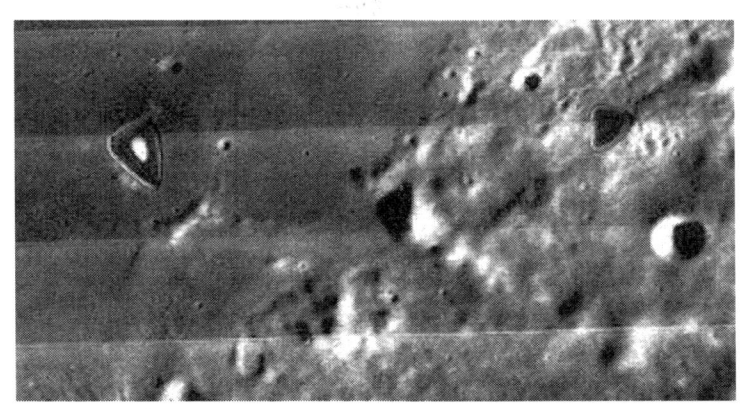

IV-128-H1

    Throughout the Lunar Orbiter Photographic Atlas pictures one can find other similar alien structures. In the L.O. IV-128 there is a crater next to a larger crater or another "alien hanger" - depending on how one looks at it. This is one Mr Steckling seems to have over looked.

    The warehouse hanger appears to be another one of those "ponds" according to Steckling, one that he likes to flaunt as alien in origin. Comparing the above frame to the larger Medium framed 157M the artifact hunter can see a larger section of the same area, but on a larger scale. Close inspection reveals that the alien hanger 'blob' is missing! The blob hanger disappeared! It appears in one frame and NOT the other! Do we have movable hangers? Hangers ¼ mile in size that are portable?

Stellar Sterility

Where did it go? Well, it just moved somewhere else and we do not know how. Closer inspection reveals that there are hundreds of these 'hanger' blobs in many of the NASA negatives. The aliens must have airports, hangers and adjacent swimming pools everywhere on the Moon. 157-H-3 appears to have a few splatter blobs, and 157M also has TWO separate horizontal emulsion patterns. Comparing these emulsion blobs with the hanger reveals that it is a recurring stray emulsion splatter. It has something to do with the development system on the orbiter.

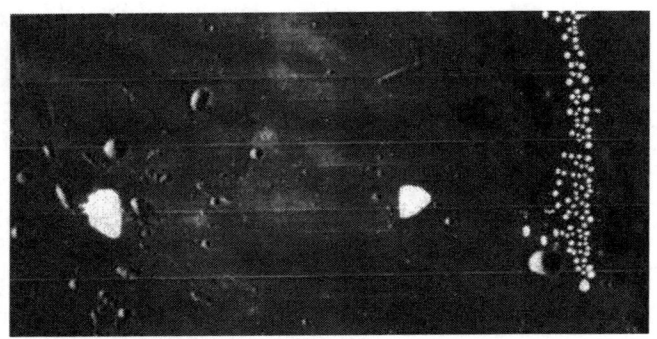

IV-114-H1
Steckling's "White Ponds"
Notice the supposed "clouds" to the right.

Another interesting feature about these blobs or alien hangers is that under magnification, enhancement and advanced focusing, the blob-hangers are more focused than the local lunar terrain surrounding them! This therefore shows us, by association, that the hanger or pond is just a BLOB!

A photograph is a photograph and each has a uniform focus. Nothing can be more focused than any other thing in the picture. Notice how the hanger sits fine focused upon the surface of the photo as an

Stellar Sterility

independent additional "thing" contrary to the less focused lunar surface features. Everything is slightly blurred except the blobs.

The same type of splatter blobs can also be found in Steckling's other "pond" photos in LO IV-128-H-3, which should actually be frame number "1." Fred said the structures are irrigation ponds, canals or dams. When enlarged the "ponds" display the same effect of being superimposed over one half of the raised lunar surface, which is also much less focused than the ponds! He said they are irrigation ponds situated among water bearing clefts and the rough terrain about the Crater Plato.

In all cases the "ponds" and "hangers" display the same smooth rounded and triangular pattern. Compared closely with the true splatter patter blobs of 157M one can see that Steckling FABRICATED his "pond" and "hanger" theory! Such a wild imagination! What a best seller for UFO buffs to read! From such poor photo reproductions, they DO look like California irrigation ponds! (See Plates 55, 56, 60. Revised Ed., Plates 51, 52, 56)

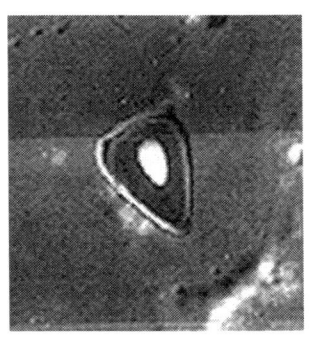

LO-IV-128-H1 Close up.
Compare photos 128-1, 126-1, and 114-1

Stellar Sterility

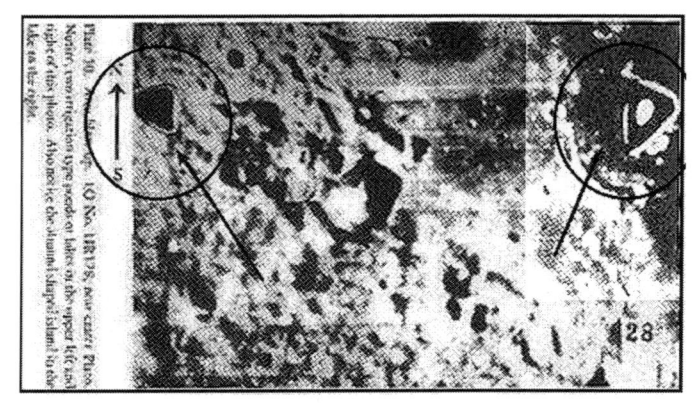

IRRIGATION PONDS

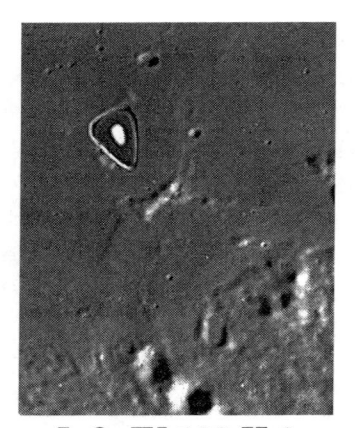

L.O. IV-128-H-3

It appears without contest, that in ALL cases the ponds and hanger images display the same finer focus with microscopic jittery edges, just as we see in the true emulsion splatter blobs in 157M photo of Crater Pluto.

So, now we know "WHAT" these blob hangers and ponds are. But, can we find more of these 'ponds' that Fred missed? Are there more to be found in the frame "1" series of photos? Yes there are plenty of them. Here

Stellar Sterility

is a list of the NASA Lunar Orbiter Photographic frames (L.O. IV "H-1" series) showing this "same" exact set of Ponds [1].

What is also interesting is the symmetry involved. It is as if all the High resolution Frame Number 1's seems to always have this splattered pond effect. The emulsion appears to re-appear over and over again. Apparently, in developing the H-1 frames they must have experienced the same "consecutive" development problem. Many even run off the edges of the photos! The consecutive examples in the Lunar Atlas show that they are a series of splatters unique to the #1 series of the frames.

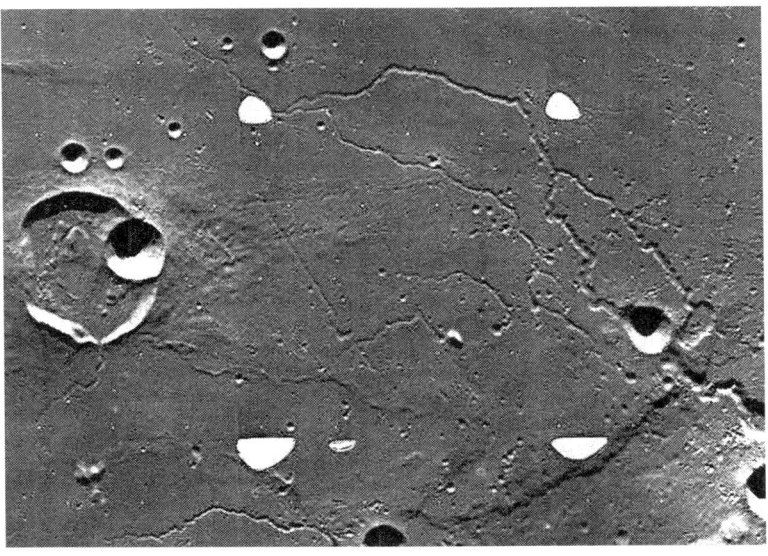

IV-151-H1
Steckling Plates 57, 58, 59, 57A

In checking the Lunar Orbiter Atlas page by page for other "ponds" reveals that there are over 57+ plates out of 404 photos that have these anomalies or

MULTIPLE POND emulsion blobs. Many have splatters and dotted 'cloud' looking patterns, and fingerprint patterns next to them, as the photos above show (See 114h1 photo).

Out of 404 photos every frame number "1" has the same TWO ponds. There are white ones over black areas and white ones over visible terrain (transparent ponds). Also clouds, which are also emulsion splatters with the fingerprint looking clouds sometimes accompanying them. After careful reviewing of all 57 plus plates and their supposed multiple "ponds", I found a single common denominator. They ALL show exactly the same number, distance and symmetry.

Compare the following set of "ponds." Look at the geometric layout of the 'White' Ponds! All have that slight triangle look with the same axis line. See my drawing about this phenomenon. What is interesting is all the axis lines have the same perspective points. The duplicated symmetry is staggering and apparently ALIEN according to Fred! Alien, because he didn't choose to reveal this repeated pattern in all the Lunar Atlas photos to his readers. I presume that he choose NOT to because it would have exposed him as a fraud!

Comparisons made of all the listed examples indicate that the ponds (the TWO ponds that always show up in every photos listed), are the SAME TWO emulsion blobs, which show up exactly the same way in ALL the photos listed! It is the same with the FOUR pond sets. This obviously presents us with one conclusion. That these sets of "ponds" are emulsion splatters from photo development processes.

When NASA photo 128-H1, is enlarged, what do we see? We see the same 'pond' anomalies as found in the previous 157-H-1 photo. Fred argues the same asinine rhetoric as before to prove his case: *"The most*

Stellar Sterility

revealing things on this picture are the two irrigation type ponds or lakes with reinforced walls. They are triangular shaped on the left side of the photo, the other on the right side, with an almond shaped island." (p.82)

Fred asked a question here, *"Is it perhaps a pump station?"* Well, NO, not really. Fred could have given us the excuse that these ponds must have put on some alien theatrical and moved as quickly as the Lunar Orbiter Camera moved to shoot them. Of course, taking a fixed position in front of the camera as the camera scanned the regions.

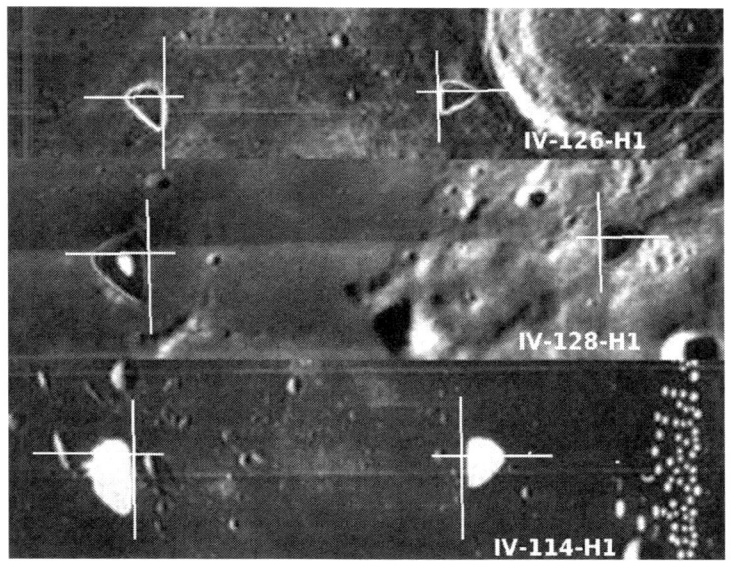

Emulsion Blob "pond" symmetry

There are simple explanations for these technical problems:

1.) The "ponds" are emulsion blobs in the development of the negatives. 2.) They are specks of

Stellar Sterility

dust on the camera lens. 3.) They are some form of light leakage into the camera that appears on all the frames numbered "H·1" They do not appear on frames H2 or H3 or any other ones.

In one photo the pond is miraculously transparent showing the fault line under the pond! Why would aliens dig a pond right through a hill?

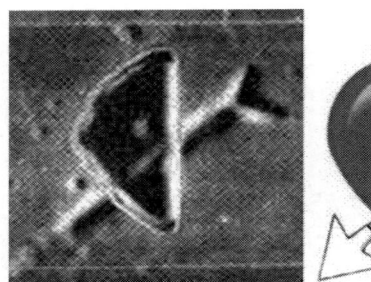

IV-139-H1

Stellar Sterility

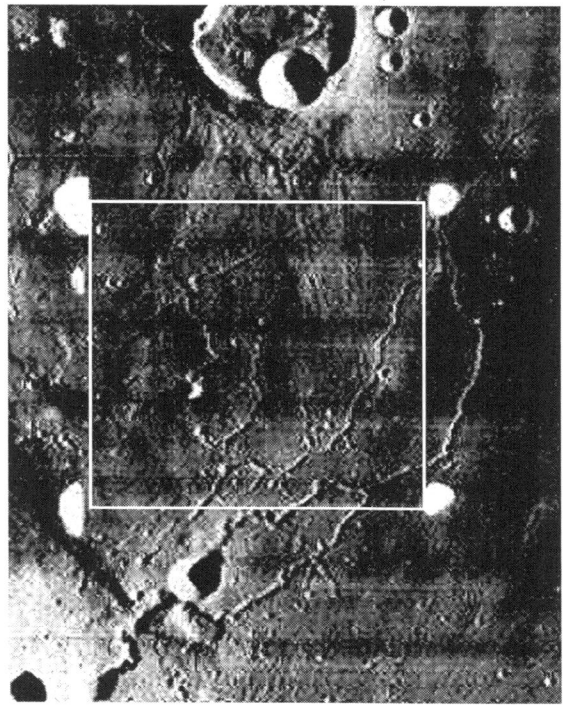

151-H-1
Duplication of the emulsion blobs show the ponds to all be the SAME set. It is the same with the set of four ponds.

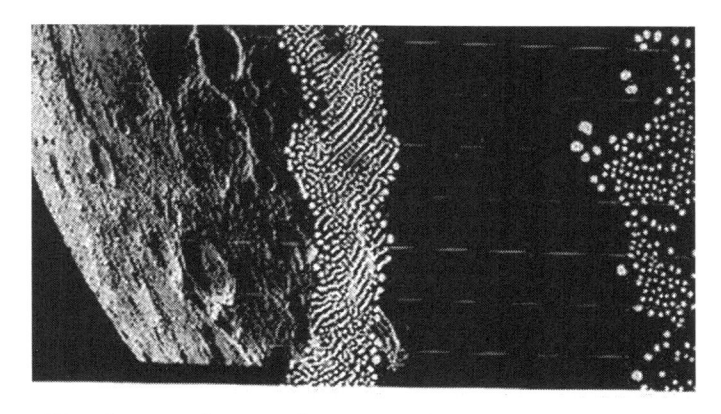

Check out these bacterial 'emulsion' splatters.

Fred uses the photo 151-H3 of Crater Krieger and the surrounding areas to prove they are advanced alien constructions. (P.101 Plate 53) There are four ponds as above shown. He said, *"To the right of the crater Krieger, notice five triangular shaped irrigation ponds, or reservoirs just like the ones on the previous pictures. Except here, the sun shines from the north just at the proper angle and altitude to reflect its light off the water, which shows up white in a black and white photo."* (p.114)

Computer enhancement reveals these are photo development errors and are not found on the lunar surface. The blob is sharper than the blurred Lunar surface, which contradicts LO photographic focus consistency. The ponds have more clarity than the lunar surface, which means to me, that the camera did not shot them from off the lunar surface at all, but must have picked up this added phenomenon from inside the camera or during the development of the negatives. Again, they only appear on the high resolution frames "H1" and not the Medium resolution shots.

4
CLOUDS OVER CRATER VITELLO
L0-V-168 M

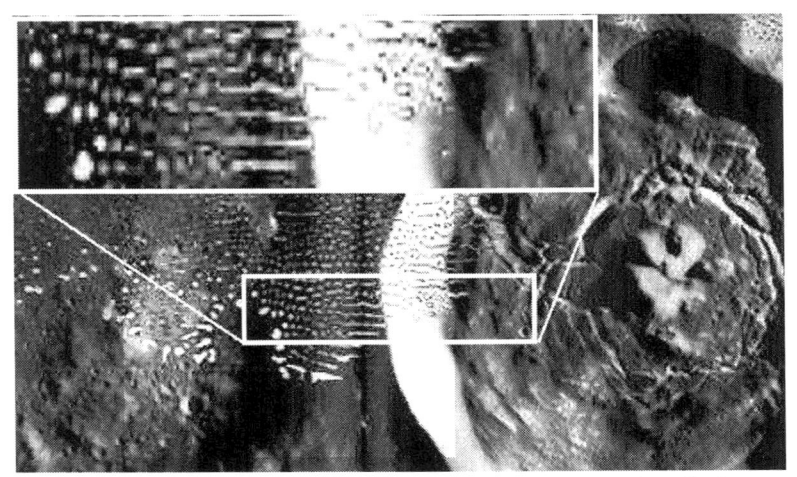

Notice that the "clouds" run toward the right side off the lunar surface!

Fred's next bold attempt to establish an alternative Lunar Atmospheric Science is with LOV-168-M), his photo 45, page 77. Crater Vitello is shown here with "clouds" floating over the sky above and across the Crater Rim. Fred said, *"Notice cirrus type clouds left of crater rim. Also notice platform in the center of the crater."* (p.77)

Cirrus type or Macheral type clouds? What a fantastic scientific observation to throw off the ignorant lunar meteorologist. HOLY MACHERAL! This photo is fishy to me! A real fisherman's tale, too!

All Fred gives us here is a "chopped" photo. Now what difference does this make? Is this just a cropped

Stellar Sterility

version to fit the book for publishing? NOPE! He crops the picture to hide a very important exposure (pun intended) of his fakery.

A little investigation into the complete NASA negative reveals that these "clouds" are miraculous clouds and not Macheral clouds, and that they are not clouds at all, but is obvious emulsion blotches, blobs, splats and splatters. These so-called clouds "miraculously" float right off the edge of the negative and into the dark unexposed area of the film proving it is all a development flaw. There are hundreds of them on the Lunar Orbiter photos!

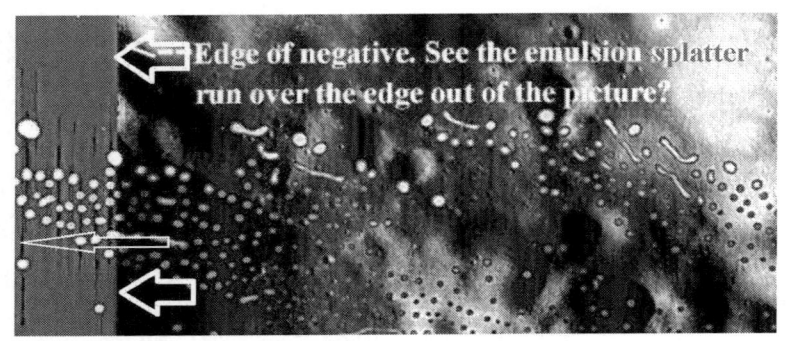

Emulsion splatter runs off edge of picture.

Now, why the lie? Why the big bold-faced misrepresentation of the facts and the hiding of what the negative really shows? Fred said he had access to them! Why did he assume they were clouds when they are blatant emulsion problems? A BEST SELLER! Hell man, it was the early 1980's and everyone was into UFO's and Aliens, just as they still are today! And so, THE BOOK! "We Discovered Alien Bases on the Moon."

Well, we have discovered alien interpretations of the Moon alien to what the negatives really reveal. The

Stellar Sterility

Lunar Orbiter V photos show that there are at least 18 such "cloud" photos and probably more if one takes the time to search for them. [2]

Why are there so many selenography researchers hell-bent to lie to the public? Answer: To promote the 'Alien Agenda" and hide the boring truth, discredit religion and real science, propagate hyped up sensational misinformation and promote book publishing. Not everyone in the 1980's could afford to buy up all the NASA negatives to verify the truth. Maybe the public does not want the truth and they prefer to want aliens to exist. Maybe they WANT to believe nonsense and mythologies because it's fun, easy to believe and does not demand responsibility. Maybe they want fraudulent soothsaying evidence to pacify their fears of a silent universe. Maybe they are not happy with a simple God and simple religion of hope and prefer a more romantic story that aliens are coming to save them from atomic destruction. Maybe people are just stupid, uneducated and gullible! It is the opinion of many decent scientists that the alien presence myth offers an alternative hope evasive of human responsibility and void of future judgment, and that it is a replacement for traditional theology.

Human nature gets real bored with the mundane reality of life, a life void of mysteries, suspense and the magical. There is a part of human nature that loves the horrible, the outrageous, alien invasions, threats and doomsday scenarios. Myth brings relief to daily drudgery, work and utilitarianism. Science fiction seems to release stress and open doors to what is not natural to human existence. The fantastic is necessary if there is no true god to turn to. The extraterrestrial alien savior is the new cosmological answer to the human dilemma. Nevertheless, some of the public may actually be interested to know that they are being lied to, if they could see the truth about these photographs.

The truth can be stranger than fiction. Moreover, what is stranger than this moon monuments mad fiction is the fact that the majority has swallowed the E.T. hook line and sinker. It would appear that human nature prefers the lie over the truth, for the truth always brings human responsibility.

It is easy to see to see how people can be deceived. But to lie and deceive the public for self-gain is intolerable! The only theory is "THEY" can do it and get away with it with immunity. They, as well as the government can also make a lot of money at it too.

Nevertheless, there are those who disapprove of such deception, if for anything, for the sake of the truth. Hence, the purpose of this "Moon Monuments Madness" book is to expose the fake alien habitation of our solar system.

5.
CRATER DOMES AND ALIEN HOUSING PROJECTS

Now, what else have these myth-makers found? CRATER DOMES, little lunar houses for aliens to hide in, weird work worm-hole lava tubes and meaningless mining operations have been some of the many mythical inventions fabricated by the spin doctors.

LO-IV-187-H2
Take a look at this "dome" sitting on the edge of a crater.

Stellar Sterility

Mr Steckling again has found some more interesting alien artifacts in the Orientale basin area. About the highly fractured hummocky terrain there are numerous odd shaped domed mounds, some of which have fractures on their crests. Fred thinks these are alien structures and finds what appears (and I mean "appears") to be a mound or blob sitting on half of a crater.

One of the most interesting "domes" found by Fred (and every other NASA geologist) is located at 85 degrees W X 15 N. Mr. Steckling points out that these pictures also show examples of atmospheric clouds and alien cigar shaped vehicles.

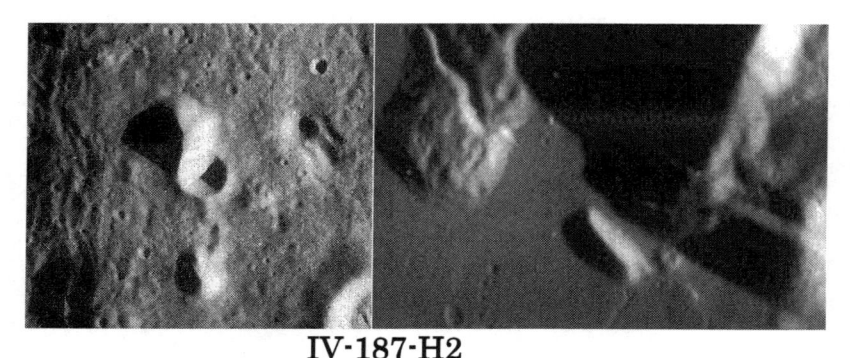

IV-187-H2
The crater mound "dome" and another cylinder shaped dome.

In Plate 66, he said that, in *"The Alpine Valley area of the Moon...* [We may] *Notice (a) large cloud-like object covering half of the central crater."* In Plate 67, Fred insists that in an area blow-up of LO IV-187-H2, we will undoubtedly notice an *"oval shaped object, UFO or cloud, one mile east of the large crater.* And that we will also see *"More clouds are in the upper right plus three domes (two northwest of the crater)."*

Stellar Sterility

These so called lunar domes are only examples of natural lunar morphological processes. Geologically speaking, domes can be naturally made. These natural mounds show up in other locations in the Lunar Orbiter photos. They are common in this area and are part of an inner blocky ring encircling the basin at the same radius. This particular IV-187-H2 *"crater mound"* (above) indicates an isolated peak surrounded by or protruding through this unit. (Shultz. p. 478). The crater rim "dome" has at least one other possible explanation besides the crater ejecta theory. It is the "fumarole" or gas vent theory.

In Plate 67a, which is LO IV Photo 187-H2, he continues to suggest that *"we can see a large cigar shaped object above the crater floor."* (p. 120-121).

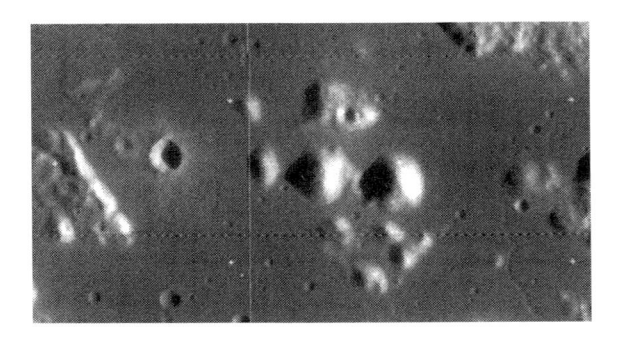

Stellar Sterility

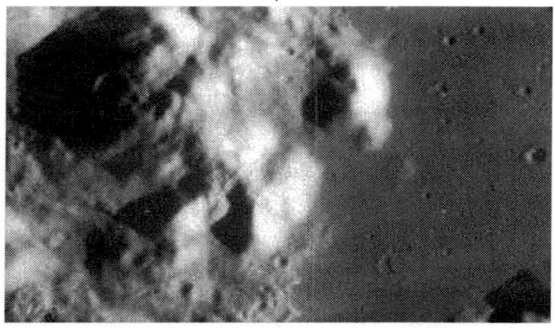

Other "clouds" or "domes" found in this area.

A close-up reveals no alien activity around the site to suggest that the mound is 'artificial'. Actually, the area is less chaotic than other locations and should not be a surprise to a lunar geologist.

The following illustrations demonstrate how an impacting meteorite can strike the lunar surface causing volcanic disturbances, release gas "fumaroles" and spew out lunar ejecta regolith materials, thus creating a lunar regolith mound to pile up on one side. Photo IV-187-H2 admittedly presents an interesting geological event, yet nevertheless, a naturally created one.

A fumarole (Latin *fumus*, smoke) is an opening in a planet's crust, often in the neighborhood of volcanoes, which emits steam and gases such as carbon dioxide, sulfur dioxide, hydrogen chloride, and hydrogen sulfide. Fumaroles may occur along tiny cracks or long fissures, in chaotic clusters or fields, and on the surfaces of lava flows and thick deposits of pyroclastic flows. A fumarole field is an area of thermal springs and gas vents where magma or hot igneous rocks at shallow depth are releasing gases or interacting with groundwater. The formation called Home Plate at Gusev Crater on Mars which was examined by the

Mars Exploration Rover (MER) Spirit is suspected to be the eroded remains of an ancient and extinct fumarole. [R.V.Morris, S.W. Squyres, et al. "The Hydrothermal System at Home Plate in Gusev Crater, Mars". *Lunar & Planetary Science* XXXIX(2008)].

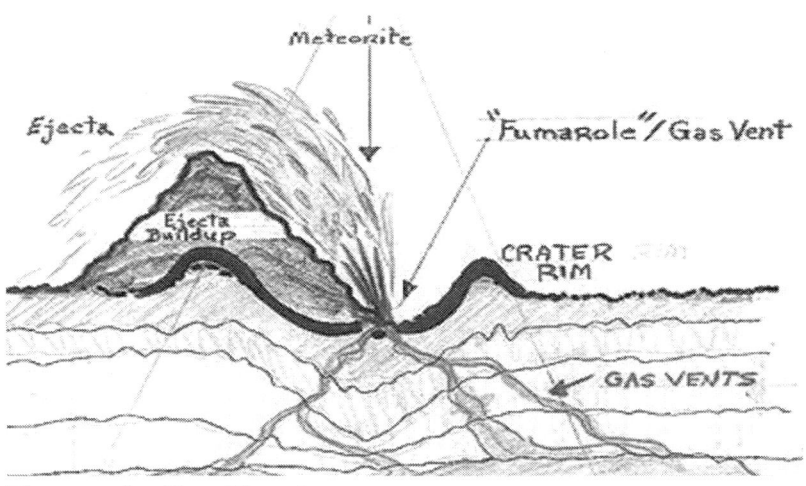

Fumarole Gas Venting

Stellar Sterility

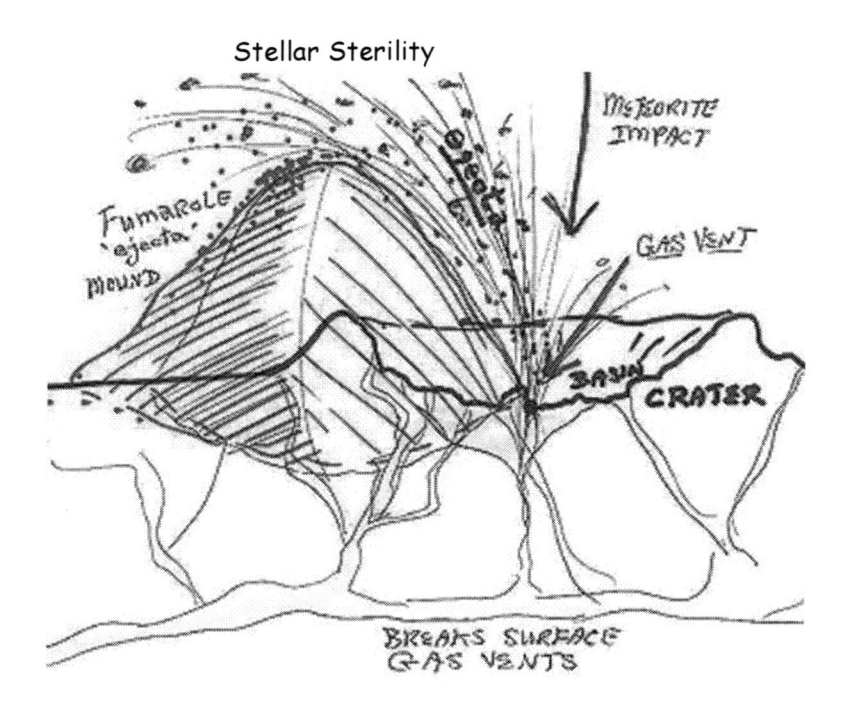

6
FRED'S DAM'D CRATER WALLS AND RILLE RIVERS

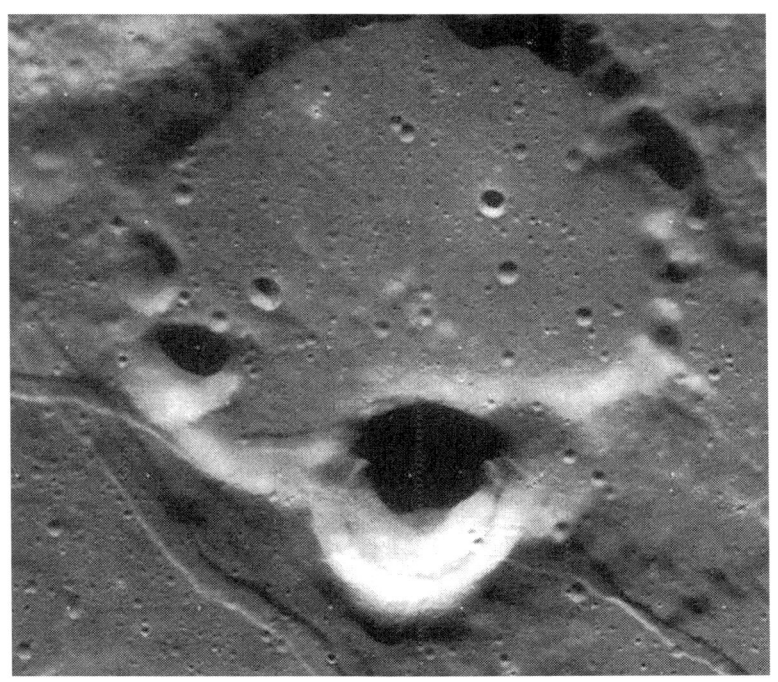

LO-IV-161-H3
Close-up of the crater with dam walls.
(Plates 71 and 72, p. 126-127)

Steckling's problem of wrongly numbering his NASA negatives seems to be an on-going problem. This misled me for a while, for about six months, looking aimlessly for this location on the wrong frame. The Lunar Orbiter IV photos gave me a much clearer picture of the surrounding areas. I found the correct frame and also found no need to waste any more time or money on NASA Negatives. I found the problem!

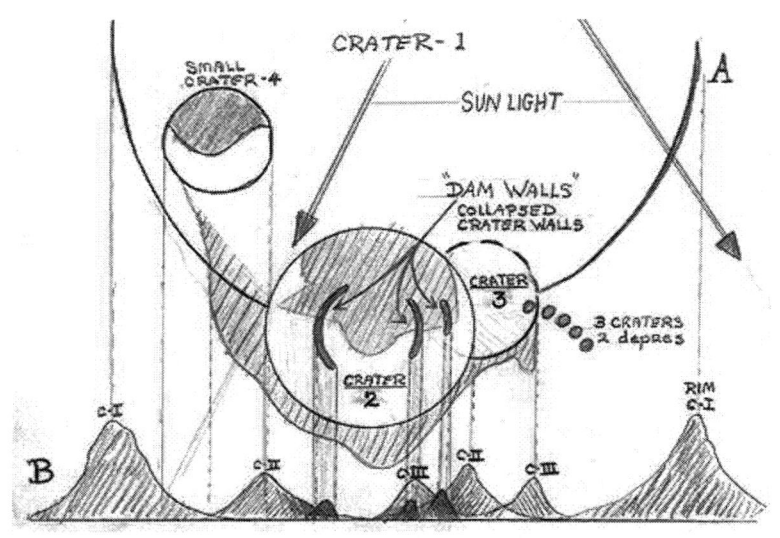

CRATER DEVELOPMENT

Fred said, that when the Moon had water, the aliens converted Damoiseau-D Crater (161-H-3) into a gigantic reservoir by building what 'appears' to be dam-like walls inside the crater. Again, Steckling misses the frame number calling it IV-161-H1, while it is actually the high resolution frame H3. This is the crater Damoiseau-D.

Fred imagines he sees a "smoke pillar" drifting from west to east, clouds floating over the crater's edge, a long platform entering a hanger, various assorted constructions, platforms and a larger platform extending over a riverbed (p.102). Besides all this selenology silliness, he sees two more double walls, more riverbeds, what looks like dam walls and five

Stellar Sterility

evenly spaced constructions, *"much like pump stations"* (p.103).

Well, I admit there are many signs of erosion in the area, but does erosion necessitate alien activity, artificial dams, drainage canals and pumping stations? Of course not. All of these things can be explained by natural lunar geological and morphological processes.

All of Fred's photos, including this one, are blurry and undecipherable. No one can really tell much about the details in his book and I am sure he knew this! One's imagination can run wild with blurry photos. The same goes with cloud formations. People see animals in cloud formations all the time. But, in the latter case, sane people know that there are no real animals in the clouds except for air plane pilots, birds and high flying insects and maybe a hail-storm of frogs.

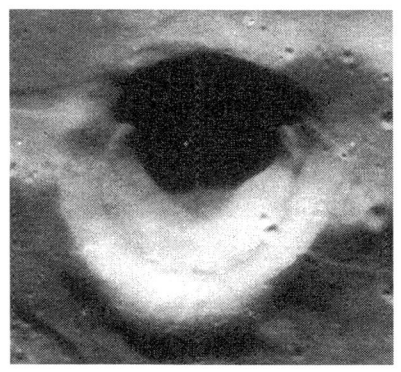

Enlargement of crater Damoiseau D

Once released, the untrained imagination, prompted by the whorish ramblings of Mr Steckling will humbly follow and believe anything. Imagination derives from "image" or 'icon', which is the basis of 'idolatry', a false representation. With this in mind, anyone can see how a person reading this cheap

Stellar Sterility

balderdash can contract all kinds of adulterated ideas and absurdities flying through their mind.

So, Fred has found, in (Plate 70), many proofs to alien activity. He has smoke pillars, clouds, platforms, constructions, rivers, pump stations and dams. After reviewing frame IV-161-H3, it will be found to be true: The "smoke Pillar" is an old double crater rim highlight from intense sun light. The clouds over the crater rim is sun light at same angle as inside the opposite wall of the crater.

The platforms and hangers are more intense sun highlights burning their way around and over the lunar hills or from the photographic glare when the picture was taken. The constructions are crater rim ejecta, highlighted and shadowed with light playing on the surface. The riverbeds may very well be some sort of ancient flow either by water, lava or volcanic erosion and surface fracturing. The pump Stations or five objects described by Fred are three craters and two mounds or depressions casting shadows. The dam walls inside the crater are two opposing crater wall collapses. Other craters show this kind of double crater rims. One is usually larger with the smaller one are inside the larger.

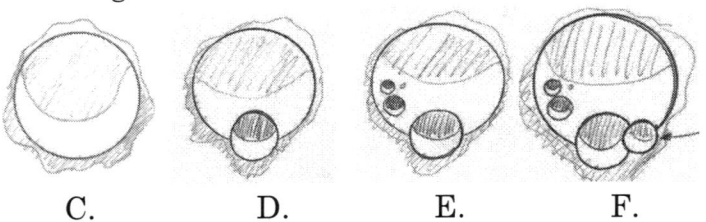

C.   D.   E.   F.

The crater's formation from multiple impacts
    C. Older original crater
    D. Smaller secondary impact crater
    E. Other minor impact craters
    F. More impacts

COMPLETE CRATER LAYOUT DRAWING

Notice the progression or history of the development of the crater through crater impacts. Crater C is the original crater. Crater D adds a secondary crater that sits on the previous crater rim. E and F add other minor craters overlapping the original crater. These multiple crater impacts cause crater rim collapses, thus creating what Fred calls dam walls inside the crater.

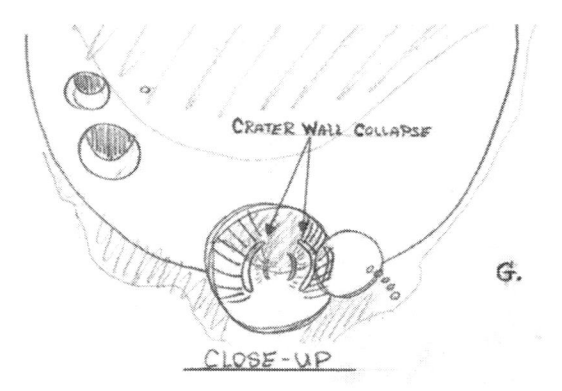

Crater wall collapse

Stellar Sterility

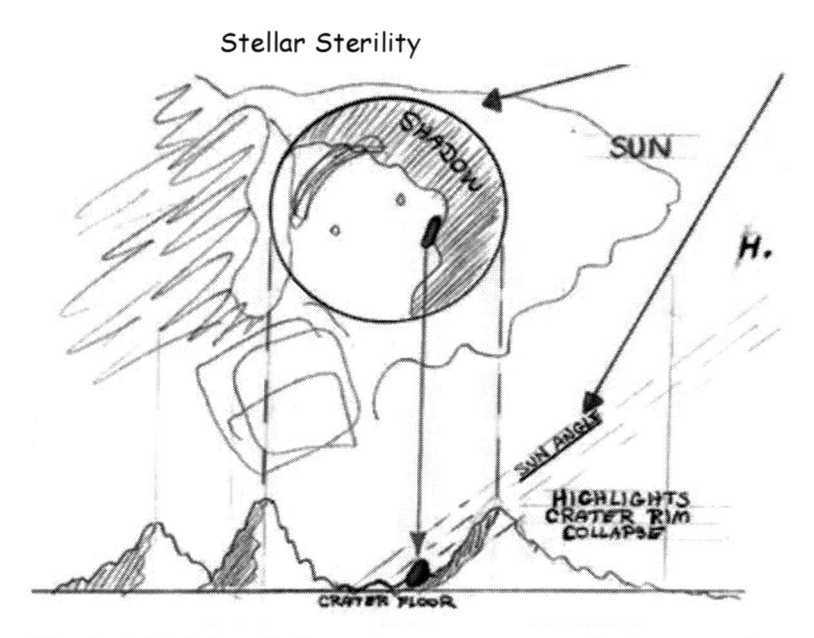

The bright "cigar" shape is now apparently just another chunk of collapsed crater Rim. Here is what the crater geology is showing.

7
CIGARS ON THE MOON

Craters with hidden UFO's? Cigar shaped alien space craft? This is what Fred sees in this particular NASA frame. In the center of Crater Romer there hides an alien UFO. As Mr Steckling said, *"Notice the large platform and long object on the center mountain."* (p.161).

AIRFORCE BOMBER FOUND in a Moon crater?

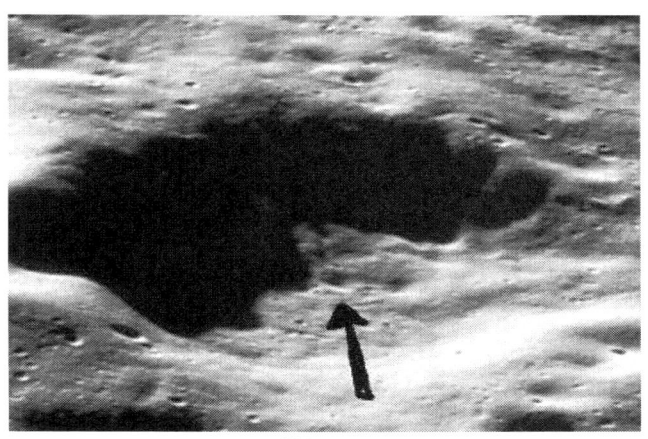

Stellar Sterility
AS16-118-18923   A Crater Cigar!

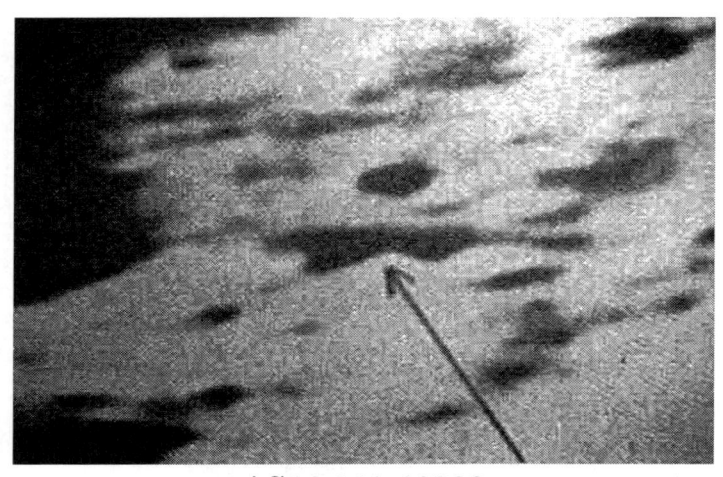

AS16-118-18923
Fred's Close-up of "Cigar UFO"

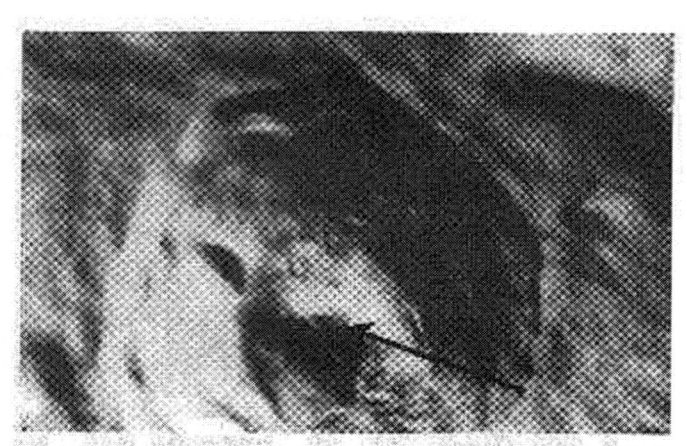

Plate 110. LO IV Photo No. LO IV 89H3. The Crater Romer. Notice large platform and long object on center mountain.

IV-73-H3
Plate 110. The Crater Romer: Notice the large platform and long object on the center mountain.

Stellar Sterility

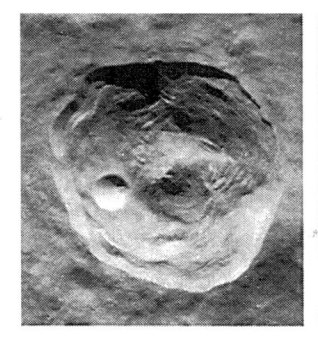

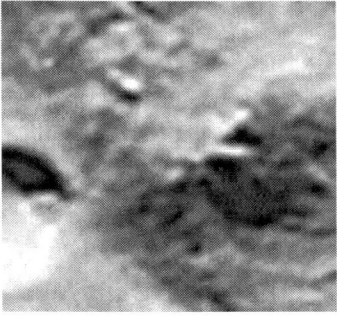

Fred lists Crater Romer as on Lunar Photo IV-089-H3 and locates the alien space craft on the central peak (arrow Y). But Crater Romer is located in the Lunar Orbiter photo IV-073-H3. Photo 89-H3 lists three craters: Crater Alfraganus, Delambre and Zoller. Crater Romer is NOT located on this frame. These craters too have lots of central debris in them. Almost every crater has some of kind of odd boulder arrangements and central pile of dirt.

Crater Romer surely has debris in it as the picture shows. But alien artifacts? If left up to this kind of scholarship, every crater has UFO's in them.

Returning to the original location, the UFO surly looks convincing! Yet, when on closer examination, rather than a little air plane or UFO it is actually a pile of crater debris. In the center of the crater is a pile of boulders and rocks forming a single cone-shaped central peak that actually may have a summit pit?

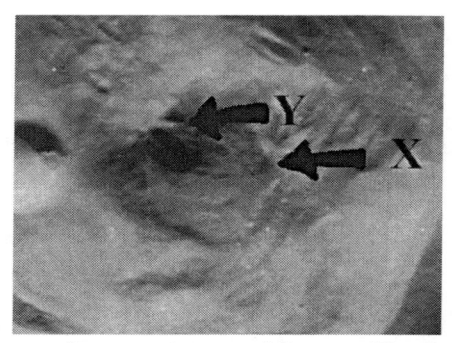

A closer cleaner picture of Romer. Fred's Ref:
L.O. IV 089-H3 Should be frame IV-073-H3

Another view of IV-089-H3
and other crater examples of interior debris.

Shultz said (p. 100), "*The peal comprises the entire crater floor and crests 1.8 km above its base to the east* [i.e. between X and Z along the crest] *but is about 2 km below the eastern crater rim. Thus the three-dimensional view has been greatly enhanced. Adjacent to its west slope is a ringed depression (arrow X) that might represent a collapsed form.*" This collapsed area can be seen within the whitened area

Stellar Sterility

where the two arrows are pointing. At "Z" is where Fred sees a UFO.

On closer inspection, the little UFO is actually some over-hanging rocks that receive a highlight from the sun while below is the shadow area.

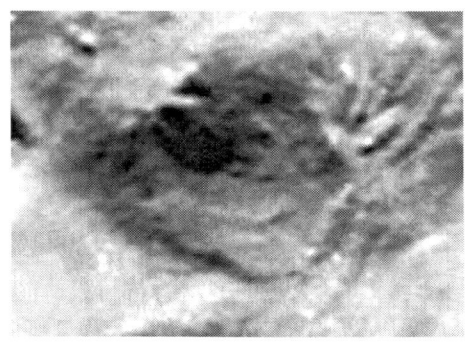

Overhanging highlight area seen in these Apollo frames.

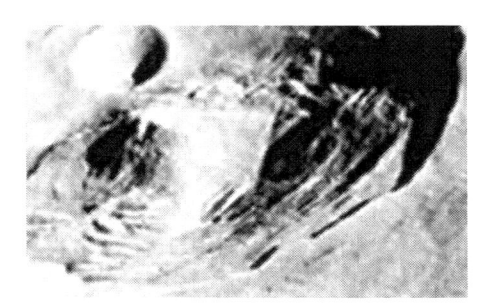

AS17-P-2293

Other crater "cigars" and alien oddities are found in almost all of Fred's photos, as if alien spaceships are just lying around all over the lunar surface. Maybe they are just rocks, boulders and crater ejecta?

Stellar Sterility

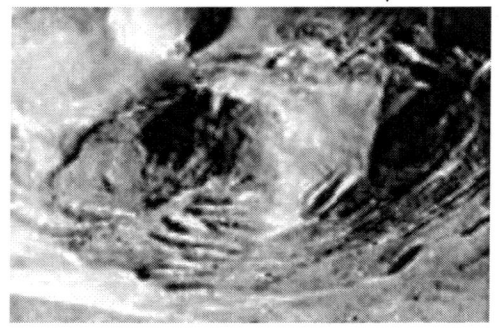

AS17-P-2295

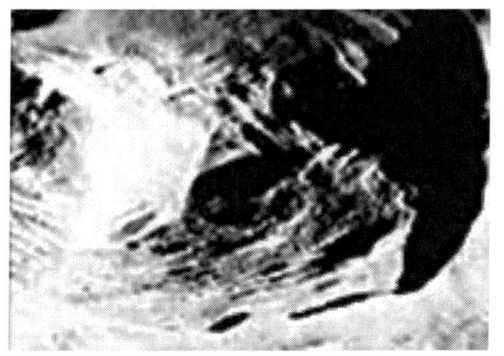

AS17-P-2298

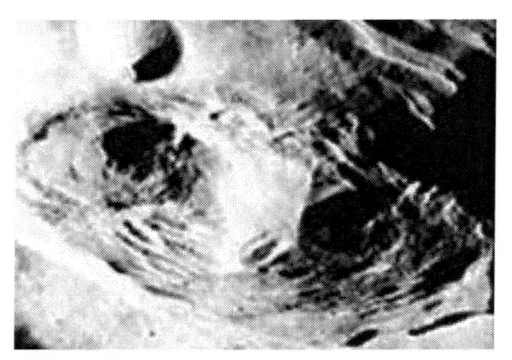

AS17-P-2300

Stellar Sterility

Try to find other cigar shaped items if you can. Fred found them! We have shown a few (below) that are very obvious and one is used for the front cover of Mr Steckling's book.

AS16-19238
Looks like part of the Orbiter?
Background to Fred's front Cover

8
ALIEN LISTENING DEVICES

IV-169-H1
Bull's Eye! Fresh new crater inside older crater!

Besides crater walls, collapsed rims, and flying cigars notice Mr. Steckling's alien made crater device. He said, *"See the double crater in the larger Humboldt Crater. This one looks artificial, Mining or?* Well what is it? He never says what it is actually. Maybe it is just a double crater!

Of course, Fred would have us consider that they are giant alien speakers or listening devices similar to the Arecibo Observatory. Here are some other craters that depict speaker-like shapes. Unlike craters formed by endogenetic (volcanic) processes, double craters may be formed by rare coincidences of a meteorite striking

Stellar Sterility

in the middle of an older crater. This is implied by the subdued nature of the outer wall compared with the fresh "crisp" sharp edge rim of the central feature. See IV-169-H1 photo frame.

Geologists tell us these weird looking double craters are examples of natural lunar phenomenon, where crater rims have collapsed down inside. Another possibility is that two meteorites struck the same spot one after the other in the same way that multiple meteorites strike successively in a row forming crater chains. Still another is that a new meteorite struck an older one as implied by the newness of the inner crater rim compared to the older outer crater rim. Yes, they look weird, alien and artificially created, yet they are only the product of natural crater processes.

Fresh crater impacts in area of older surface in area of Fred's double crater

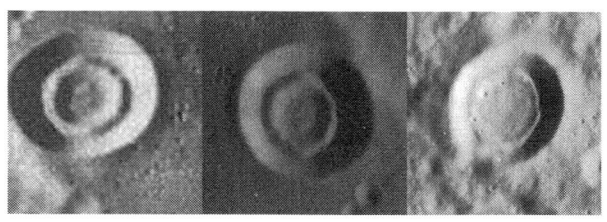

IV-125-H1   IV-119-H3   IV-156-H3

9
PLANETARY PIES
THE "PIE-CUT" MOUND

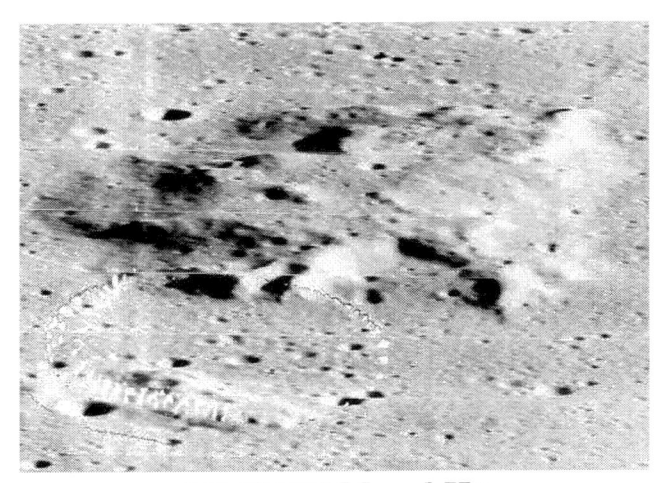

LO-II-213-M and H2

Lunar Orbiter II photo 213-M is a very beautiful and romantic lunarscape shot. This is a great site that excites the wondering imagination. This is Steckling's "Pie-Cut" mound site. His most accurate pie-hole said, that the *"hill, left of center, has been cut out like a piece of cake. Above the hill appears large construction casting a shadow to the left. To the right notice several perfectly cut holes or craters and two oval objects on the rim."*

After careful study of the original frames, there will be found no constructions casting shadows but for some shadowed craters, no perfectly cut drilled holes accept fresh crater impacts and no oval objects on any crater rims, though one can see a play of highlights against shadows all over the frame.

Stellar Sterility

Also, one can see little white equidistant dots dividing the scanned area. Some of these white dots hit exactly on some crater rims causing what looks like UFO's and "alien objects." If one takes the complete frame and follows the white dots, they proceed all the way up the frame and into dark space, perfectly divided according to the distance between the scans. Wow, what a "piece of cake!"

Pie shaped cuts or landslides play a common role in lunar surface features. So do cracks, gas vents, double and triple craters, crater chains, multi-rim craters and extremely odd shaped boulders. This pie-cut is nothing other than a collapsed mound. At closer range in a good zoom in on the mound, one can see that the pie-cut thief has left some residue at the inner edge.

The Lunar Orbiter Atlas "The Moon as Viewed by Lunar Orbiter," has a large reproduction of this pie-cut on page 122. A drawing will show this land slide effect we are talking about.

The left side appears collapsed and debris has slid down, somewhat obscured by the heavy sunlight hitting it. The right side is in shadow. Collapsing mounds do not always create rounded openings, edges and cliffs, etc. This is what makes this formation unique.

Stellar Sterility

The composition of the mound determines the differences. In this case, the mound could have been struck by an early meteorite and then affected this way by lunar erosion. Microscopic meteorite bombardment over long periods of time could have caused secondary erosion patterns, such as the pie-cut effect. Even adjacent land flow could have pulled away and removed debris. The apparently visible erosion effects as seen in 213's medium shot and in 213-H2 evidence this.

Fred's perfectly cut pie hole and the two oval objects can be easily explained, without having to assume alien influences. Heavy shadow and light play a major role in such odd looking land masses as this pie cut. Shadow and light playing off round boulders! Yes, boulders can be round! Even square and rectangular! A good whack by a small meteorite can crack one even into a triangle shaped rock!

"PIE MOUND"

The reason the crater holes look like the ends of large tubes sticking out of the lunar surface is because they are in extreme light and shadow. The rim's details are obscured either by excessive shadow, which the

Stellar Sterility

Moon has plenty of or by highlights that tend to bleach out the details on the sunlit sides in the photos. All that would be left in this case would be the sharp rims that have no delineation as to the oblique crater walls and regolith soil.

Under more balanced lighting conditions and by a different camera and sun angle, the crater(s) would look the same as all the millions of other craters!

The funny white "finger print" image is a development process disfigurement. At about 10:00 O'clock, Northwest in the photo is actually another pie cut mound, yet not as visible as Fred's mound. There is land flow around the area with newer crater impacts.

There are at least five large craters and numerous smaller ones within the pie-cut area. There is much lunar regolith disturbance. Now that we have shut the pie-hole of Fred's alien pie-cut mound blab and have shown it to be nothing but a mound of dirt, we may proceed to Fred's bowl of alphabet pottage soup and other supposed alien objects.

10
CRATER CAMBELL'S ALPHABET SOUP!

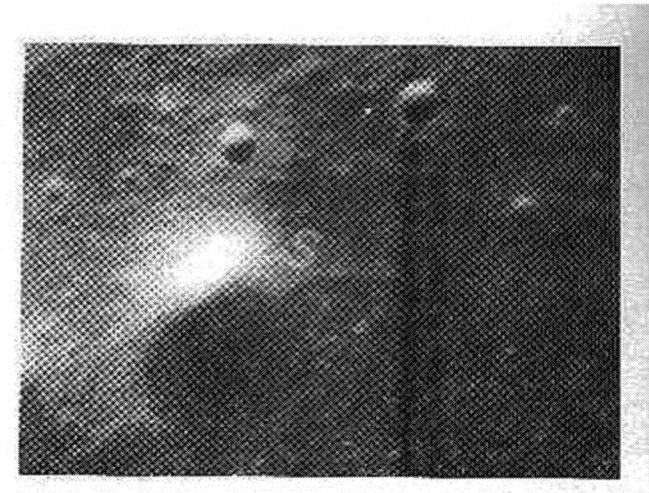

Plate 109. Apollo 14 photo No. 14-80-10439. Another letter "S". This time placed beside the crater.

AS14-80-10439

This is an example of one of Fred's "S" Letters, which he finds scattered all over the moon inside craters. This time, this "S" is not found in a crater. It is sitting next to a mound. This is a familiar "emulsion hair" caught in development of photo AS14-80-10439, which by the way cannot be located. It is not on the internet and is not in the NASA Apollo-14 photo archives. It is said to be classified? It is nowhere to be found. In fact, I cannot locate film Magazines 79 or 80. See footnotes of reference of image close to this one proving magazine 80 does exist [1]. Maybe someone has these sets of photos?

Stellar Sterility

In L.O. III photo frame 194-3, Fred sees the alphabetical letter "S" artificially inscribed within a crater. It looks like some 'signature' or locator marking for bypassing UFO's! Admittedly, this really excites the human eye and throws a real S-curve ball when first viewed. This appears to be a real genuine anomaly. Needless to say, this is not some artificially created mark, but just another explainable lunar phenomenon. A viewer will see the same effect when looking at a river from an airplane. Many rivers curve and meander like the coils of a snake in an S-shaped pattern. Well, so do crater rims when they collapse. Each end of the "S", the bottom and top being crater rims remain. The pushing in of the two collapsed rim creates the middle of the "S". They look connected and probably are from the erosion collapse of the rim. Notice that the entire rim does not have to collapse, just two opposite sides. Other crater rim materials can easily slide down creating the connecting parts of the so-called "S" shaped letter.

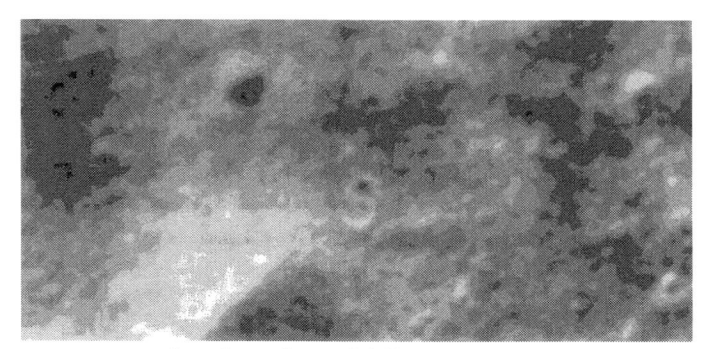

Another shot of AS14-80-10439
(http://keithlaney.net/ApolloOrbitalimages/AS14http://keithlaney.net/ApolloOrbitalimages/AS14)

Stellar Sterility

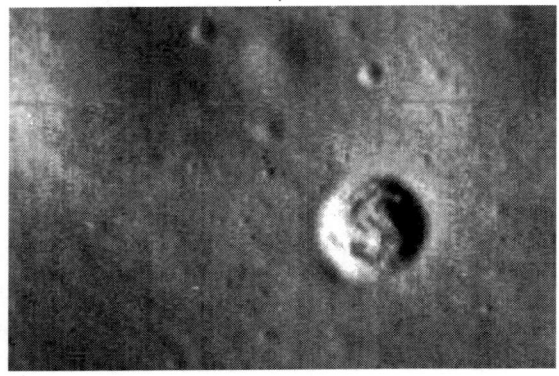

III-194-H3
Fred's loony alphabetical letter "S"
Plate 120, p.141

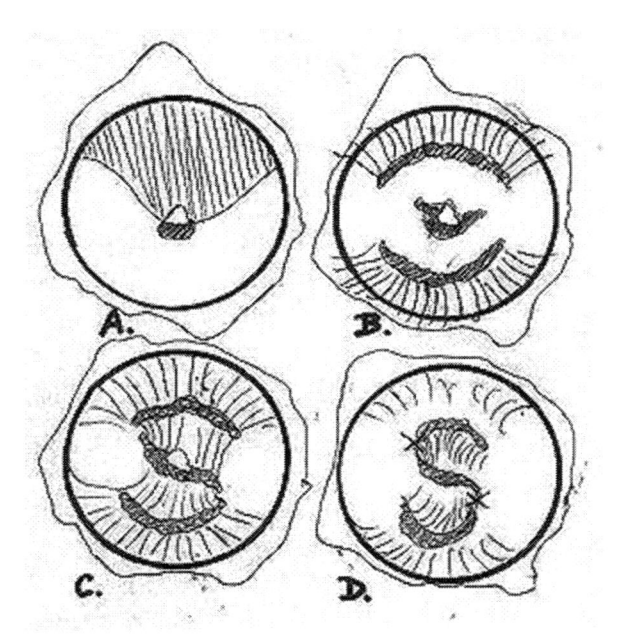

Explanation of "S" shape development inside craters.

If Fred would have studied other crater interiors, he would have to assume aliens were learning the English alphabet or something! Crater floor features

Stellar Sterility

can appear to look like many things to alien artifact hunters. When you add numerous geological events together, such as crater rim collapses, double impact craters, double impact crater rim collapses, boulders, cracks, fractures, shadows and highlights you can have aliens trying to rewrite Shakespeare!

Many craters have this "letter" effect phenomenon. After craters are formed, regolith sometimes, and in most cases, pours back into the crater from the rims. This creates mounds and interior rim ridges, scarps and other funny looking piles of "twists and turns" that would appear from a distance to be "letters." Some blurriness does not hurt in adding more grammar to the effect.

Fred found other consonants such as the letter "R" half-hidden in a crater shadow on the back side of the Moon. See p.140, plate 119 and plate 122.

The Letter "S" seems to pop up in many craters, usually smaller ones. Why so many small craters and not the bigger ones? Why not crater Plato? How about Crater Tycho? The drawings we have just seen are a series of examples of how an "S" can be formed inside a crater. The impact leaves a small mound or ridge in the middle of the crater.

Crater rims symmetrically slide inwards as they collapse causing the top and bottom of the "S" to form. The collapsing debris forms uplift in soil forming the middle of the "S" and a ridge connecting the top and bottom of the "S."

Stellar Sterility

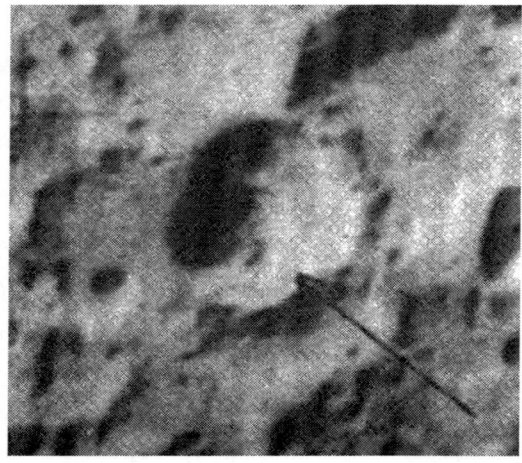

(Fred does not give us a photo number for this frame)

Using fuzzy versions of Lunar Orbiter photos one cannot help but see what one wants to see. Yet, there are details enough through crater comparison to illustrate how shapes like this can form. With slightly blurred images, shadows and highlights illusions emerge causing many factious features to appear. If the details are smoothed out around the regolith mounds the sliding rims may become anything the alien artifact hunter wants them to be. The three parts of the crater collapse, the top, bottom and middle would tend to blend together under such lighting. Exposure also helps to blend the disconnected material into a perfect looking "S" shape. Thus, according to these effects some look complete, while others appear fragmented.

According to Fred's photo references from Apollo-14-80-10439, which shows a white letter "S," many NASA photos do present white streaks, hairs and dust scratches that create what might be misinterpreted as alien artifacts. They are so numerous that one might suppose that E.T's once inhabited the whole surface of the Moon. Hotels, motels, houses and mining operations seem to be everywhere according to this man's theory.

For example, just enlarge any Lunar Orbiter photo and POOF! You will have thousands of alien structures, buildings and craters with alphabet soup letters! The "S" in plate-121 is obviously a curvilinear shaped dust "hair" caught in the development of the photo.

How about BLACK letters? Can anyone find a BLACK letter? These would be harder to find, but not impossible to explain though. They would be cracks, fault lines, gas vents or even small rilles.

11
TANKS AND TOWERS
Alien Oil Refineries!

AS16-118-18918

Another, strictly speaking, Steckling hideout for alien structures is NASA Apollo-16 photo AS16-118-18918. Fred said, fuel tanks, towers and other platforms are supposedly visible in this photo. Let us take a closer look at this.

The photo is very blurred and the highlights are too intense, while the shadows are very dark and imposing upon the terrain. With a super imagination, a Stecklingite can easily discern artificial structures with very little effort. What one stuffs into the pie is what one gets out of it after it is bake in the oven of warped imagination. Fred apparently sees geometric

Stellar Sterility

and other artificial shapes created by some super technical mind, as the following fantasy drawing shows. (See drawings 1 and 2). Here is his oil refinery tanks.

Another concept drawing of Towers

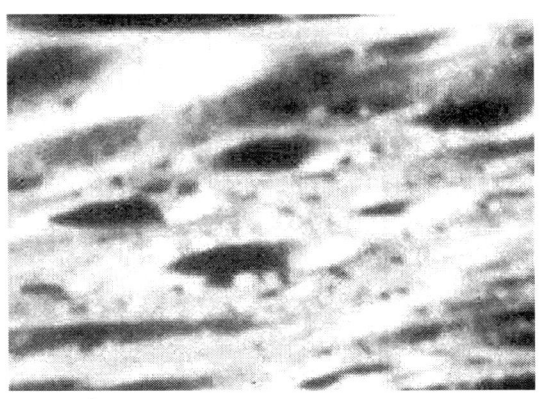

1.) Enlarged and enhanced

Stellar Sterility

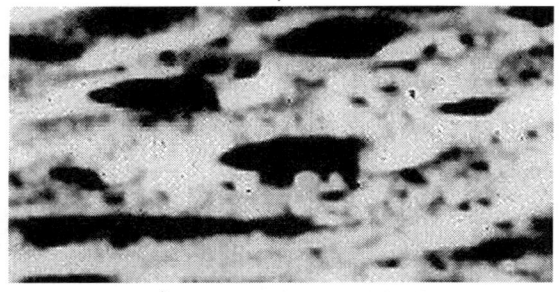

2.) Contrast added

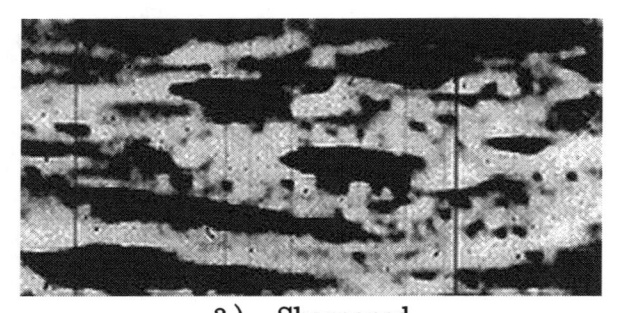

3.) Sharpened

Fred's Plates 148 and 149 (Plates 133 and 134)

Drawing- 1 See the oil tanks?

Stellar Sterility

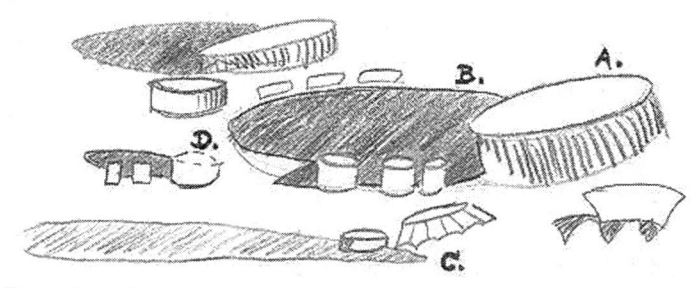

Drawing-2

Contrary to such fancy interpretation one can easily extrapolate from the photo a purely 'natural' set of lunar surface features (See drawing -1) and twist them into "tanks," "towers" and "platforms." Whereas, after looking at the photo closely, with some enhancements, drawing – 2 exposes the "A" object as not a platform at all, but the Sun baked highlight of the right side of Crater "B."

Object "B" is the termination line between the crater shadow cast by the sun upon the crater rim. Object "A" is the highlighted right side of the rim. "B" is the shadowed area. The crater to the upper left shows this light-shadow phenomenon much better, as do all the craters in this enlargement.

The little white 'geometric' obstructions around the crater rim are nothing more than what was left of the rim after it collapsed into the inside and of whatever crater ejecta was thrown out from the impact. They appear to be crater ejecta boulders. In some cases, the crater rims have collapsed, leaving "toothed" openings along the edges, which appear in blurred imaging as supposed artificial "digs."

Stellar Sterility

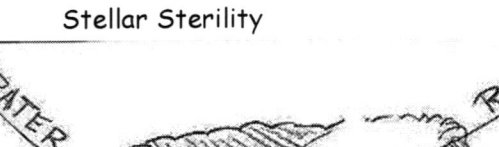

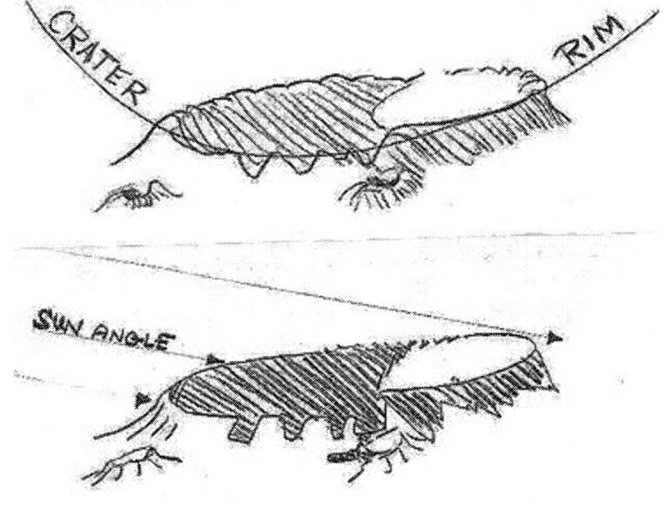

Drawing-3

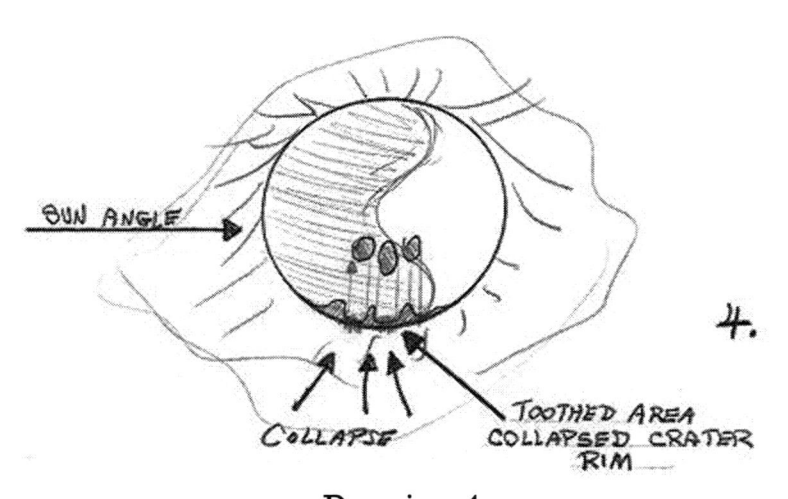

Drawing-4
Tanks, Platforms and Artificial Structures?
No! Simple crater anomalies.

Stellar Sterility

Large chunks (not regolith soil) of rocks, boulders and larger pieces of the lunar surface (that was not broken up in the impact) might cause the toothed appearance. These are some that never loosened in the collapsing of the rim.

The openings are the ones that slide down into the crater leaving 'holes' or toothed ridge areas along the rim. Crater shadows would show through, while the sun would continue to highlight the remaining crater rim producing what appears to be 'teeth.'

12
ROLLING BOULDERS
ALIEN MINING MACHINES

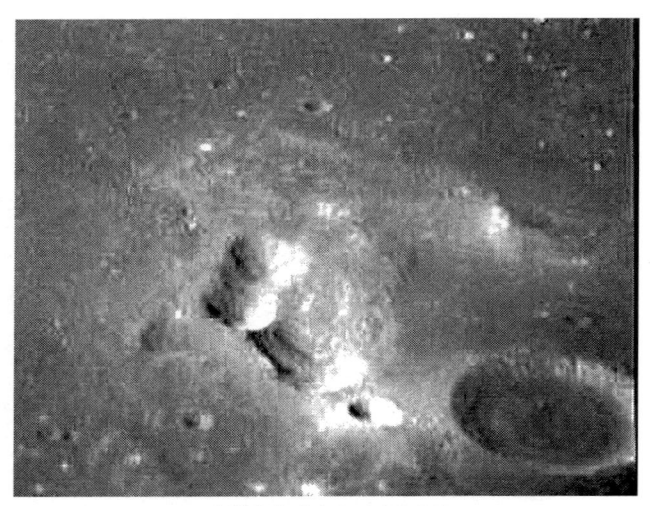

AS16-119-19067

Using alien mining operations as a premise, Fred argues for an artificial origin of these 'rolling stones'. He said the paths left behind the boulders are "tracks." These tread marks are the evidence for alien mining machines left behind after a great alien exodus from the Moon! Why they left no one knows.

Fred shows us some lunar photographs of (boulder) tracks produced by rolling rocks in the Crater Vitello as well as from other areas, insisting that *"Tracks of vehicles rolling over the ground, up and down the hills, have been photographed."* And as far as he can tell, the *"large vehicles, some seventy-five feet across, seem to*

Stellar Sterility

probe the soil for future mining possibilities. The tracks left behind by these vehicles show definite 'stitch-marks' by some form of belted vehicle."

The Apollo-16 photo 119-19067 shows a 30 mile stretch of stitched path or track left by some URO (unidentified rolling object) that tumbles along up and down the hills, and up and over the crater rim continuing down into the other side. Fred believes it must be alien because the tracks are *"identical to Plate No. 23 tracks"* (in V-168-H2). He forgets to explain how they appear out of nowhere.

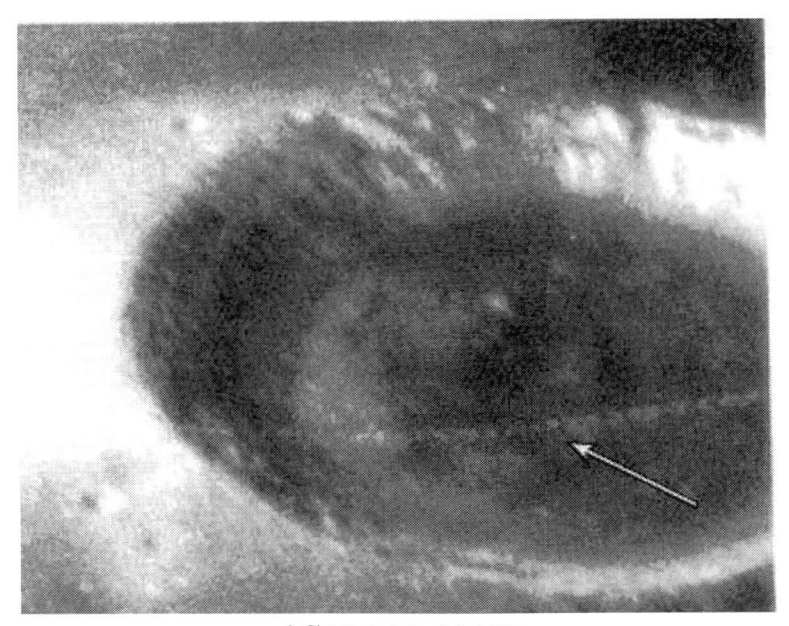

AS16-119-19067
Close-up Lunar back-side, Crater Vitello
(Fred's photo, Plate 24, p.39)

He continues to tread on our reasoning that one rolling object (boulder) measures 11 by 21m and judging from its shadow, is 11m tall. A second

Stellar Sterility

mechanical monster shows as a bright spot above the first one. (LO V-168-H2 "30.4 S, 38.5 W.) The large object leaves a definite 'stitch-mark' path caused by some form of belted vehicle (Plate 23, p. 39). The following some different shots of 168-H2. After close inspection the tractor looks nothing like a wobbling moon machine. What is seen are just boulders!

The second set of photos are of a famous boulder from the old Lunar Orbiter V shown in the 1972 research article "Boulder Tracks on the Moon and Earth" by H. J. Moore.

The track is produced by a skidding boulder in the crater Petavius. The boulder is about 68 by 81m; the track is 50 by 137m. (Frame H-36, framelet 144, 24.8 S, 59.8 E). The frame to the right is from the LRO Camera, frame No. m183131379rc_pyrd1.

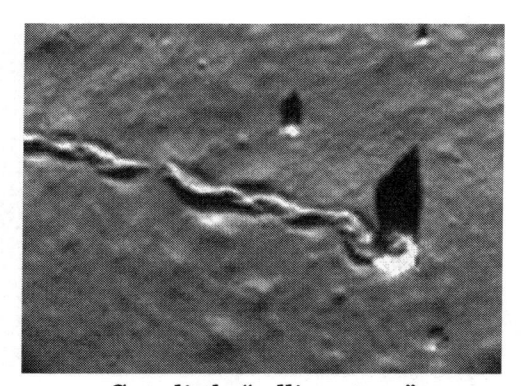

Cute little "rolling stone"

Stellar Sterility

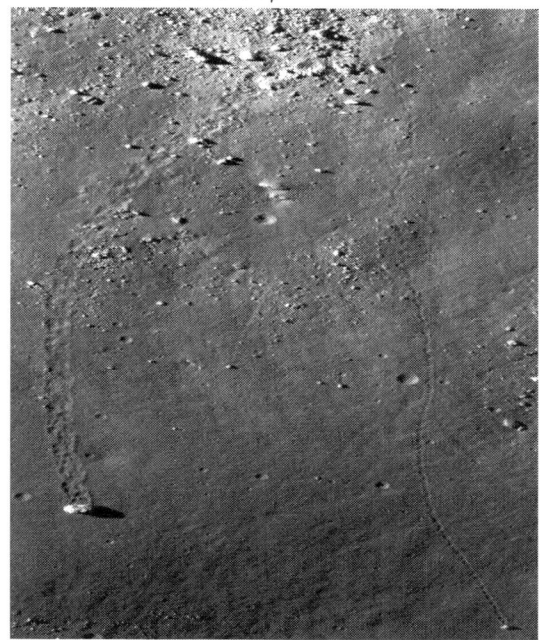

Photo: m1354337521c_pyrred
Lunar Reconnaissance Orbiter Camera photo

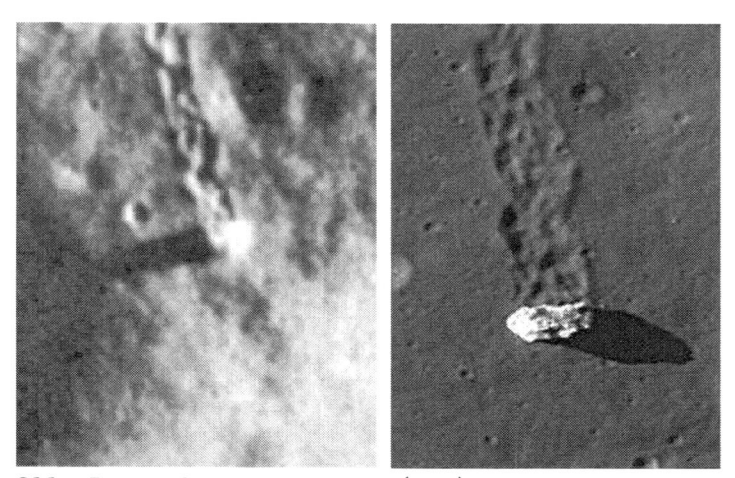

Older Lunar Orbiter V-168-H2 (Left) compared to the new
Lunar Reconnaissance Orbiter photograph above (Right)

Stellar Sterility

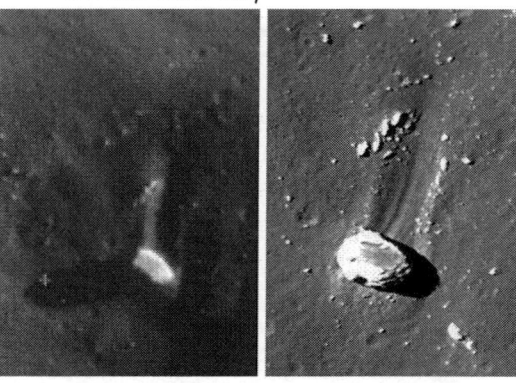

LO-V-036-H2 compared with LROC

This next track example was produced by a "walking" boulder in Scheoter's Valley. The boulder is 19 by 35m. The length of the track is almost 540m. Notice the zigzag pattern of the track and angularity of the boulder. It is located at 49.8 degrees West and 25.6 degrees North (Moore p. B171).

Fred said these tracks are "stitches" left behind by some alien "stitching" or "digging" machine. In his book, he sketches what he believes it looks like. Another Japanese web site extrapolates another design of what is causing the tracks.

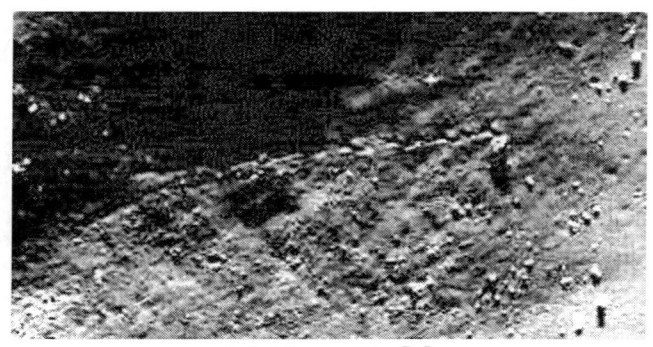

LO-V-204-H3-[2]

Stellar Sterility

Plate 24. Author's conception of what rolling objects in plates 20 and 23 may look like. A type of mining vehicle or soil testing device?

Fred's drawing of the stitching machine

Fred must "dig" inventing alternative theories, purposely defrauding his innocent readers and defying all his so-called government training by believed this 'bull' dozer theory himself. Knowing Fred and knowing the NASA materials as I do, he looked at the same photos and negatives of the traveling clouds of the Vitello area that I have and probably disregarded any further study into the anomalies.

With a bit of effort one can read all they want about the truth of traveling boulders in Moore's monograph, "BOULDER TRACKS ON THE MOON AND EARTH", in USGS Research, 1972; Survey Professional Paper 800-B, pages B165-B174.

Stellar Sterility

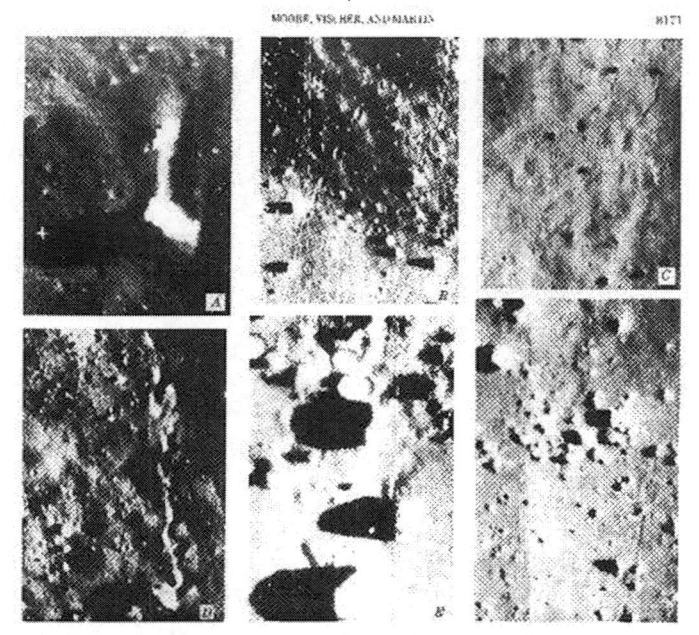

Geological Survey Research Professional Paper 800-B, p.B171.

Fig. 4. – L. O. V photos of boulder tracks showing types of tracks

In this study are examples of lunar and earth boulders rolling up and down hills. Other examples that I found are as follows.

Stellar Sterility

LO·V·063·H2

Another taken from L.O.V·036·H3·[2]

Stellar Sterility

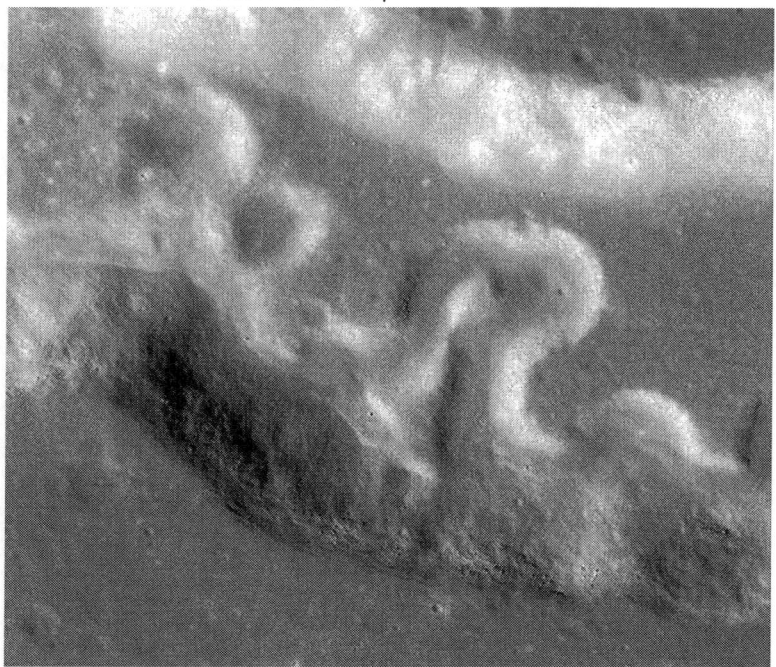

This image was captured by Lunar Reconnaissance Orbiter on August 14 at 00:18 UTC

The above picture is from Vallis Schröteri, the largest rille on the Moon, and is on the Aristarchus Plateau. The channel actually contains two rilles, a larger one 4.3 kilometers wide and 155 kilometers long, and a narrower, inner one 600 meters wide and 204 kilometers long. The main feature is called "The Cobra's head".

Notice the long track going down the sloping wall of the outer rille. It is continuous at the top of the slope, dis-continuous toward the base. Is this the track left by a boulder that first rolled, then, as it picked up speed, bounced down the slope?

Stellar Sterility

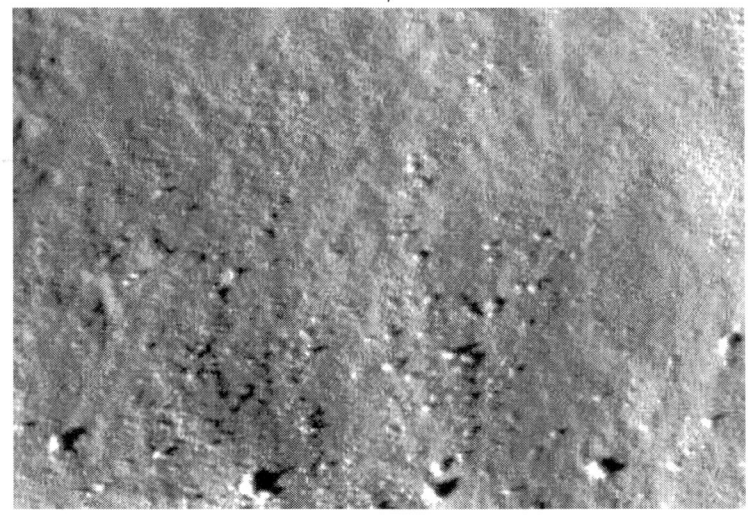

If Fred knew about Moore's article, he would have also have known they were ejecta rocks thrown out by crater impacts or some other lunar geological event. The above three writers in this monograph tell us exactly what they are and how they have done what we see in the above "traveling boulder" shots.

Here is a good shot of two boulders chugging along on the Lunar surface taken by the Apollo-17 astronauts using a telephoto lens to photograph "boulders and boulder tracks" on the North Massif at the their landing site. The largest object, making a weaving, tread-like path is about 5 meters across.

There are other lunar rolling rocks that Mr Steckling seems to have missed. Over 300 "boulder tracks" have been identified in photography taken from the lunar orbit. (See, Grolier, Moore, Martin in "Lunar Rock Tracks," 1968; and "The Moon - Boulder Tracks and Nature of Lunar Soil," 1973 by D. Reidal Pub.).

Stellar Sterility
Apollo-17 photo AS17-144-22129

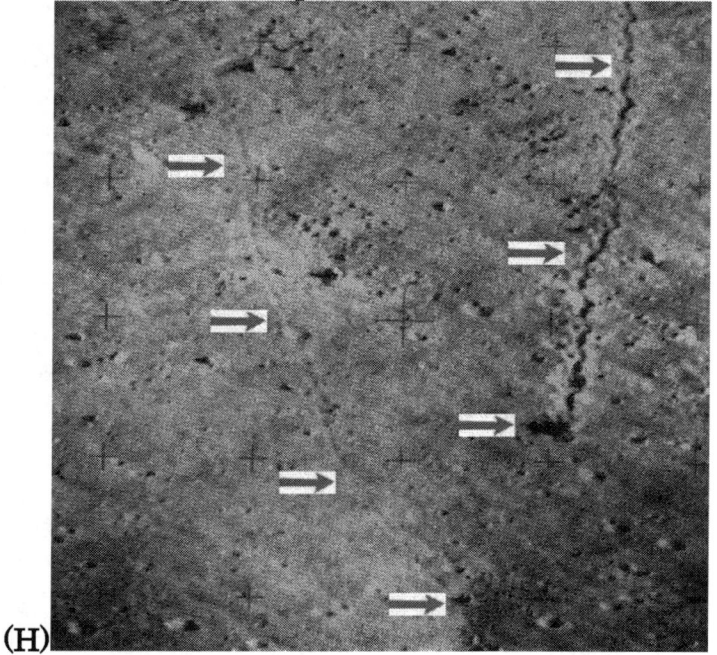

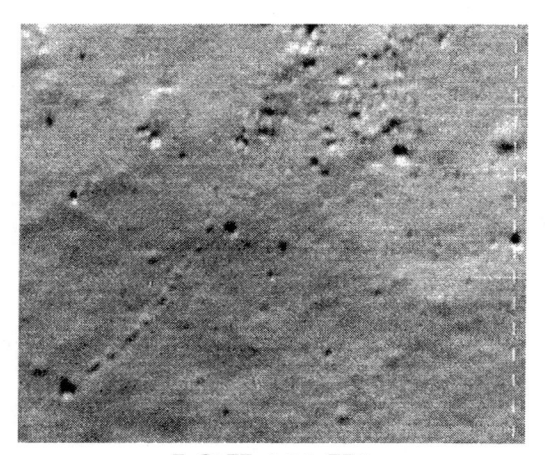

(H)

LO II-092-H1

The above frame is from Lunar Orbiter II at an altitude of 44 miles. The image is taken from frame 92, frame let 445, and has resolution is 0.98 meters/pixel. As such the large boulder that has left a trail is around 6-7 meters in diameter.

Stellar Sterility

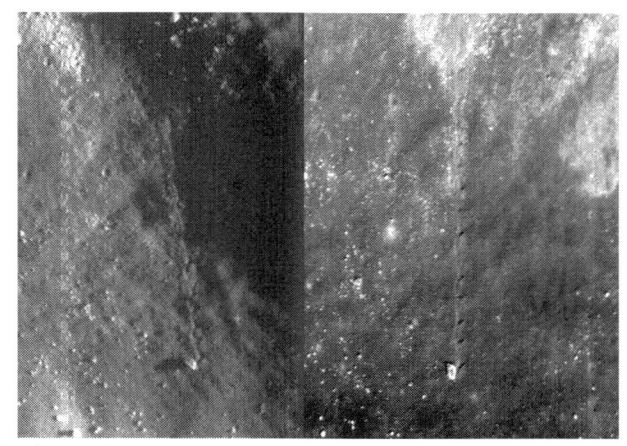

Right: LO V-204-H3 Photo of same rolling boulder as captured by LROC m188572067rc_pyrred (left). For exploratory fun in locating other Lunar "rolling stones" and "boulder tracks" see this web site:

http://the-moon.wikispaces.com/Boulder+tracks]

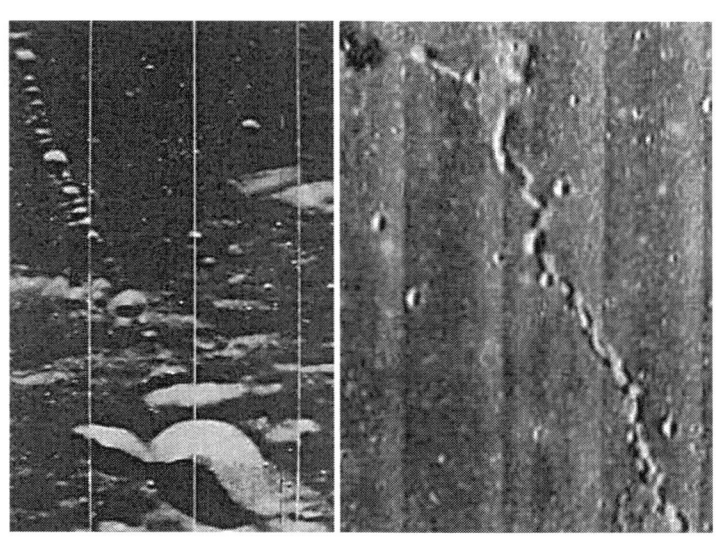

Moon as Viewed by Lunar Orbiter p.97, No. 119

13
CRATER LAKES

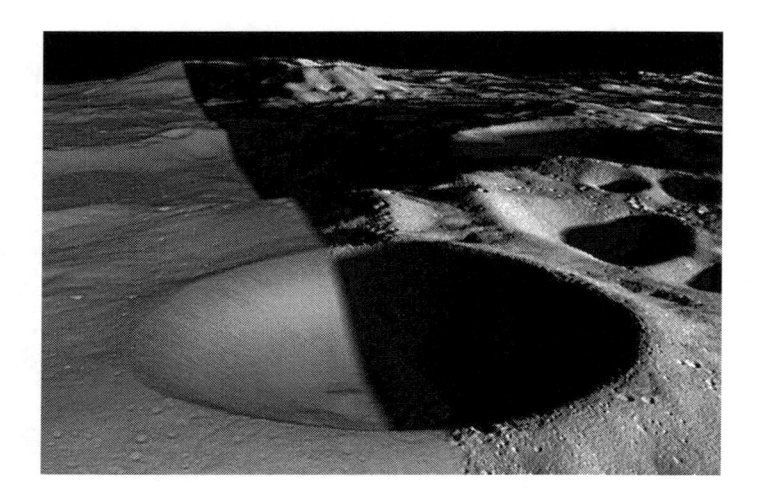

Is there water on the Moon? Fred Steckling said there was and he was right, but not to the extent he dreamed of. In a recent finding NASA said ice may reside in cold dark crater pockets, especially at the North and South poles of the Moon.

Actually, a team of NASA and university scientists using laser light from LROC laser altimeter examined the floor of Shackleton crater. They found the crater's floor is brighter than those of other nearby craters, which is consistent with the presence of small amounts of ice.

The crater is named after the Antarctic explorer Ernest Shackleton, and is two miles deep and more than 12 miles wide. Like several craters at the moon's south pole, the small tilt of the lunar spin axis means Shackleton crater's interior is permanently dark and

Stellar Sterility

therefore extremely cold. (Eddie Wrenn, "More Water on the Moon."

[Ref. www.dailymail.co.uk/sciencetech/article-2162505]

The South Pole is not the only place where water might be located, but the north Polar Regions offer equal hope.

Shackleton, (12.5 mile-diameter) permanently shadowed crater near lunar South Pole,. Possible an ice coating at its base

The North Polar Region is said to contain scattered pockets of ice and water, and a lot more than previously calculated. One estimate by NASA is that there is some 600 million metric tons concealed away in about 40 craters. This is not much, compared to Fred's estimate of "oceans" of water, but enough to supply a large city the size of Seattle for about three years. Water appears to be turning up everywhere on the moon. Not bad considering it was once thought to be bone dry. The latest discovery from the Indian Chandrayaan-1 lunar orbiter found 40 craters each containing water ice at least 2 meters deep.

Unfortunately, these small traces of water on the moon do not mean we are to start buying up water-front crater properties. Nevertheless, Mr. Steckling believes there is huge amounts of water, and not just water, but plants, trees, vegetables, lichens and mosses, and with all the dark spots at the poles, let's not forget the possibility of mushrooms!

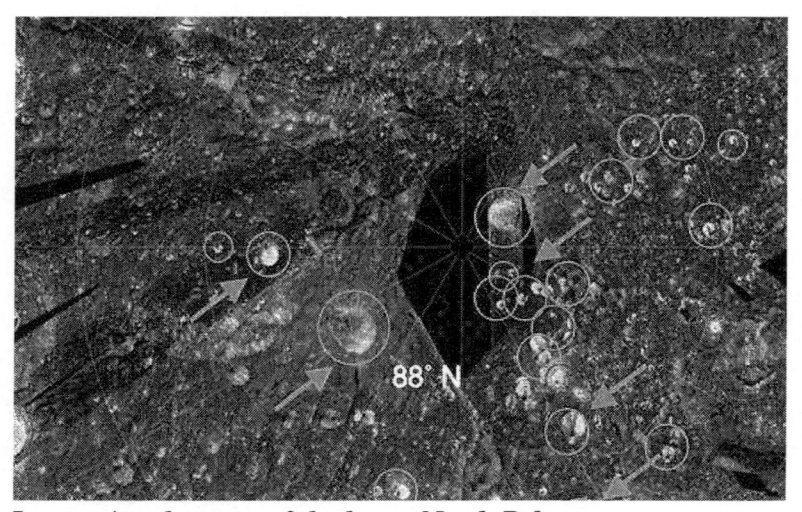

Image: A radar map of the lunar North Pole.
Craters circled believed to contain frozen water (NASA)

This is good news for those supporting the alien occupation of the Moon. It would mean humans will occupy the Moon one day and have plenty of water to drink too!

Before NASA ever proved these traces of water, Fred and other scientists speculated about huge

Stellar Sterility

reservoirs of water, such as can be seen in the crater Tsiolkovsky on the back side of the Moon (LO IV photos. Picture below, bottom left). Fred said it is loaded with water and is actually a giant lake! "*Notice the large "lake" and the smaller ones nearby.*"

Fred said, "*Clouds can be seen north of the lake. Objects appear to line the North West shoreline.*" Obviously Fred does not know about crater fill-in by lava flows. The crater is very old and appears to be filled in with a flat surface material. He said NASA does not know what the material is. Well, of course, this is Apollo-8, which never landed nor took Mare samples.

He said, "*NASA states that the black mass in this crater is unknown, but that it must originate from the interior of the Moon. It obviously appears to be water, but since this would upset most of the scientific world, the best thing is to wait as long as possible to reveal that*" (Page 109-110).

From a long distance and at the right angle and lighting, it can look like a lake as he said Astronaut Aldrin mentions: "*When I looked at Tsiolkovsky Crater, it reminded me of a mountain lake with a quiet surface and with a small island in the middle*" (page 110).

Stellar Sterility

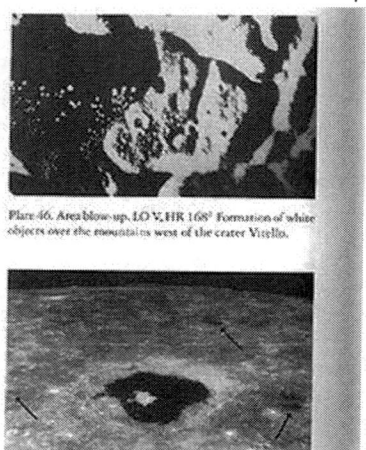

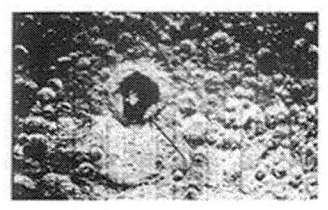

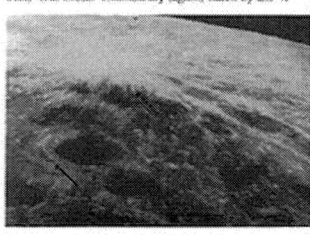

Bottom Left: Apollo-8 photo of the crater. Top Right:
Tsiolkovsky Crater taken by Lunar Orbiter IV. Bottom Right:
Apollo-8 photo of a crater he calls a
"lake" on the back side of the Moon close
to the North pole. Bottom Right: More crater lakes.

Apollo-8 Photo

14
TSIOLKOWSKY CRATER WATER DROUGHT

Close up of Tsiolkovsky Crater

Fred argues that these are lakes built by aliens for water storage. They left them openly visible for us to see when we were to ever go to the Moon!

Here is a close up of Tsiolkovsky crater. I wonder what those small holes are in the surface of the water. Let's take a closer look? Wow, more small impact craters! There is no water. Looks like a complete drought to me. Maybe we are looking at a dry sea bed?

Stellar Sterility

AS17-139-21302HR
Tsiolkovsky Crater floor littered with small impact craters

Maybe they are floating craters? Holes in the water or alien boats? Or, maybe they are just what they look like, craters in the floor of Tsiolkovsky, peppering the surface like gun shots.

It is amazing how a so-called scientist, Mr. Steckling, avoids close up imaging and enhancements. It's all the same blurry medium shots that are printed out of focus. Anyone can take a photograph of a crater, snapped from a medium or long range distance and make out all kinds of things.

Alternately, when you ZOOM in and see the details, all you see are craters, "regolith", boulders, rocks, dust, fault lines, rilles, cracks, as well as mountains and valleys, and what looks like areas where water or lava once flowed. Other NASA photos show craters that look like lakes: See AS8-12-2209 and AS8-12-2196.

An atmosphere was once on the Moon! Fred said that water was once on the Moon in huge quantities!

Stellar Sterility

Maybe so. Even NASA has made such claims. But an atmosphere that still exists? Fred said, NASA is telling lies to throw off the public, because it would scare us! That they want to hide the truth about lunar rivers, lakes and ponds. Such tremendous knowledge would destroy our present thinking.

Fred decides to show us, lakes, rivers and ponds to debunk NASA, who said there are no such lunar features. Yes, there may well be water, ice and other frozen gases, but as to gushing rivers, water falls, oceans and plants? Let's take a trip rather into the hollow earth and search for giant subterranean mushrooms!

The World's Largest Mushroom!

15
MOLEHILLS, MUSHROOMS AND MICROBES

Here is a big long emulsion splatter running across the whole photo. Surely it looks like some great storm crossing over the lunar surface causing the source of lunar water.

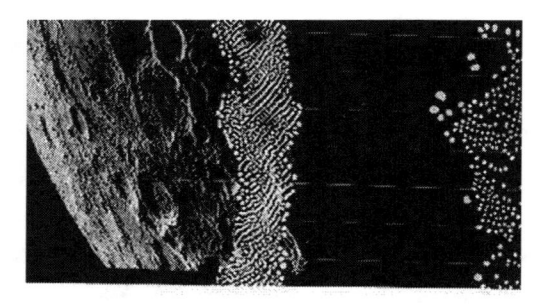

L0-V-65-M Largest Microbial Colony in Outer space

Emulsion splatters, people! Look at the evidence for yourself and stop swallowing exoquack theories for truth. Without investigation and verification you can end up believing anything.

What about space born viruses and big huge bacteria-like "things" floating over the Moon? More emulsion problems? Look at this shot from Lunar Orbiter V-22-M. [See Lunar Orbiter Atlas, plate667].

They sure look like viruses floating around the Moon. If one were to see these in a fanciful alien theory book, without proper reference, he would believe they were. Actually, the internet is loaded with hundreds of web sites claiming all sorts of science fiction theories.

Stellar Sterility

The higher resolution the images get, the closer and smaller artifact hunters search and thus, the smaller the artifacts become.

This is no great wonder to the true scientist who knows that as higher resolution photos are obtained the more the blurry out of focus older photos are debunked of having any alien evidence.

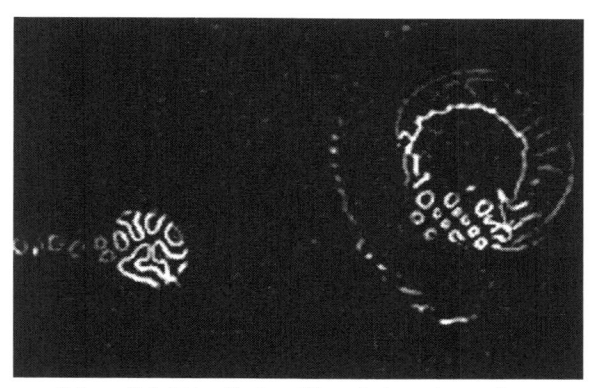

More Multicellular Emulsion Amoebas

Mysterious Alien Flying "rod" Spy Device?
Be careful what you read and believe now days!

Stellar Sterility

This easy to believe syndrome is even more wide spread now days with alien surveillance "rods" flying around the earth at light speed! Ever hear of "rods"? Well, they are weird looking spiral devices sent here by aliens or they are micro-size alien creatures of some kind. No one really knows. Maybe they are photographic glitches?

In the next section we will deal with more lunar laughables with the studies done by Mr. Swaney in "Objects on the Moon."

SECTION-4

JACK SWANEY'S
"OBJECTS ON THE MOON"
(Fate Magazine, June 1982-84)
Intermediate Idiotic Incidentals of Intergalactic Garbage

1. MARE CRISIUM CONTRAPTION

LEVER STICKING OUT OF THE MOON
Located at 50 degrees E. Longitude, 15 degrees N. Latitude.

Jack Swaney, of Fate Magazine, decided to jump in on the extra-terrestrial artifact bandwagon in a series of published articles. As an amateur astronomer and fan of Fred Steckling and George Leonard, Jack decided to be nimble, and was quick to secure his fame, fate and fortune by finding some more alien junk overlooked by his predecessors. There was plenty of ET debris and manufactured material missed by them to be found, lending further credence to the alien presence lunacy, and leading money to the bank account.

In scanning the moon through telescopic observations, Mr. Swaney was aware of objects and formations so bizarre, as to beg description. For example, he found a 45-mile long rectangular pit or housing on the western shore of the Sea of Crisis. This was something so amazing, obvious and apparent, that one cannot help but ask why previous artifact hunters did not see it. The blindness to such bizarre formations begs the question how they could have so easily overlooked them.

This particular object has a large lever sticking out of it. The rod and shaft, comprising this contraption *"shimmers with a metallic gleam."* If this thing actually is a lever, Swaney said, then "weird connotations come to mind." He asks, *"Who or what could actuate such a gigantic thing and what does it do?"* Jack offers the above drawing (rather than a photograph) as demonstrable of the artifact.

One may ask why Jack did not offer any photos of such a gigantic and most visible object. Is the mechanical contraption found in any NASA photos? Checking the coordinates above we turn to the Lunar Orbiter photos for verification. LO-IV-191-H3 shows a nice shot of the whole Mare Crisium area. Swaney published his drawing upside down, with south pointing up to northeast. If we rotate the frame 191-H3 to fit, we see that the only shoreline object that could fit his description is within the yellow circle.

Stellar Sterility

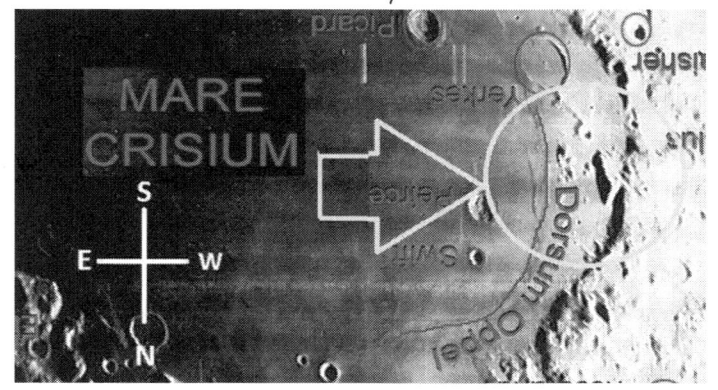

LO-IV-191-H3
Mare Crisium "Lever" object supposedly parked on western shore area.

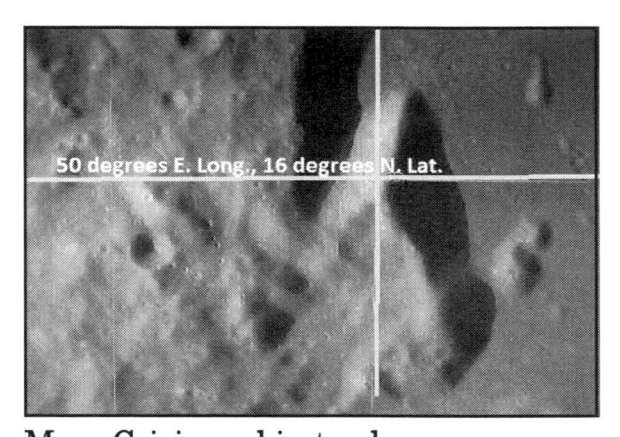

Mare Crisium object - close-up
LROC   Act-React   Quickmap:
[http://target.lroc.asu.edu/da/qmap.htmlhttp://target.lroc.asu.edu/da/qmap.html]

2. ALIEN CRISIS ON NORTH SHORE

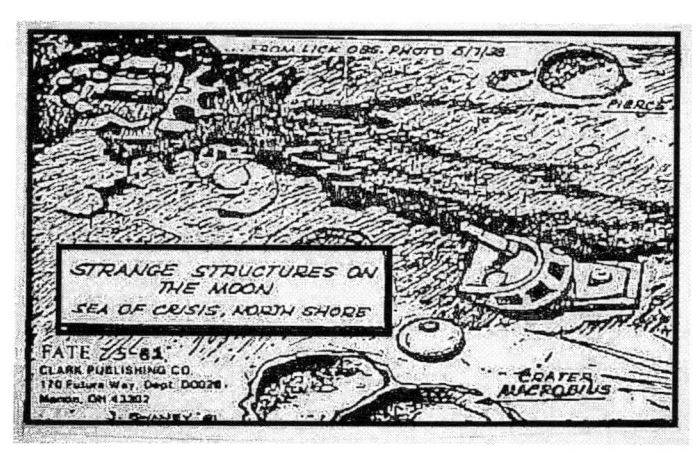

Strange Structures on the Moon in Mare Crisium
(Swaney drawing of Lick Observatory photo 5/7/1938)

Persons who look at the moon through telescopes sometimes see formations that resemble artificial buildings and constructions unlike anything known on earth, said Swaney. One area he points to is the north shore of Mare Crisium, where a small round spot sits just above right center on the moon's face. He said there are a number of them sitting on the southern and eastern shores, and two on the northern sector. South of Crater Macrobius *"stands a quaint round structure with a peculiar slant, measuring 25 to 30 miles in diameter."* He describes the object as *"a cannon"* that *"protrudes from its topside at the back, while half a dozen openings ring the front."* Below Crater Cleomedes about 100 miles east is another anomalous formation 18-miles-long, rectangular in shape like a building that has six silver-white spheres sitting on its

Stellar Sterility

roof. Again he draws a picture rather than give us a photo.

Jack proposes that the government has covered up the truth of these objects by suppressing older observatory photos predating 1950 and has actually confiscated them. Persons searching NASA photos for these structures will be inevitably disappointment, if they are expecting to find anything. The truth is that anyone looking for Mr. Jack's lunar candlesticks and sand castles will be disappointed no matter what photos they dig up or what telescope they look through. Yet, at least one photo still exists from Lick Observatory, from which Mr. Swaney drew his illustration above.

A comparison of NASA photos will solve the crisis of the missing mystery photos and the vanished alien constructions. They are not there now. They were not there before and they will not be there any time in the future – they are fictions of the imagination of the writer of the Fate magazine articles.

The following photos show the area between crater Macrobius and the smaller crater named Peirce. The objects drawn between Macrobius and Peirce are interpretations of the shadows and sun highlights playing off the surface features of the moon. It seems the "cannon" with window openings has moved on to another location. As for the 18 mile size building 100 miles south and east of Cleomedes, it too must have been relocated somewhere else. The red square in the following photo outlines the closest thing that looks like what he is describing – a lunar highland ejecta area between craters.

Stellar Sterility

LROC Act-React Quickmap, northwest Mare Crisium, an area between
Craters Macrobius and Peirce.

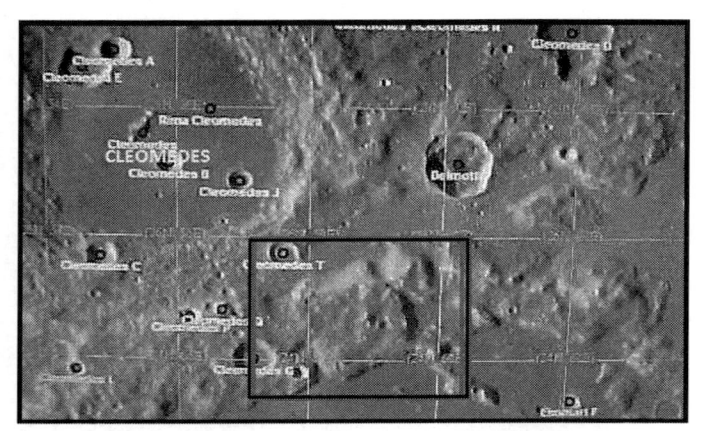

Crater Cleomedes area showing no building.
LROC Act-React Quickmap

3. THE GIGANTIC BLOCK IN ENDYMION

WRECKED OR DISCARDED artifact on the eastern rim of Crater Endymion at 54 degrees N. Lat., and 60 degrees E. Long.

In the northeast quadrant of the moon lies a strange rectangular junk heap well over 50 miles in length. It's a double-decked ocean liner size square block, tilted on its side, giving the impression of a beached whale washed up on shore. In this case, it is a mechanical crater critter resting on its side, moon bathing. It has four brilliant white domes, two of which seem to be up on poles and a circular superstructure on the upper deck.

The above monumental heap pile and mechanical mishap depicts what Swaney said is evidence to as long-concealed truth regarding our solar system's recent history, things our scientists refuse to talk about, which happened some 12,000 years ago. Swaney said, this was coincidentally contemporary with the disappearance of Atlantis. Jack claims the evidence is

Stellar Sterility

surfacing in support of ancient alien technologies and historians will have to give an account – I.e. rewrite the books, restructure religion to accommodate aliens and redirect humanity away from godly truth. It is easy to deny the existence of this technology said Swaney. Yet, if considered, it is not easy to pin point its demise.

One explanation, given by Fate Magazine, is ancient Greek mythology. The Greeks tell of a prehistoric period called the "golden Age," when the gods ruled the heavens and earth. The myths tell us that a colony of renegades known as the Titans invaded this solar system with intent to conquer. The ancient Atlanteans fought against them to no avail. The Titans used nuclear weapons and defeated the Atlanteans. They blasted away at the smaller planets and blew up Mars, destroyed the planet between Mars and Jupiter, and disrupted some of the gas planets like Uranus, capsizing it the way we see it today.

Based on the false assumption and pseudo-science of the atomic bomb explanation of crater development, the above (exaggerated) interpretation would seem reasonable. Nevertheless, having some reasonable shadow of a doubt, there must be some other reasonable scientific explanation.

The science of geology is a mystery to most people, who would not know a conch shell horn from a piccolo. Geology explains exactly what happened. The above mechanical mirage is a byproduct of light and shadow. This drawing was made from observatory photos or from direct telescopic observation, because none of the NASA photos reveal such nonsense. The article does not reference any lunar photo number. Nevertheless, if the object was there in the past, it is not there now.

Stellar Sterility

The LROC Act-React Quickmap shows the location of what may be mistaken for the above drawing. Improper lighting and shadow play can easily form artificial looking constructions. The observer can imagine giant size monstrosities from this trick of light and shadow, and if careful, the sizes can range into the hundreds of miles.

The above square shaped colossal size contraption has to be hundreds of miles long and exceeds the limits of sane engineering, even for alien cultures. It seems the further away the alien presence artifact hunter is the larger the artifact. Conversely, the closer the snoop finds himself, the smaller the alien artifact. The only common denominator would be the blurriness in hiding the true nature of the object.

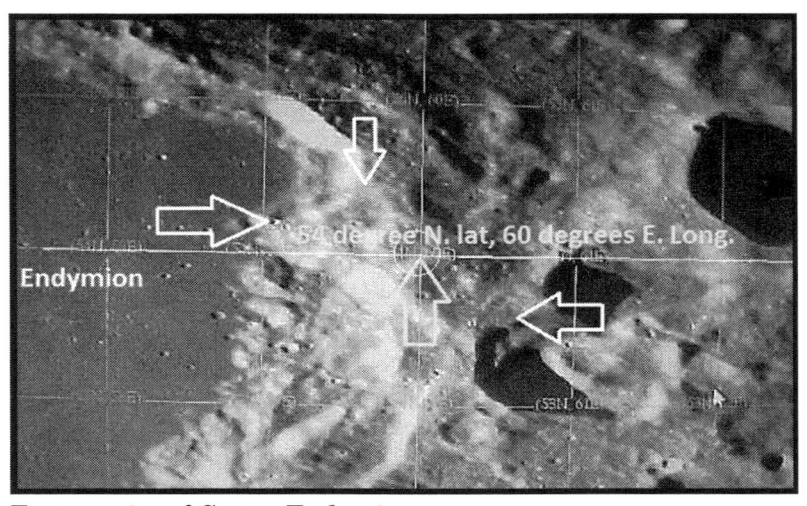

Eastern rim of Crater Endymion

4.   ROBOTIC BUNNY IN CRATER CLAVIUS

Kewpie Doll lying on southeast wall of Crater Clavius
(Location: 60 degrees S. Lat., 8 degrees W. Long.)

No, this is not a funny looking robotic gadget or some other alien claptrap. The weird shape looks like what Swaney and Felix Bach said is a *"discarded kewpie doll."* Wow! Giant Chatty Cathy's for giant aliens! The illustrious reporters declare they have seen this surrealistic toddler sitting close to the South Pole on the wall of Crater Clavius, cocked at an angle as if knocked off its base. Some kewpie dolls have support bases. This one has a base attached to its bottom. This bizarre doll-baby has a metallic sheen like dull aluminum and is visible in some observatory photos of the moon.

Nevertheless, one should not hold their breath in finding the kewpie kwickly. Most of the photos have been touched up, repainted or confiscated. Do the NASA photos expose this E.T. infant in any of the millions of pictures taken of the moon? LO-IV-118-H3

Stellar Sterility

and 131-H1 do not show much supporting evidence. Neither does frame 124-H1 and neither do the 10,000 plus other NASA photos.

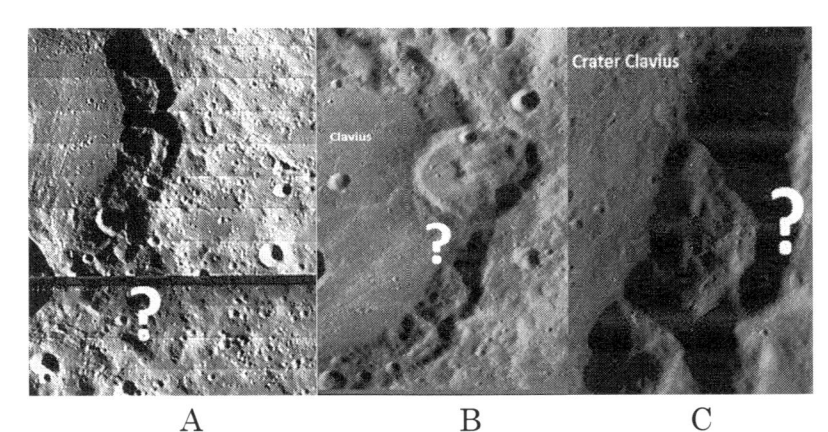

   A   B   C
LO-IV-118-H3   b. LO-IV-124-H1   c. LO-IV-131-H1
Southeast corner rim of Crater Clavius

5. THE RASCALLY RABBIT IN THE EAST

Rabbit Ear machine east of Sea of Nectaris
(Located at 25 degrees S. Lat, 45 degrees E. Long.)

Hi Ho the Dario, a hunting we will go and to the east quadrant of the moon too! We are hunting "rabbits" in the east or one-half of an ocean liner's prop. A rather bizarre object is supposedly in the moon's east quadrant, a little east of Mare Nectaris close to Crater Theophilis. Drawn from telescopic observations, the mechanical rabbit appears a dazzling white, almost fluorescent in brilliance.

The above photographic hair hunters insist that when it is near the shadow line, it is far and away (by far) the most immediately visible feature on the lunar surface. They speculate that, just like a real rabbit, the ears of this mountain size Lagomorpha-leporidae is

Stellar Sterility

intended to move, because they are mounted on a vertical shaft, which rises above the ground.

This above interpretation is a "hair" bit above ridiculous, as photo references do not show any such hair-brain rabbit device. Actually, it is by far the most elusive rabbit on the moon. None of the NASA photographs catch this bunny tale and reveal only highlighted lunar craters, regolith and shadows. This is another wild hair story and goose chase among the craters as the following photos suggests. Nowhere do we find this White Rabbit.

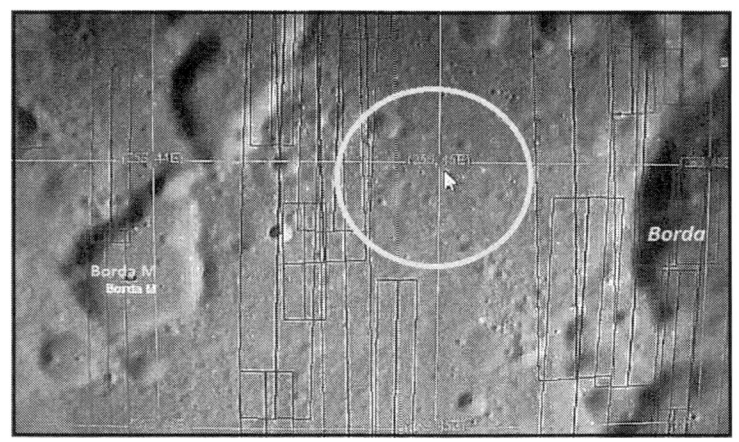

LROC Act-React Quickmap of area southeast of crater Theophilis.

Location of the elusive white rabbit.

Stellar Sterility

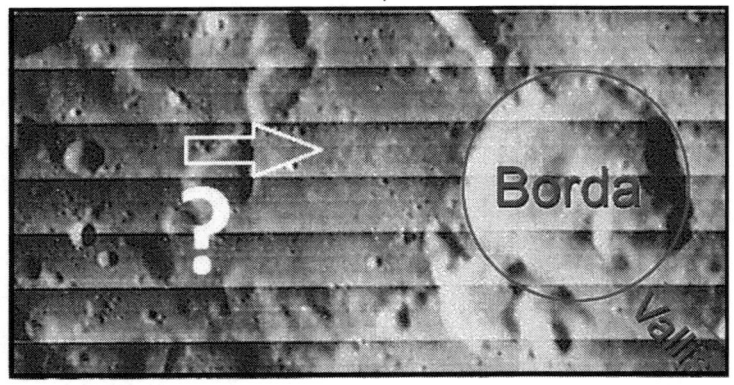

LO-IV-065-H1
Rabbit location E. and S. of Mare Nectar and west of Crater Borda.

6. MONUMENTAL MOUNTAINS OF MOON METAL

Piles of malformed metal heaped up on the south tip of the Caucasus Mountains, along the edge of the Sea of Serenity. Location: 35 degrees N. Lat., 8 degrees E. Long

442

Stellar Sterility

Our friends at Fate are on another wild salvage hunt and this time for scrap metal. In the moon's northern hemisphere (northeast quadrant) resides a junk pile location littered with twisted metal plates, broken constructions and bizarre chunks of machinery, stretching for 200 miles within a lunar formation named the "Caucasus Mountains." It is supposedly the amateur astronomer's favorite telescopic view for its dazzling piles of shinny geometric shaped chunks of rubbish. It is a real junk pile hill for possible ULO's ("unidentified lunar objects").

The Caucasus Mountains are in the Sea of Serenity, which lies southwest of Crater Calippus. Rukl's Atlas in map 13, p. 53 demonstrates a good shot of this location. Other good shots of "Junk Pile Hill" are in NASA photos LO-IV-103-H1, H2 and 098-H1. It is up to the reader to decide the location of this pile of fantastic garbage. The area is steeped with what looks like alien junk and just about any lunar-graphical location will expose the stuff.

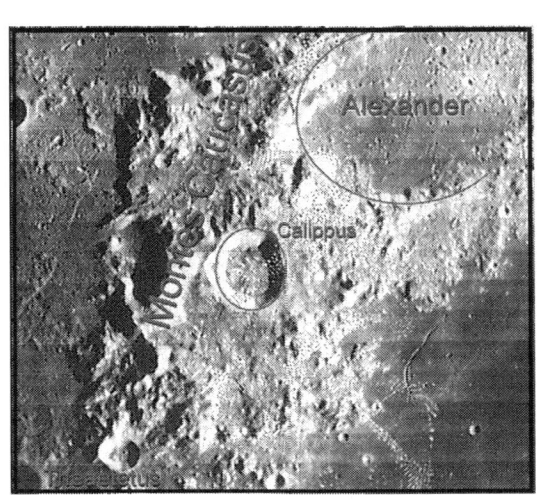

LO-IV-103-H2

Stellar Sterility

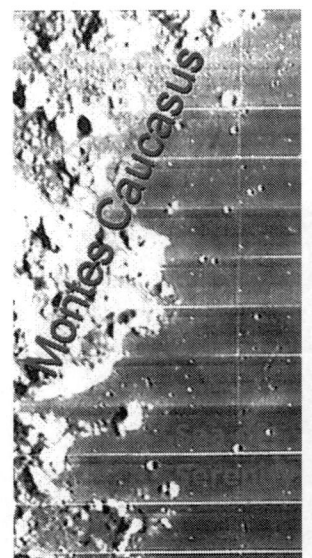

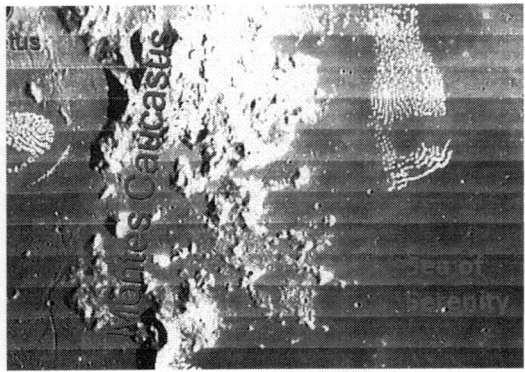

LO-IV-098-H1   LO-IV-103-H1
(Lunar Orbiter Photographic Atlas of the Moon)

7. CRANES AND BOOMS OF JULIUS CAESAR

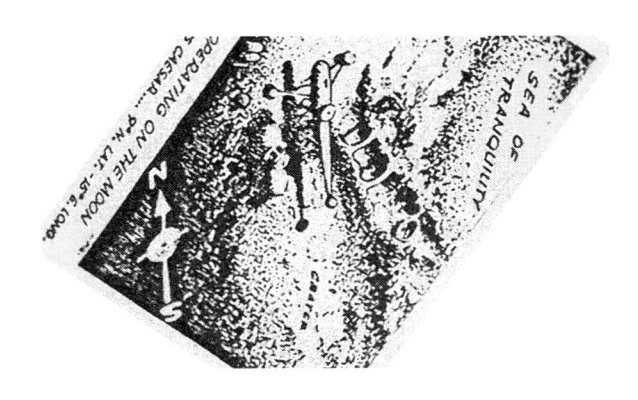

A 25-mile long crane or boom is proportionally unthinkable and it is unthinkable that people should believe such hog wash. The above is a sketch of some machine made in April 1982 by the Fate observers, when the moon was only six days old. The writer pontificates that it is located at the north rim of the Crater Julius Caesar and easily viewed at 9 degrees N. Lat., -15 degrees E. Longitude, on the western edge of Mare Tranquillitatis. The ball-tipped arm is usually orientated north to south, but on occasion reorients itself east to west. In other words, it moves! Maybe it is light sensitive and changes shape and location according to the Sun?

Stellar Sterility

LO-IV-090-H2 Crater Julius Caesar

No rotating weather vane. Just a mountain with dark shadows.

Previously, in 1980, the astronomical weather vane rotated 180 degrees, the ball-tip riding above the semicircle of structures in the foreground. The blab continues with braces that apparently steady the front end of the boom, which are ahead of the pivot point. This suggests they have to be unconnected before it makes the 180-degree swing and there is no one up there to do that. Another gigantic moving "thing" lies 75 miles west close to Crater Boscovitch.

SECTION-6
RICHARD HOAGLAND'S MONUMENTS MADNESS

A POLEMIC DISCUSSION OF
RICHARD HOAGLAND'S
"THE MONUMENTS OF MARS:
A CITY ON THE EDGE OF WHERE EVER"

Mr. Richard Hoagland

The most prevalent and modern equivalent of George Leonard and Fred Steckling is Mr Richard C. Hoagland, self-educated [1a] scientist and famed author of "The Monuments of Mars: A City of the Edge of Forever," and "The Enterprise Mission" web site. Mr Hoagland is a former space science museum curator; a former NASA consultant and during the historic Apollo Missions to the Moon, was science advisor to Walter Cronkite and CBS News. At present he hides behind a web site with no direct phone line, runs the country giving talks and lectures and sometimes visits NASA and other government agencies promoting the godless alien presence mythos.

Following the earlier tradition of Lunar surface alien trivia, Mr Hoagland thinks he sees extremely complex mathematical and hyper dimensional geometry in such things as the colossal *Face* on Mars and surrounding rock piles of Cydonia. He also sees this strange type of alien stuff scattered about the lunar surface of our moon. For instance, Richard asks a rhetorical question, surly not a real question (for he really believes in alien artifacts), that *"If a complex, hyper dimensional, spacefaring culture built, and then abandoned, the monumental ruins at Cydonia,* [then] *what other artifacts, on what other worlds, for what hyper dimensional purposes* [and he does not really know], *might also have been left across the solar system, including on earth's Moon?"*

The Mars Face was [at that time] a sure thing, but the lunar evidences were a real nut to crack. Richard was reluctant to dive into the Sea of Tranquility and disturb all the previous questions of crack-pots, for he was still without [and still without to this day] *"some kind of indication as to where to search, to even begin to pour through the literally millions of NASA close-up images taken of the Moon."* To him it was all but hopeless, but to others it can only take maybe a dozen photos to prove or disprove the alien artifact urban legend.

No matter, Richard persisted and *"Only after beginning a* [caught, choke, strain to not laugh] *serious investigation into a possible hyper dimensional influence between Earth and Moon, did we also inevitably begin consideration of a crucial corollary: the possibility for ancient ruins on the Moon, left by*

Stellar Sterility

those who long ago left us this new physics." The problem he said though, was where to find them. Strain as he did and so easy as it came to be, he then began to find more alien artifacts than George and Fred could have ever found. Almost every photographic scratch, glitch, dust particle and especially dangling crystal (development scratch) on the photos became an alien artifact.

Unfortunately for Richard, as it was for his predecessors – thanks to Richard's expose', these artifacts are as well fictional inventions created from hyper dimensional, delusional thinking – not much different than the hyper-two-dimensional thinking of the two previous researchers.

Of course, it took Richard some time to overcome the lack of "real" evidence, supposedly found by previous researchers: i.e. George Leonard and Fred Steckling, whom he purposely criticizes and debunks for proposing that emulsion development processes were alien artifacts, to come to grips with and supposedly prove that tons of photographic pixel distortions are rather the real alien artifacts! One thing Richard did maintain in common with his debunked predecessors, in establishing an alien presence, was the same old light and shadow play on lunar surface features that deceived Leonard, Steckling and many others.

The Mars Face is a prime example of this deception. What was once a real face and the foundation to a whole theory has recently been proved with higher resolution photos to be a giant mountain size pile of Martian rocks. Thus, it is this writers opinion that everything that follows the Mars Face as evidence for aliens is just more advanced piles of Martian regolithic rhetoric.

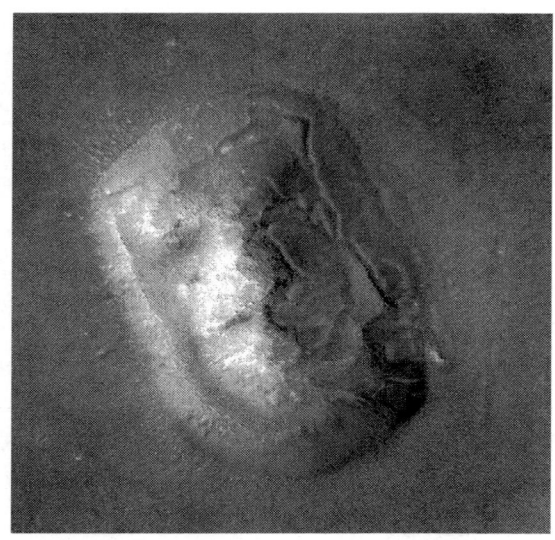

LROC Photo of the famous hyped up "Face" on Mars

The Mars Face (illusion) borders the plains of Acidalia Planitia and the Arabia Terra highlands. [1] The area includes the Mars regions: "*Cydonia Mensae*", an area of flat-topped mesa-like features, "*Cydonia Colles,*" a region of small hills or knobs and "*Cydonia Labyrinthus,*" a complex of intersecting valleys. [2]

This is where Richard goes monumentally mad. His in-depth studies of lunar and Martian alien artifacts overshadows the combined researches done in previous decades. In fact, this is exactly what he has done, combined all previous studies (minus the "water-droplets" and "pull-apart marks") adding further a hodgepodge of much new material — mythologies, fictions, imaginative inventions, typologies and abstract concepts, all gift wrapped in super scientific mumbo-jumbo. The consequences of this man's work has led to a global wide pandemic of alien artifact finds by what we may call Hoaglandites.

Stellar Sterility

According to Hoagland and his cult followers the Moon is littered more now than when Leonard bent the first lights of Selenography studies. After bleeding the Man in the Moon dry, and to escape boredom, many E.T. hunters jump this new train to other planets in search of the fabulous.

Mars has been found to house new alien landscapes for mad monument hunters. The scientific realm, especially the space programs of our nations are cluttered and mingled with this alien pandemonium of non-fact and nonsense. I suppose that the portion that does not believe is more than happy with all the publicity and has rather turned the other cheek caring less about debunking it. What a loss of publicity and funding! So, why bother?

Mr Hoagland's supposed revolutionary concepts, stemming from the Cydonia region, suggests an astrophysical origin and role of our Moon, and the possibility of finding alien artifacts. He sees remnants of an alien culture that once thrived in this area and supposes that Mars was once a global cosmopolis of buildings, pyramids, super highways and advance glass tube tunnel systems. He postulates that if a hyper-dimensional space-faring culture built the Cydonia Complex, then we all must ask, "*What other artifacts... might also have been left across the solar system, including the earth's Moon?*" [3] We will find out the answer in the remainder of this compendium.

To answer the mute question of where these Martians went to next, after cratering Mars beyond livable conditions, we must turn to the Moon, where we see they also cratered the lunar surface and disappeared. We dare not ask where in Hell they went after the Moon. But that might be a good place to start looking! With a full knowledge of Earth history and a total lack of earthly alien artifacts to suggest that Earth is where they went to next, Richard turns our attention as well as our stomachs (without much mention of Earth) to the next best location – the Moon.

Of course, the moon! Where else can one play with people's imaginations, twist landscapes to beef-up the alien presence cult and create a super income for himself as well as advance public interest in NASA? As far as NASA and Uncle Sam are concerned, they have their Mickey Mouse now, thanks to Hoagland. Who would ever dream of flushing Mickey down the toilet in favor of raw dead dry science?

Richard has only three reasons to choose the Moon. Each is perfectly strong enough to deceive the ignorant masses. First, the Moon is preferable, because the Earth has no super technology, "super" physical enough to prove an alien presence in early human history. What it has is only suggestive and not at all tangible. There are no huge colossal crystal glass domed cities, no mile high alien monuments or mysteriously advanced non-human scientific devices.

All that we do have are vague references to a high technology in our ancient past that any culture could develop within four to five thousand years of scientific evolution. Hence, ancient advanced cutting marks, sonic drilling and other weird and unexplainable artifacts are only so many evidences for advanced human engineering.

There is nothing on earth that cannot be explained by human development if one is patient enough to look and avoid jumping to alien conclusions – give or take a few mutilated cows or destroyed corn crops – and why only cows and corn crops, and not chickens and cabbage we will leave for another volume.

Stellar Sterility

Nevertheless, when it comes to "evidences" found 250,000 miles away, shot with not so fine photographic equipment, anything can be wizard into proving an alien presence. Shadows, highlights, warped surface features and other natural lunar-physical anomalies are easily contorted into alien manufactures.

Now, unknown to the common alien Oopart hunter, just the opposite can be proved, which is even truer, when the same materials are exposed to closer inspection. Domes, machines, towers, pillars, obelisks are easily seen to be mounds of dirt, rocks, boulders and other naturally developed selenological features.

Secondly, up until recently, the lunar surface had been studied in more detail than the Martian surface. Since the earth was void of alien Ooparts, and Mars was still obscured by low resolution photography, the Moon was the only best choice. It was far away, mysterious and yet close enough and available to study. Since the Viking missions, the Moon though was abandoned in favor of the more romantic new world of Mars.

Coincidentally, as Martian photography advanced to what it is today, and as lunar photography developed even further in resolution, the direction of people's consciousness moved away from the Moon, and for good reasons too. Lunar photography is even higher in resolution than the Martian photography, coming from the Lunar Reconnaissance Orbiter, and lays waste to any attempt to discover any alien artifacts, as we have seen.

We shall see in the following studies that the advancement in photographic resolution of lunar surface features has exposed all this alien artifact trash as pure nonsense, and all the theoretical nonsense as pure trash. However one prefers to mock it matters not, it is now being proved with Mars. As the photography has gotten better, so also has the exposure of Martian artifact nonsense.

A grand example that is unfortunate for Mr Hoagland (and he should have never pressed the issue of demanding finer detailed photos of Cydonia) is the Face on Mars. As we have mentioned above in the most recent photographic missions, the Face is proved to be what real science predicted – a giant pile of Mars rocks.

We shall not choke these pages full with all the double-talk of Mr Hoagland and his followers, as they try to explain the blatant evidence against the Martian face. The Internet is full of Richard's expos-fact-o sophistry. In fact it has become a sewer of debates, most favoring the Hoaglandites, for people prefer the fabulous over the facts.

Getting back to the original point: The Moon in past times was closer and offered more detail, yet still vague in resolution to really decide for sure the reality of alien artifacts. Hence, we now understand Richard's quick distracting disconnection from his predecessors.

These studies offered (at that time) the necessary *"conclusive evidence of contemporary extraterrestrial bases on the Moon."* [See footnote 3] This was the one and only and last chance to seriously investigate and embed the alien presence syndrome into people's minds, and thus prove Hoagland's so-called *"hyper-dimensional influence between the Earth and the Moon,"* and of course Mars.

The only thing missing were the facts or artifacts and exactly *"where to find them."* [See again, footnote 3] Since they were NOT good enough, though they were good enough as a starting point, Richard decided to replace the old theories with his new physics.

Stellar Sterility

With such a vast surface area to look over, Richard decided to pick up where the others left off. The most convenient *"jumping off spot"* were the previous studies done by such alien Oopart hunters as Fred Steckling [4] and George Leonard. [5] After some lengthy digging through all the *"non-stop comedy of errors, misidentified photographic defects and the general mishmash of malillustrated misinformation"* [6] of these forerunners of alien artifacts, Richard decided that it was his turn to advance the non-stop comedy of errors, misidentified photographic defects, pixel distortions and the general mishmash of malillustrated misinformation with new examples, not only from the Moon but from Mars.

Though he *"looked hard... but could see nothing"* of what they were claiming, it was not long before he looked hard and began to see something. To him, just as it was for his predecessors, something was very *obvious*, yet easily missed, if one did not have a hyper-dimensional mathematical eye to see.

Of course, anyone can develop the hyper-dimensional arithmetic eye, after days of staring at blurry photographs of dead dry boring surface features. It is not much different than gawking at Earl Grey leaves and prattling fancies about one's health and welfare, after noon time tea and crumpets.

To save the reader their eye-sight and not belabor the subject beyond sensibility, we shall reduce the Martian and lunar piles of Hoagland hog-wash to a short study, picking only the most obvious and favored, as well as rediculous examples of the new evidences.

1
UKERT CRATER BASEBALL DIAMOND

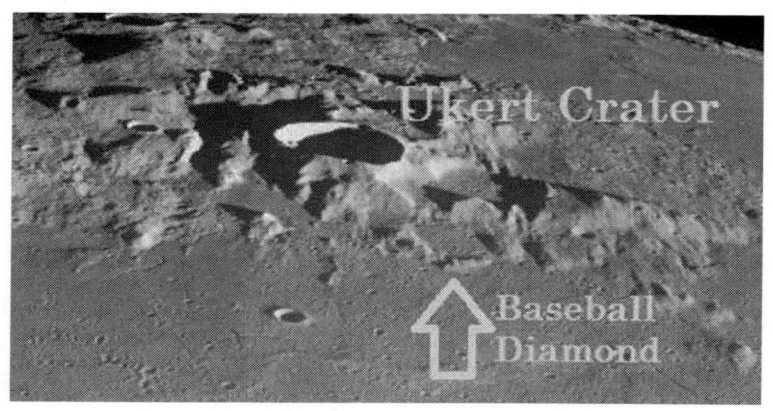

AS10-32-4819

Marvelously, after analyzing thousands of blurry, dead scientifically dry photographic lunar features, it was not long before Richard's eyes fell upon a most interesting formation immediately adjacent to and somewhat southwest of Ukert Crater. It was NASA Apollo-10 photo AS10-32-4819 that attracted the attention of Richard and it was the best choice and finest starting point for a most obvious and potential lunar alien artifact that could begin proving the alien presence on the Moon. What he found was an almost perfect square baseball diamond shaped looking plot of dirt.

From this default in clear vision followed the delusions found in photo frame number 4822. Just as was conjured from the Cydonia ruins with the Face on Mars, Ukert showed possible evidence for a similar type of "hyper-dimensional *grid*" and thus potential proof of a "hyper-dimensional *alien biological and cultural connection.*"

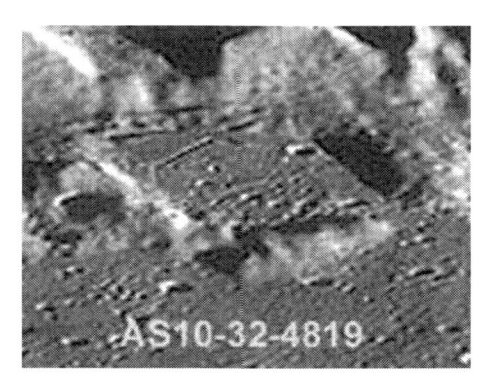

The Ukert Crater Baseball Diamond Field

This was a find as grandiose and colossal as the Giza pyramid of Egypt, the Temple of Baalbek, and the ruins of Teotihuacan. It was equal to the famous Seven Wonders of the Ancient World! This was an ancient alien city of the edge of forever or for sure, the edge of Ukert. It appeared long ago abandoned and forgotten. Yet, something was there! What was it that caught the attention of Richard's eyes? It was a little *"geometrically marked square"* next to a big crater.

Baseball Field from Top View

Stellar Sterility

With heart in mouth, an apple in his throat and a pen in his hand, Rich began to analyze this remarkable compact, bluntly obvious geometric arrangement that was leaping off the page at him, and him alone. For, after some further attention by this writer of the same photo, and after what seemed to be an eternity of looking, nothing could be seen to verify such claims of alien biological cultural connections but a square plot of dirt created by possible crater ejecta materials.

Nevertheless, Richard persisted and with ruler in hand, and some scientific thought, or what he thought was scientific thought, he applied to this tiny geographical region of highly fascinating, but highly anomalous organized dirt some super mathematical calculations. Richard points his own Leonardean magic wand at absolutely nothing and instantly transformed this regolithic nothing into a megalithic something. A close examination from different angles from all the relevant photos of Ukert revealed to Richard a magnificent alien construction, whereas the same study repeated by this writer revealed absolutely nothing. See As10-32-4813 through 4822. [8]

Further study with a ruler in hand, a calculator, a micrometer, a spectra-meter, an oscilloscope, an old but usable electro diagnostic violet ray wand, some Flemish, English and French ell, fathom, feet and cubit tape measures, this writer found some odd and offbeat "moot distances," yet coincidentally similar to the smoots (364.4 smooths) the length of the Harvard Bridge.

Yet, all attempts to translate this anomaly into varied other important measurements – megalithic yards, bloits, pyramid and British inches, sheppey, a few Mickey's multiplied by the factor of 1000, furmans, and a few attempts of Potrzebie, cowznofski, vreeble and some hoo and hah – failed to reveal any further so-called Hoagland super hyper-dimensional geometrical equations. There was nothing whatsoever akin to the supposed geometry of Cydonia. There was no stroke of alien Oopart luck revealing Lego type constructed buildings, no electromagnetic hovercraft freeways, no mechanical stairs, steps, walkways, glass domes or used abandoned UFOs. There was only a crater and a funny little (somewhat) square baseball field diamond shaped plot of crater ejecta.

2
RETURN OF THE UKERTIANS

With some extra careful observations, beyond the abilities of Richard, one can notice other supposed obvious and blatant evidences of alien activity or what appears to resemble alien "Ooparts," "Oops," and "Oops, it's busted and left behind" equipment. Glaring at Lunar Orbiter-III photos for hours on end, one cannot help but see, even when hallucinating, what Hoagland missed by not looking hard enough. The aliens apparently never left or they came back and are still using the area. There are five or six small white orbital shaped UFOs floating about the crater!

Interesting little blobs. Maybe the aliens are back observing or salvaging their old properties? Surly with the help of their little fleet of "water-droplets" and "pull-apart marks" floating in zero gravity emulsion development fluids, and at 1/6th gravity (located to the upper right middle in the crater) hovering over Rime

Stellar Sterility

Hyginous, the aliens will have no problems rebuilding the city – or at least a baseball team reunion in the diamond square. What of the huge squadron of UFOs to the East?

We are not sure, said people like Hoagland, so they might well be the reconnaissance missions of smaller scout ships hovering south of the larger assembly of the crater blob crafts, and directly north the group of nine crafts running their mission over Ukert City area. Richard! You missed this. These are not emulsion splatters. Because, if you were to snap another picture of the area today, we'll bet you all the tea in China that they are not there anymore – they have moved on to another location – actually, the Lunar Orbiter Photographic Atlas of the Moon is replete with examples of these "things" floating all over the Moon. They seem to be there one minute and then not the next: Now you see them; now you don't.

Fleet of UFO's or Space amebas? or Emusion splatters?

Toward Ukert Crater & the baseball field.

Moreover, if you look at the most recent photos of the same area taken by the Lunar Orbiter Reconnaissance Mission, they are not there either.

Stellar Sterility

Poor Richard's Almanac does not stop here. Richey moves on in his fabulous researches for E.T's and artifacts way northward from the Square to what he calls "*the best yet to come.*" Oh, I can't wait!

From the square or Hoagland's way-up-stairs "L-7" to the upper right "*lies a diagonal array of brilliantly reflected hills now suspected... to be the shattered ruins of several ancient, environmentally enclosed 'acrologies' (city-sized architectural ecologies) on the Moon.*" Beyond arcology row "*lies an equally remarkable 'rectilinear ground pattern' eerily reminiscent of an array of city-sized blocks... (And) anomalous vertical 'reflections' and curiously translucent 'panels' scattered throughout the area,*" [8] with "*overall, striking 'grid-like' 3-D rectilinear geometry... anomalous curved 'pipe-like' structures... (And) the elevated 'glass-like, geometric lunar construction'... Sinus Medii dome, (of) 'wavy shower-glass veiling'... 'Of irregular vertical lineation's' (of) 'optically reflective lenses' dangling and hovering over the city area.*" After reading and prepping ourselves with Gulliver's Travels and dangerously straining our eyes, we could very well see cross-eyed, "*a shattered crystalline city on the Moon.*"

With all this hyper-dimensional flap and blab, and the potential evidence of alien UFO scouting activities of the little white blobs, the Ukertians may be returning to claim their old land grants. But, little white blobs and floating orbs are not the only things dangling around the skies of Ukert.

3
DANGLING CRYSTALS OF UKERT CITY

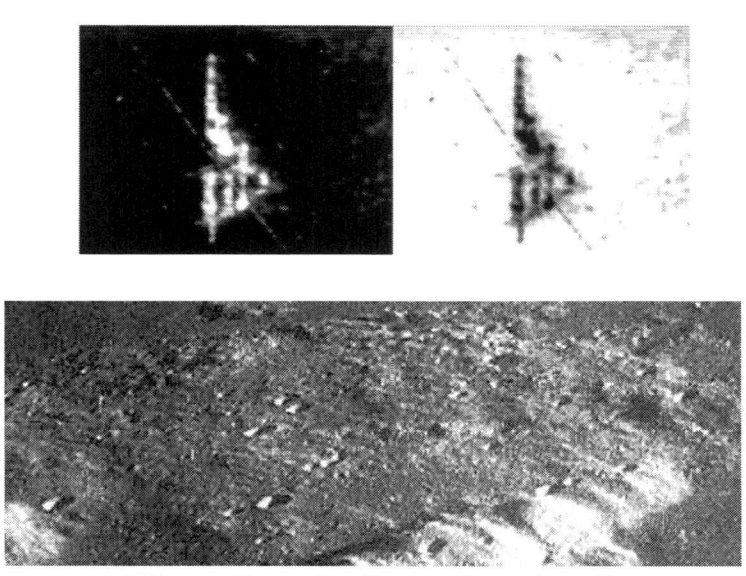

Ukert City, the Lunar Los Angeles

In the Crystal City of Ukert (and it does not matter where one looks in the Lunar Orbiter photo III-85M) one can find what appears to be "crystalline structures" supposedly hanging above the lunar surface, within a *"geometrical grid-like"* system of decayed glass dome support beams and lattices. Now take a deep breath! Our fiction writer said, they all seem to be irregular shaped and unnatural to normal or natural lunar geological structures. Zooming in deeper, both into Mr Hoagland's imagination and into the finer details of the photographic emulsion of frame III-85M, we find what looks like these crystals of wire-like, hair-like bright reflective shapes sticking out

Stellar Sterility

like a sore eye hovering over truly square shaped double craters. Here are a few enlargements of the photo and some drawings extrapolated from these high resolution enhancements of the frame, along with some other newly exposed examples demonstrating E.T presence.

When looked at properly, as development processes, and explained as such, it becomes evident that they are not above the Lunar surface, but actually scattered along the plain of the surface of the photo itself. Without taking the hair-like object in context of the whole surface of the frame, it is an optical illusion and deceives one into thinking it is on the Moon.

When taken in context of the frame surface, and when hundreds of other examples are located, it becomes apparent that it is one of many photo development Ooparts or artifacts, even possibly dust on the camera lens, or some other intrusive element. Compare these drawings with the following visual optical illusions.

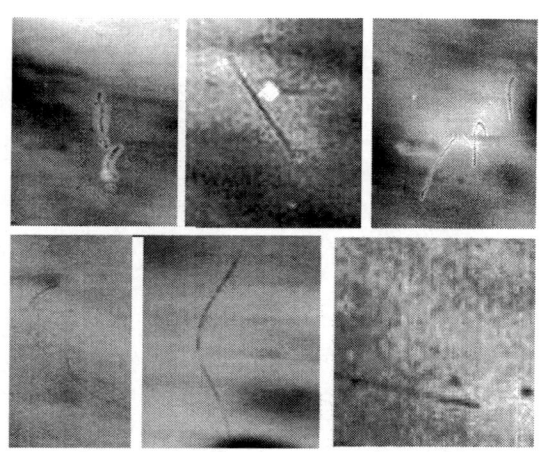

Hairs, lines, scratches and dangling crystals

Stellar Sterility

DOUBLE CRATERS

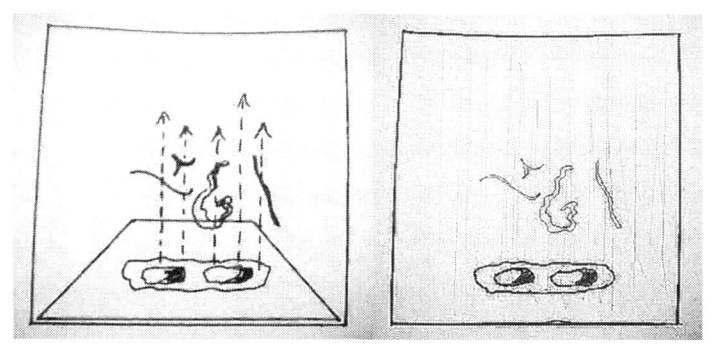

The first drawing demonstrates the way Hoagland wants us to see the artifacts. The second drawing shows the way we should look at them.

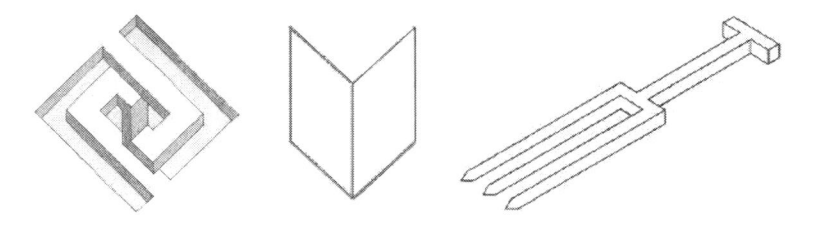

The best example of this eye trick is the middle drawing of the above collection of deceptive constructs. Ask yourself: Is the book open to a page inside the book or are you looking at the cover and spine on the outside? Now turn back and look at the so called crystal slivers and alien artifacts. Is this really a chunk of crystal glass from some ancient domed structure or some other type of alien building material "suspended above" the lunar surface?

Are they really dangling crystals of glass or plastic hanging from geometrically shaped structural support beams? Yes, said Richard Hoagland. But, this is only a hair-brain comic book idea postulated (or hatched) in his overly imaginative mind. No, they are no more real artifacts than the ones pointed out by George and Fred. And they are not dangling in mid-air over the moon. This is an illusion

From Richard's descriptions it would seem more than an idea and actually an alien artifact. Richard is not spinning a yarn or blowing fantastic glass bubbles. He, maybe unlike the first two yarn spinners, actually believes what he is describing. As George Leonard made dirt digging machines out of boulders and as Fred Steckling made ponds and lakes out of emulsion splatters and crater floors, so Mr Hoagland overly imagines and hyper exaggerates misidentified emulsion errors, pixel distortions making Ukert monuments out of lunar molehills.

This is not just some alien hairy crystal hanging above the Moon, but actually a "hair" or "dust" particle recorded in the photographic development when the Lunar Orbiter processed the picture and transmitted it to earth.

GLASS SPHERE ENCLOSING THE MOON?

Believe it or not, there is more than one layer of debris artifacts in the process before final printout on

Stellar Sterility

earth. There is camera lens dust, micro-fine cosmic dust particles inside the Lunar Orbiter, the tiny straightations or lines in the machines processing of the prints and then there is possibly the scan lines in transmission, not to mention the dust on earth that could have been imprinted on the original copies made for the different Space agencies. You have the NSSDC, JPL and a few others.

After studying the lunar photographs this writer found many other hair-like fine slivers, scratches, blobs, cuts, scan lines and anomalous looking imprints on thousands of frames in thousands of areas on the Moon. Just review one of the many NASA published lunar atlases.

It seems that if Hoagland is right (and he is not), the entire Moon should, for all intents and purposes, have been covered with one gigantic glass sphere. Hmm? Interesting. A glass sphere housing a small planetary "moon" inside! But, it will burden anyone's sanity to find even one humongous colossal support beam column that would have held up such a structure. I can understand maybe the almost complete destruction and decay of the glass dome(s) and even the smaller support beams, but not one larger one has been found! We do see one qualifying as the only possible support beam - the famous Shard, which we shall discuss later. But, it is not big enough.

Needless to say, getting back to Earth and common sense, frame LO-III-85M, which we have been discussing, has uncountable so-called alien hair-like glass crystals, dust particles and processing scratches. We really need not look any further than this frame. This one is enough to keep the bewitched busy for months. There is no need to waste thousands of dollars buying NASA negatives to verify what is being said here.

Getting back to the above optical illusion, the particles or debris artifacts always appear flat to the real surface of the photograph. It seems that no matter which little debris spot one chooses to look at, not one has proper perspective to it. The particles always appear flat to the surface of the photo.

The crystals seem to be found in almost every place you look and much higher than normal for any crystal dome. They can be found even in the corners of the photos, some few off the photo area and others right in the middle. They all seem to run parallel to each other all along the surface. Close inspection leads to other fantastic evidences.

5
UKERT CRATER TRIANGLE

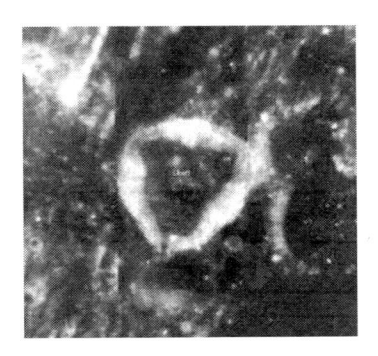

Attached to this array of geometrically shaped objects or what looks like a broken conk shell is Ukert Crater. Ukert is a lunar impact crater that lies on a strip of rugged ground between Mare Vaporum to the north and Sinus Medii in the south. It is located to the north-northwest of the crater Triesnecker and northeast of the crater pair of Pallas and Murchison. The outer rim of this crater is not quite circular, with outward bulges to the north and the east. The interior floor is irregular in places, with a central ridge running from crater midpoint down to the southern wall. It is the one previous discussed as having the little group of UFO scout ships – the cluster of little white blobs. It is a nice little crater not uncommon, but somewhat unlike many others. It has a "tetrahedral triangle" shape inside its crater walls.

Unlike other craters that have various objects, alphabetical letters and multiple rims, Ukert is almost geometrically perfect. It is perfect enough for Mr Hoagland to Heehaw all kinds of occult significances.

He scribbles page after page of mysterious mathematical formulas, from Euclidean triangles, to a six-pointed star, the solvability of the quintic, and the trisection of angles, divisibility of instances and theories of alien technology.

Now, after all this pseudo-scientific gibberish we are presented with some simple minded alien baseball diamond shaped field and a funny triangle inside a crater. Obviously, since aliens do not need entertainment nor know anything about sports, we can sterilize all lunar finds of any super hyper-dimensional, cultural entertainment and strictly "sport" the idea of a completely sterile scientific hyper-dimensional purpose. The question now is what does such an anomalous looking square field mean to this alien presence?

CRATER FORMATION AND WEIRD FEATURES

The Ukert Crater triangle and the Diamond Field, along with the busted conk shell fragments are not the only location of odd shaped formations. Other crater gaming fields and odd shaped craters can be found such as the triangular shaped one to the bottom right of Ukert. These natural selenological structures can be found almost anywhere on the Moon. There is a natural explanation for every one of these if the student takes time to think reasonable rather than invent artifices.

The natural explanation of Ukert is crater wall deterioration and collapse, where the higher reflectivity of the lunar soil contrasts with the shadows forming what looks like some artificial alien geometric shape. One of the main images of Ukert is from the Clementine spacecraft. It shows that the sides and the vertices are not linear and as sharp as the Lunar Orbiter photos as Hoagland would have us believe.

Stellar Sterility

This higher resolution photo reveals the triangular anomaly to be natural to lunar geophysical processes. There is nothing bazaar about Ukert except the light-shadow anomaly caused by the combination of crater development shape, high and low reflective lunar soils and sun angles.

A meteorite strikes the lunar surface creating a crater, the walls collapse creating interior crater wall landslides exposing highly reflective materials. In this case the illusion of a triangle is formed by the uneven land slide of only three sides. With the lunar soil of a higher reflectivity, the walls would highlight brighter at lunar Noon-time obviously explaining the supposed artificial looking shape.

The triangle makes its appearance in the optics of telescopes at full Moon (high noon at its location on the lunar surface [9] creating a 16-mile-wide dark equilateral triangle within the crater floor. Spaced 120 degrees apart around the edge are three brighter areas that when connected create the triangle.

Considering the shape and construction of the crater rim and the reflective properties of the crater walls, as well as the brightness of the sun as it rises, it is not difficult to see how the mysterious triangle develops at high noon. The more the light brightens as the sun rises the brighter the three crater rim peaks expand.

Without much consideration of alternatives and a total disregard for real scientific analysis of the crater, Richard fabricates alien "redundant geometry." Obviously a hallmark of his proposed "tetrahedral geometry" -- a two dimensional representation of a three-dimensional, double tetrahedron inscribed within a sphere identical to the geometry found in the Cydonia Complex on Mars.

Stellar Sterility

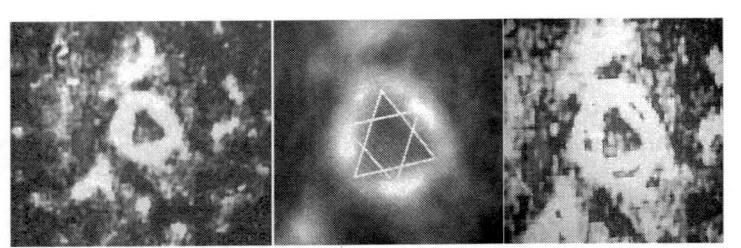

DEVELOPMENT OF HOAGLAND'S TETRAHEDRAL TRIANGLE AND SIX POINTED STAR

6
CRYSTAL DOME OF SINUS MEDII

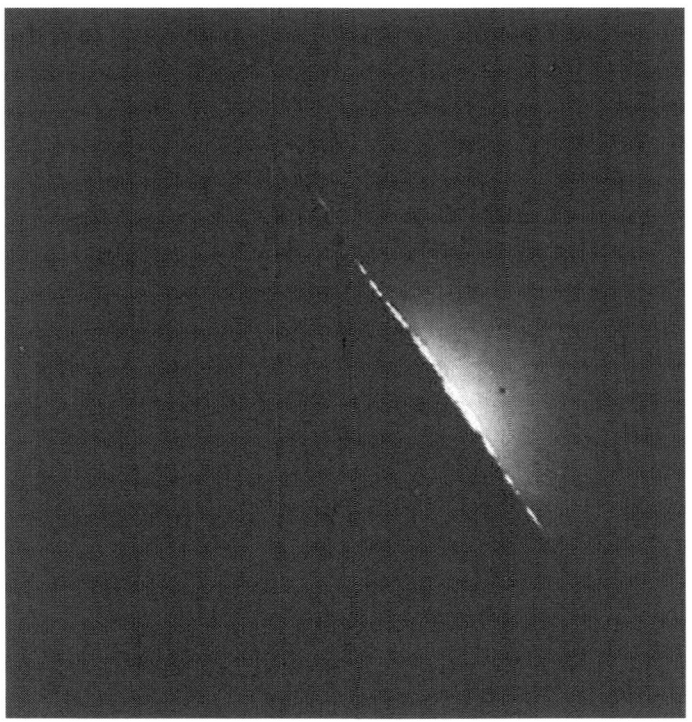

SURVEYOR-6, Photo-7 NASA 67-H-1642
November 24, 1967
Sunlight diffracted at Moon's limb as seen in Surveyor VI
picture of the horizon west of spacecraft.

Richard's search for enlarged pixel distorted crystalline-like structures migrates to the landing site of the Surveyor Missions, particularly Surveyor-6 and subsequently Surveyor-7. The Surveyor-6 spacecraft was the fourth of the Surveyor series to successfully achieve a soft landing on the moon, obtain post landing

Stellar Sterility

television pictures, determine the abundance of the chemical elements in the lunar soil, obtain touchdown dynamics data, obtain thermal and radar reflectivity data, and conduct a Vernier engine erosion experiment. It carried a television camera, a small bar magnet attached to one footpad, and an alpha-scattering instrument as well as the necessary engineering equipment. It landed on November 10, 1967, in Sinus Medii, 0.49 deg in latitude and 1.40 deg West longitude - the center of the moon's visible hemisphere.

This spacecraft accomplished all planned objectives and also performed a successful 'hop' rising approximately 4 m and moving laterally about 2.5 m to a new location on the lunar surface. The successful completion of this mission satisfied the Surveyor program's obligation to the Apollo project. On November 24, 1967, the spacecraft was shut down for the 2 week lunar night. Contact was made on December 14, 1967, but no useful data were obtained. [10]

Yet, Richard said, it did send back some fantastic photos of what he believes is an ancient alien crystal dome situated in Sinus Medii. He said of the following photo: *"This may be a photograph of the extraordinary glass dome covering the region of the moon known as Sinus Medii. It was taken by the unmanned Surveyor 6 on November 24, 1967, one hour after sunset."* [10]

This photograph (NSA7-100) of the sun's corona can be seen in the NASA technical report 32-1262, August 15, 1968. The view angle is to the west toward the western part of Sinus Medii. It shows the sun's corona reflecting upwards from below the horizon and through what alien Oopart hunters say is a grid-like glass structure of numerous objects and dome support beams dangling above the surface

Stellar Sterility

Most scientists suggest that this effect is caused by micro particles from the surface trapped in electrostatic clouds hovering over the area, and that when the sun shines through it, it gives off a reflective effect exposing a very light thin electrostatic cloud atmosphere, and a very thin one at that. [10] Mr Hoagland said this is what is left of an alien glass dome.

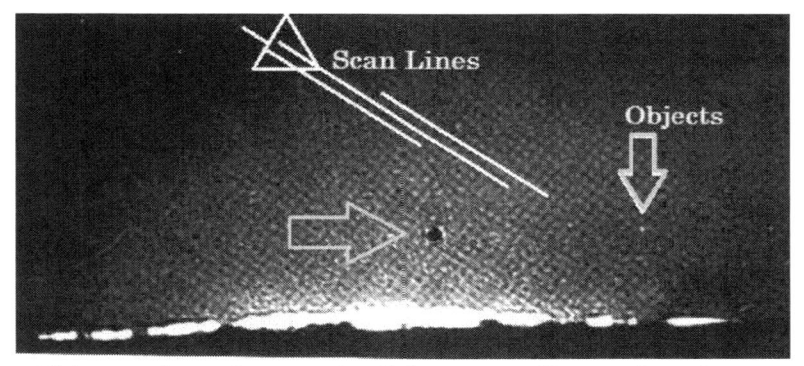

This may be a photograph of the extraordinary glass dome covering the region of the moon known as Sinus Medii. It was taken by the unmanned Surveyor 6 on November 24, 1967, one hour after sunset. - R. Hoagland

Other Surveyor photographs, such as Surveyor 7, seem to suggest the same scientific explanation. Surveyor 7 was the fifth and final spacecraft of the Surveyor series to achieve a lunar soft landing. The objectives for this mission were to: perform a lunar soft landing (in an area well removed from the maria to provide a type of terrain photography and lunar sample significantly different from those of other surveyor missions); obtain post landing TV pictures; determine the relative abundances of chemical elements; manipulate the lunar material; [5] obtain touchdown dynamics data; and, obtain thermal and radar reflectivity data.

This spacecraft was similar in design to the previous Surveyors, but it carried more scientific equipment including a television camera with polarizing filters, a surface sampler, bar magnets on two footpads, two horseshoe magnets on the surface scoop, and auxiliary mirrors. Of the auxiliary mirrors, three were used to observe areas below the spacecraft, one to provide stereoscopic views of the surface sampler area, and seven to show lunar material deposited on the spacecraft.

The spacecraft landed on the lunar surface on January 10, 1968, on the outer rim of the crater Tycho. Operations of the spacecraft began shortly after the soft landing and were terminated on January 26, 1968, 80 hours after sunset. Operations on the second lunar day occurred from February 12 to 21, 1968. The mission objectives were fully satisfied by the spacecraft operations. [11]

7
SINUS MEDII DOME DEBUNKED

The above Surveyor 6 frame (NAS7-100 photo number 7) of the Sun's corona can be seem in the NASA Technical report 32-1262. [12] The view angle is to the west toward the western part of Sinus Medii. The photo shows the sun's corona reflecting upwards from below the horizon and through what dome hunters believe to be a lunar atmosphere and what they say is a grid-like glass structure of numerous objects (emulsion blobs) and fragmented structures suspended above the lunar surface (See arrows). These anomalous structures and objects are located to the right (west) of the so-called "Tower" and "Shard."

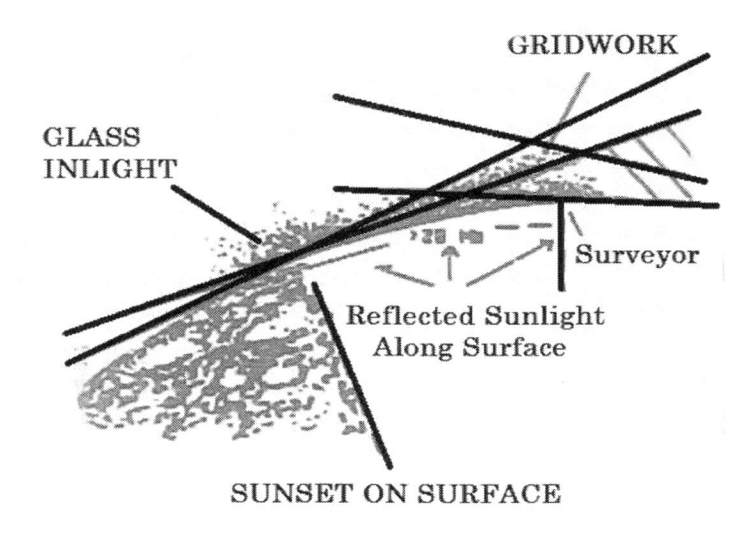

GRIDWORK

GLASS INLIGHT

Surveyor

Reflected Sunlight Along Surface

SUNSET ON SURFACE

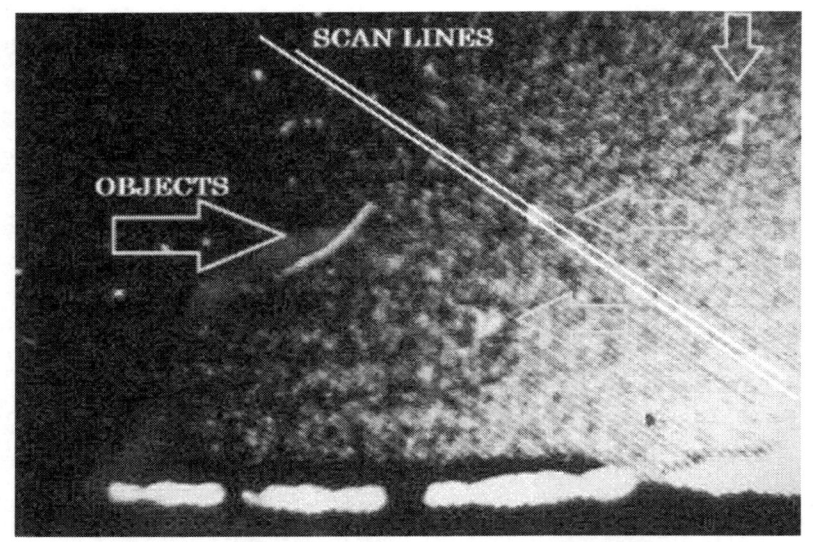

(Left) CHART SHOWING REFLECTION ANGLE
(Right) OBJECTS ON LEFT SIDE

The big difference with this location and series of photos is that they were taken with the light coming from the front from below the visible horizon in what is called *"forward scattered light"* unlike most all other mission photos of the same area, where the sun light is behind the camera causing *"back scattered light."*

In other words, rather than photographing these so-called enigmatic geometric structures in back scattered light, the camera obtained these photos of sun light radiating through some kind of atmosphere in front of and toward the camera. The purpose of these type of shots were to record and study any possible light-scattered properties in the space above the lunar surface caused by the solar corona.

The right side of the photo shows this anomalous reflective material very clearly. Mr Cornet points in his publication that *"This material becomes less illuminated by the sun's corona the further away the structures are located from the axis of the sun and corona, proving its existence."* It is interesting how Mr Cornet replaces the word "material" with the word "structures" leading the reader to believe it is foreign and alien to the Moon.

To the left side of the photo can be seen what appears to be horizontal bands of light reflecting materials and a bright band of bead-like reflective materials along the horizon. Mr Cornet said these anomalies *"cannot be explained as forward refraction from suspended Moon dust."* They, therefore, must be *"some sort of lens-like system of glass-work at the base of a geodesic dome,"* where sunlight is reflecting through rather than off of surface materials. [13]

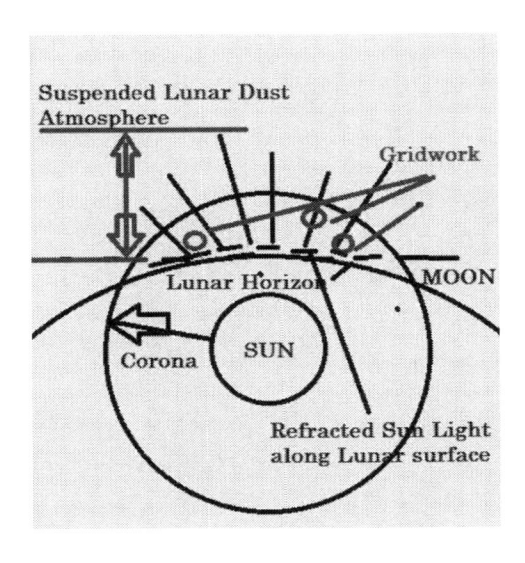

According to conventional NASA sources the view angle of this photo shows intense light from the Sun reflecting off the surface material on the horizon — the bead-like effect reveals the Lunar atmosphere. The grainy-like reflective materials supposedly suspended above the surface and supposedly indicative of some kind of alien built structural dome can be explained by a number of theories.

The large chunk-like objects are probably particles of dust adhering to the camera lens after Surveyor 6 landed. The finer atmospheric refracting crystal-like material is nothing more than lunar dust and other remnant gases of the lunar atmosphere — and the moon does have a very weak atmosphere. The term "atmosphere" is a loose term designating the very minimal gases, particles and electrostatics in the space above the lunar surface. [15]

Nevertheless, Richard Hoagland would have us believe that this material cannot possibly be native lunar dust particles in the atmosphere, because certain other Apollo dust sample experiments eliminated *"suspended lunar dust particles: as a probable cause of this intense forward light scattering effect."* Thus, the Enterprise team of alien artifact hunters concludes that this phenomenon is *"caused by forward-refracted sunlight: myriad lens-like images of actual inner corona of the Sun, being literally bent over 20 miles around the sharp curvature of the Lunar horizon... caused by the presence of some kind of glass-like structure."*

This of course is only implied and is not proven by independent observations of similar "glass-like" stuff photographed by Lunar Orbiters and Apollo Mission cameras from the same area. There are many ways to explain the floating debris in this photo. Other than the super fine cosmic dust particles floating in the static electricity filled space above the surface there are camera lens dust [15], gases, scan line interferences and the low resolution blurring of the camera. The Sun corona reflects through the lunar dust and gases causing the refractive effect. Then, the image is seen through the camera lens that is obviously less than spotless, the transmission lines and the emulsion patterns when developing the earth based negatives.

Richard asks the question how this reflective effect could take pace in an airless atmosphere implying that the space above the lunar surface is a total empty vacuum and void of all gases. Apparently he denies the scientific reports of the lunar missions that tell otherwise. Gas itself reflects and scatters light.

It was once believed that the moons of other planets including our own moon had no atmospheres. Since the Apollo and other planetary missions, measurements have shown that most of these moons are surrounded by a very thin region of molecules, which can almost be called an atmosphere.

Are there gases on the Moon? Yes, according to recent finds. The weak lunar atmosphere may come from a couple of sources. One source is out gassing where gases are released from deep within the Moon's interior.

The Moon has been found to have abundant amounts of nitrogen, carbon dioxide, carbon monoxide and rare gases such as radon. Another source of atmospheric particles are molecules, which are loosened from the surface when other molecules from space hit the ground. [16] Other noxious gases compose the hot air and cigar smoke within the sitting room of Mr Hoagland and his think tank.

It has been suggested that this Surveyor 6 horizon-glow (HG) is sunlight, which is forward-scattered by dust grains present in a tenuous cloud formed temporarily just above sharp sunlight/shadow boundaries in the terminator zone. Electrically charged grains could be levitated into the cloud by intense electrostatic fields extending across the sunlight/shadow boundaries.

Detailed analysis of the HG absolute luminance, temporal decay and morphology confirm the cloud model. The levitation mechanism must eject more particles per unit time into the cloud than could micro meteorites. Electrostatic transport is probably the dominant local transport mechanism of lunar surface fines. [10]

This would be the lunar version of Rayleigh scattering, which is *"the elastic scattering of light or other electromagnetic radiation by particles much smaller than the wavelength of the light. The particles may be individual atoms or molecules. It can occur when light travels through transparent solids and liquids, but is most prominently seen in gases. Rayleigh scattering results from the electric polarizability of the particles. The oscillating electric field of a light wave acts on the charges within a particle, causing them to move at the same frequency. The particle therefore becomes a small radiating dipole whose radiation we see as scattered light."* [17]

Needless to say, the moon does have enough atmosphere to cause some weak scattering of light, especially when viewed at very low angles over long distances, such as in the Surveyor photos. Here are

Stellar Sterility

some more examples or horizon shots showing this effect.

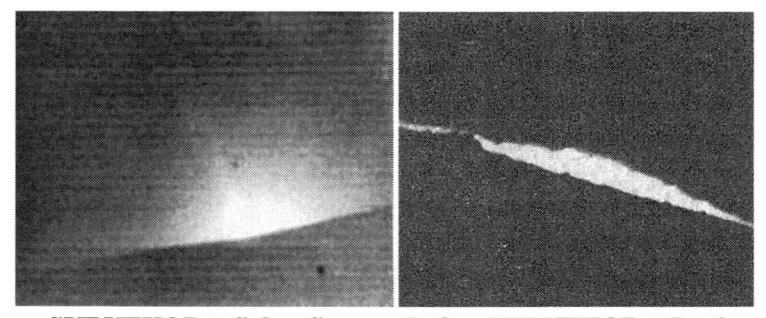

SURVEYOR 1 Solar Corona Spike. SURVEYOR 7 P64b

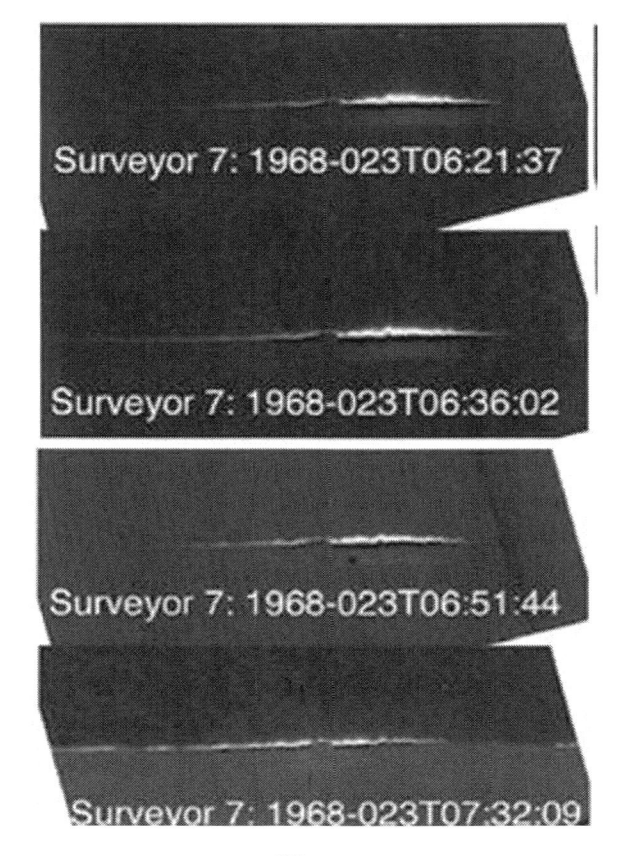

Surveyor 7: 1968-023T06:21:37

Surveyor 7: 1968-023T06:36:02

Surveyor 7: 1968-023T06:51:44

Surveyor 7: 1968-023T07:32:09

8
THE "SHARD" LOCATION

FAMOUS SHARD PHOTO (Enlargement)
STRUCTURE OF THE "SHARD"
LO-III-84M Famous shot of Ukert Crater area.
LO-IV-101-H3, 102-H1, 108-H3, 109-H1, 12-M

Apollo-10 photographs 4854-4856 show camera looking west at the terminator (Lunar surface sunrise line) from above the eastern side of Sinus Medii. Photo 4822 is looking northeast across Ukert Crater towards the northern edge of Sinus Medii. [See AS10-32-4822, 4854, 4855 and 4856]

The Lunar Orbiter photo II-84M and the above Apollo shots (4854-4856) supposedly show Hoagland's glass-like crystal structures suspended in the Lunar sky along with the alien Tower and Cube mounted on top peaking just above the horizon. The Apollo photos unfortunately for Richard do not show the Tower, Cube

Stellar Sterility

nor the Shard and reference can only be made to the LO-III-84M in the following pictures.

Photo 84M was shot by Lunar Orbiter-III towards the south-west side of Sinus Medii from about 30-miles altitude with the horizon at about 256-miles. The Shard and Tower are approximately 230-miles and 200-miles respectively. Orientation of this shot is 45 degrees to the south of the Apollo-10 photos.

LO-III-84M Ukert Crater (left)
Looking towards the south-west side of Sinus Medii.

Other good location photos are LO-IV-108-H3 the Rima Flammarion and Crater Mosting -A area. Locator line "C" passes through four possible locations (A, B, C, and D) for the Shard.

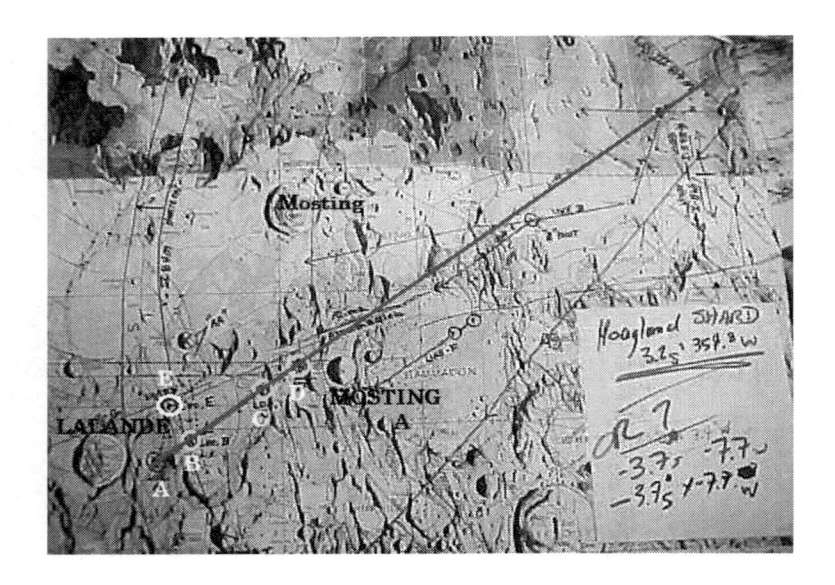

An enlarged and enhanced section of the photo reveals the Shard is an apparently real lunar geological structure sitting out in the middle of the plains area, possibly next to Mosting-A Crater, or beyond westward. As for the Tower and Cube that Richard mentions, it is identified as a faint emulsion blob splatter stemming from the huge splatter line running across and above the horizon. Many of the blobs are compared to show a similarity between the blobs and the Tower Cube. It is most probably an emulsion blob as it too demonstrates the same exaggerated pixel "crystalline structures" as the other emulsion blobs.

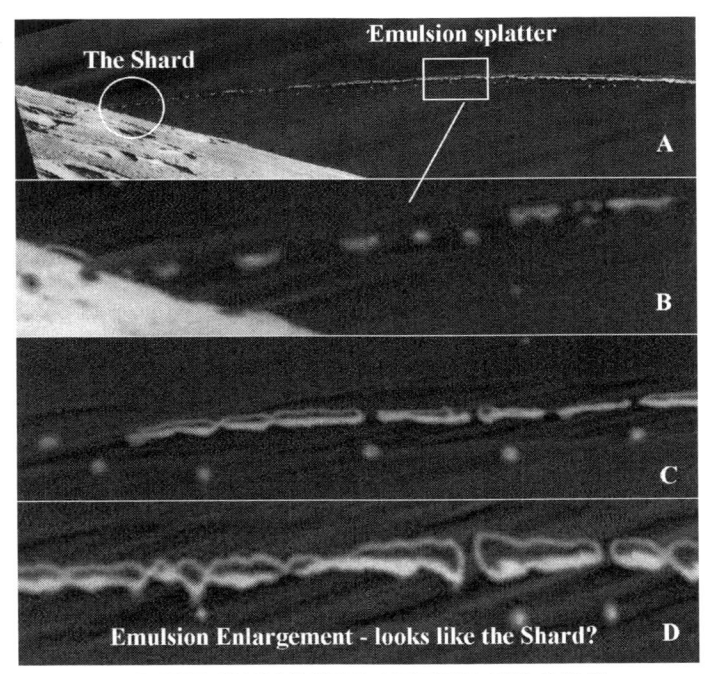

EMULSION SPLATTER BLOBS

After analyzing detailed enhancements of the Tower, Cube and other anomalous objects for so-called "suspended glass-like structures," "support beams" and "alien architectures" all that can be derived from eye torturing study are developmental emulsion splatters, blobs, scratches, smears, and (with Hoagland's photos) over enhanced, digitally exaggerated pixel distortions. In fact, almost every Lunar Orbiter photo of a horizon background can be over exposed, extremely enhanced beyond normal enhancement and made to expose so-called alien dangling crystal structures. LO-III-84M is loaded with them from top to bottom.

Stellar Sterility

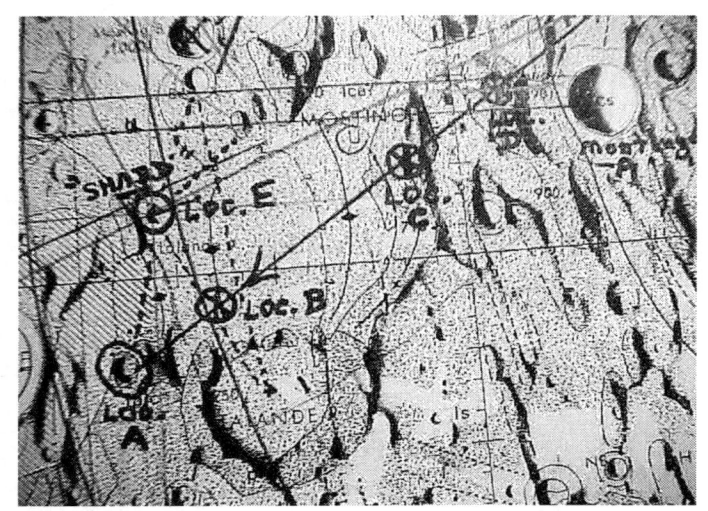

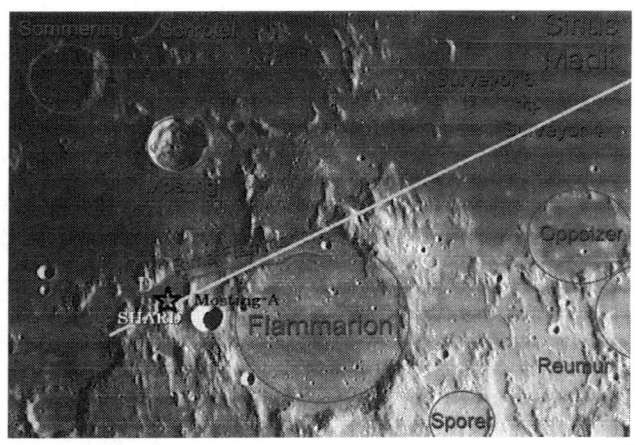

Location Lines A, B, C and D show possible Shard locations.
"A" = 4.5 degrees S, 7.7 degrees W – Hill Top

Shard Loc. "D" is closest to the camera at 3.1 degrees S, 5.8 degrees W, next to Crater Mosting-A

9
THE "SHARD"
(LO-III-84M)

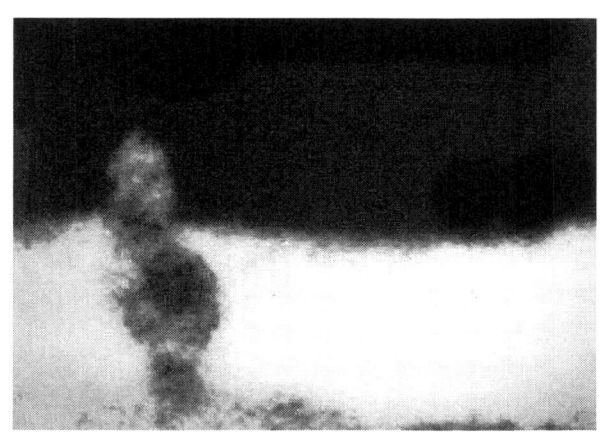

The "SHARD" (Super enlargement)

There are a number of other photos of Sinus Medii besides Surveyor 6 NAS7-100 of special interest relative to alien structures. There are those of the Lunar Orbiter Missions [18] and those from the Apollo Landing Missions [19], especially the famous "Shard" photo LO-III-84M. Sinus Medii is one of the most interesting sites for Mr Hoagland as it is supposedly the location of a great dome and many alien structures. He has found, along with the Shard, such colossal structures as the "Tower" and the "Cube."

Stellar Sterility

LO-III-84M Sinus Medii Horizon

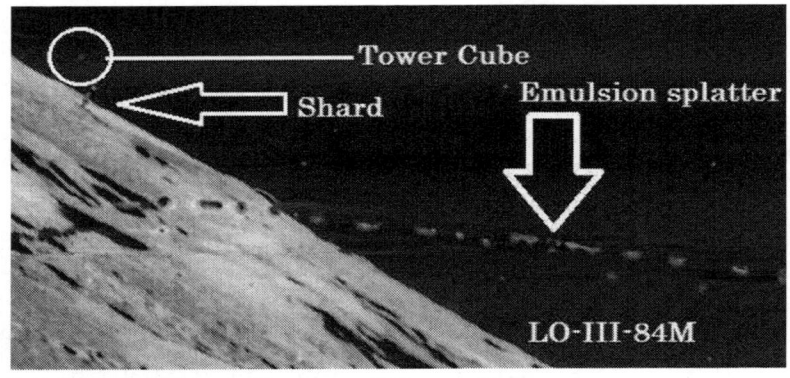

LO-III-84M Shard, Tower and Tower-Cube
Compared to emulsion splatter and emulsion blobs

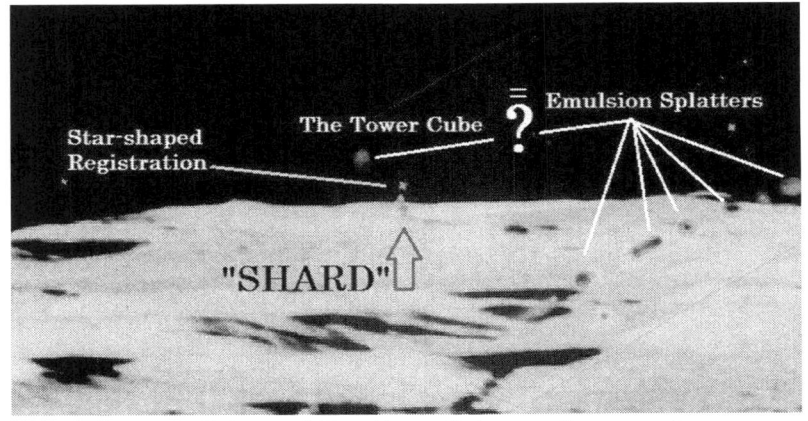

![Shard, Tower, Cube and Emulsion Splatter labeled photograph]

Shard, Tower, Cube and Emulsion Splatter Compared

The Shard is a very obvious tower-like structure that rises above the moon's surface to a possible height of over one and a half miles. It has an irregular "*spindly shape*" of regular "*geometric patterns*" and has constricted modes and a swollen internode.

Mr Cornet as well as Richard Hoagland points out that no known natural processes so far observed can explain such an object of that shape and size. As far as the construction of the Shard, its high reflectivity exposes its artificial nature and alien manufacture. "*The amount of sunlight reflecting from parts of the shard indicate a composition inconsistent with that of most natural substances. Only crystal facets and glass can reflect that much light...*" - i.e. on the Moon. The shard must therefore be "*a highly eroded remnant of some sort of artificial structure made of glass-like material.*"

The Shard is an amazingly huge object but not really made of some amazingly artificial alien material. The reflectivity of the object when compared with surrounding reflective lunar materials reveals that it is not much different in composition than the Lunar

Stellar Sterility

regolith "soil" and no more "unnatural" than any other object in its surroundings.

It is not all that odd in shape and size, though it is unique in its position and placement on the surface. Seems it is the only large object in the area. Even Mr Cornet points out other large structures in the area with the same reflectivity, though not as large as the Shard.

The Shard is shown casting a shadow revealing that this object is without a doubt real. The shadow mirrors to a great degree the Shard's unique shape. The shadow is also consistent with the local surface geology on which the shape is sitting. Hoagland said, *"It is this shadow – more than any other aspect of this object – which in the eyes of many observers solidifies its reality as an extremely anomalous, manufactured*

Stellar Sterility

lunar structure." [Martian Horizons V2, #5. p.16] This is one of the few things Richard and Mr Cornet are right about. The odd size and shape could be explained away as an emulsion blob if not for the shadow. The shadow forces one to admit it is a real lunar surface object.

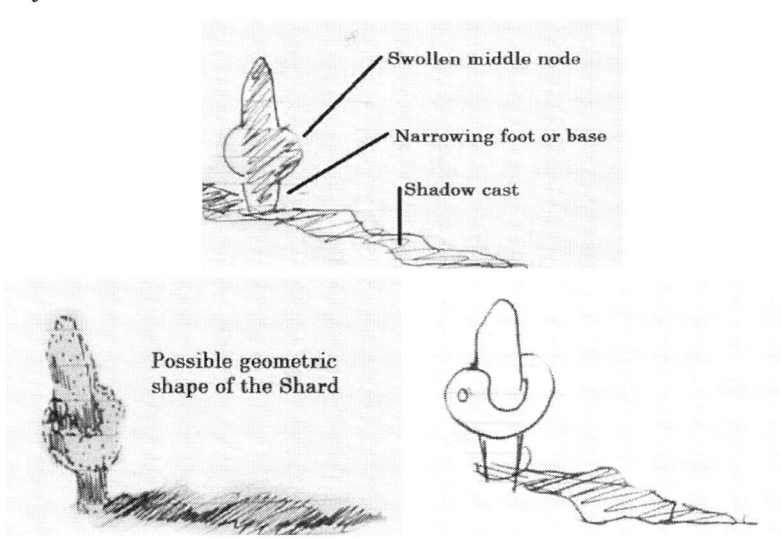

According to these two guys, the size and shape, in association with the shadow *"argues forcefully for an origin far more recent than the extremely ancient, rounded and eroded lunar landscape that surrounds it."* The shard, they continue *"somehow has managed to defy the laws of entropy that have determined those surroundings (according to geochemical and isotope age data returned by the Apollo missions): an incessant meteoric rain, which, in the absence of "recent" geologic activity, should have reduced this object to rubble indistinguishable from those of its surroundings."* [p. 17-18]

Stellar Sterility

Mr Hoagland, then, continuing on page 18, rattles off super geometric jargon describing the advanced crystalline structure of the object. Apparently, he over developed and over ultra-enhanced the shard to the point the pixels degrade into geometrical shapes, lineation's and cubical shapes revealing "*highly-eroded, highly reflective internal crystalline geometry*" with "*details of optically active vertical lineament patterns.*" [p.18]

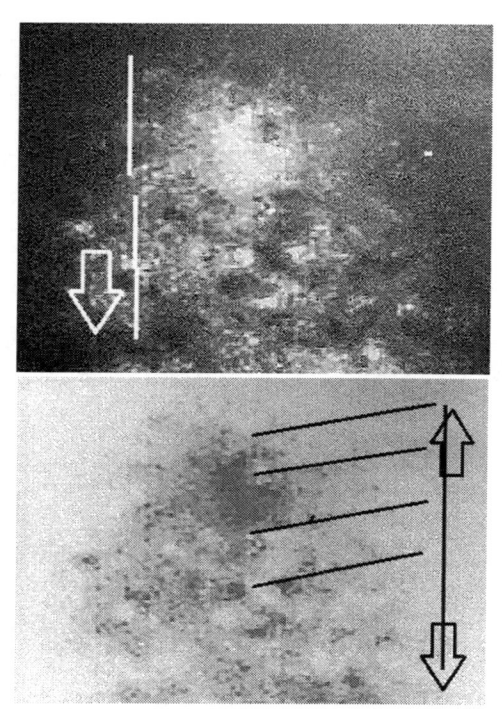

Extreme enlargement of Shard with vertical lineation pattern. Could be the pixel pattern?

10
"TOWER" AND "CUBE" LOCATION

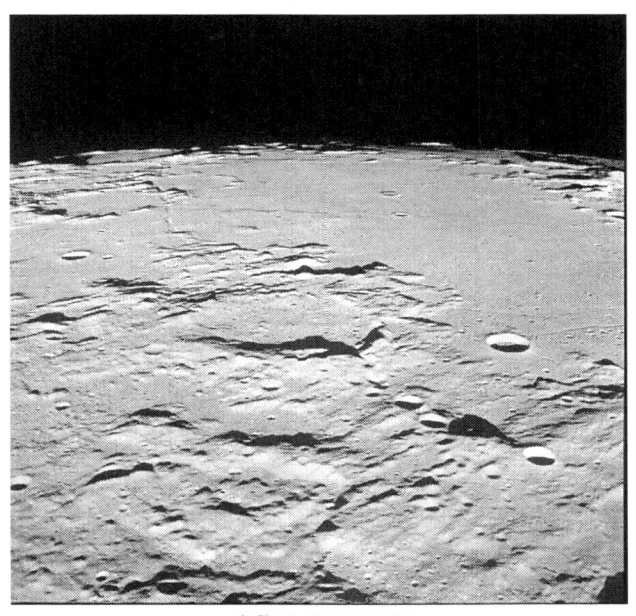

AS10-32-4855
APOLLO-10 photo looking toward Sinus Medii

The tower and Cube are seen to the left of the Shard in 84M. It looks like another emulsion blob, but Richard Hoagland said it *"appears at first to be merely a 'smudge' on the original full-frame NASA photograph – with a 'tail' extending down toward the lunar horizon."* But, it is not any other possible thing – it is not a nebulae, a gas cloud, a comet, dust, nor a distant galaxy, or any other variety of mundane deep space objects- but *"a massive 'mega-cube' hanging more than 7 miles above the Moon!"* [Martian Horizon p. 19-20] Wow! Seven miles above the lunar surface? This

Stellar Sterility

is nothing. Wait until we discuss the Castle structures that hang 35-150 miles above the surface close to Ukert.

The above photo AS10-32-4855 and the following 4856 frame show a shot looking westward of Sinus Medii over Rhaetieus, Reaumur, Oppolizer and Flammarion-T at Flammarion and to where the so-called Tower-Cube just barely peeps above the lunar horizon. It is hidden among all the other small tiny specks of defects within the photo. If one looks careful they will see a tiny white dot – the Cube.

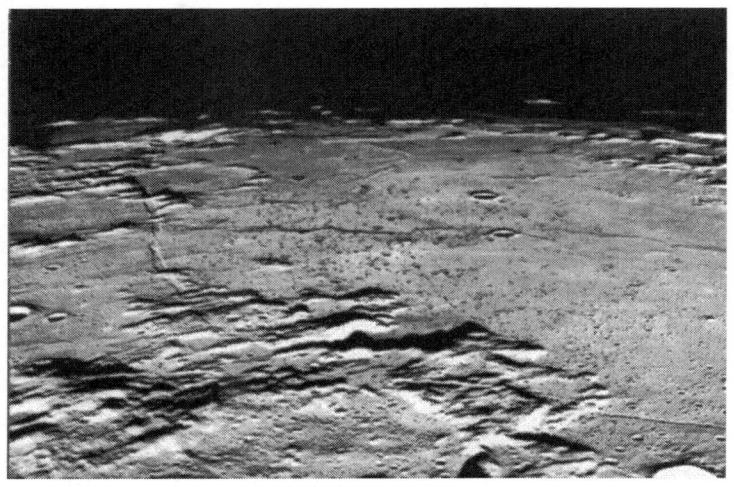

AS10-32-4856

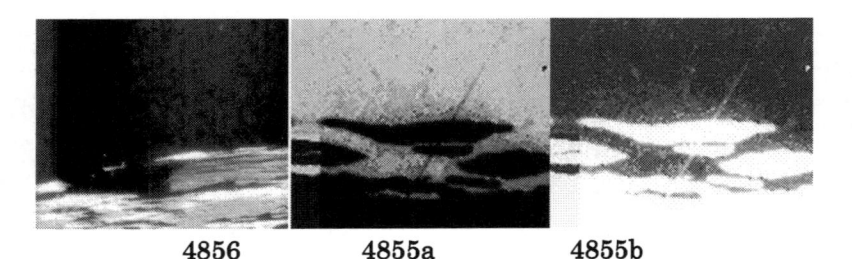

| 4856 | 4855a | 4855b |

Ultra-enlargements of Sinus Medii area of Shard, TOWER and CUBE locations

Stellar Sterility

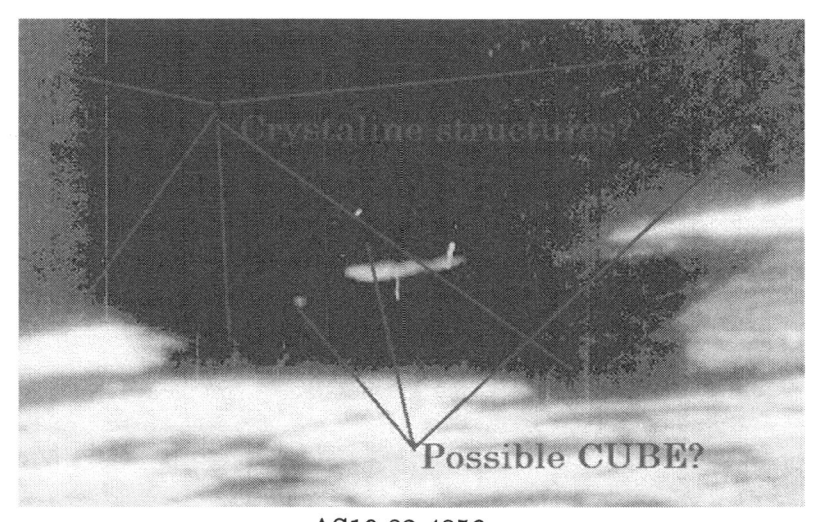

AS10-32-4856
Can you see the top of the Tower and the Cube? Take your best pixel pick!

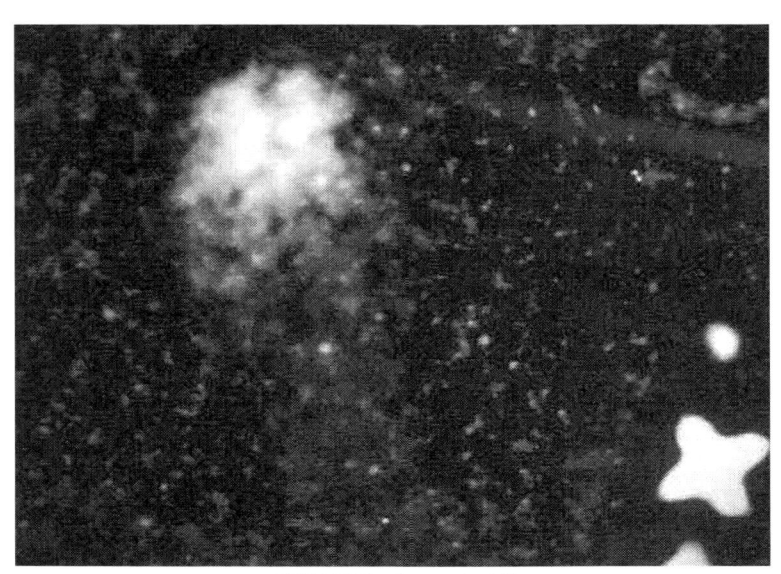

SHARD photo 84M
Tower "Cube" to left of cross-hair star and top of shard

Stellar Sterility

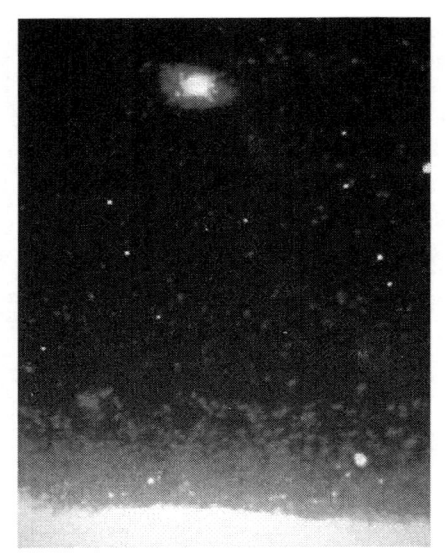

**AS10-32-4856
The CUBE**

LO-III- 84M [1]

Stellar Sterility

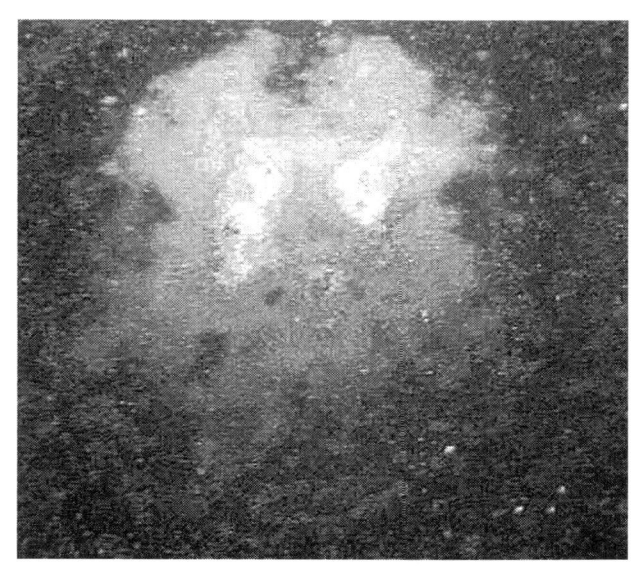

LO-III- 84M (2, 3, and 4) MORE CUBE (enhanced)

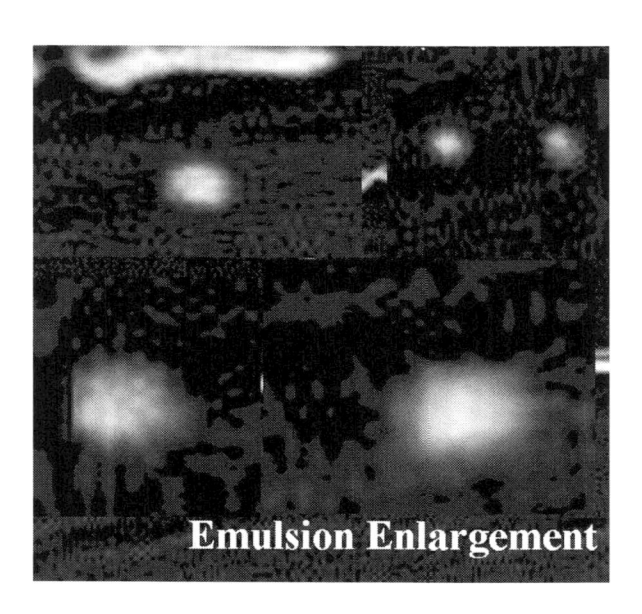

Emulsion Enlargement

11
WHERE IS THE "SHARD" ON USGS LUNAR MAP I-566 ?

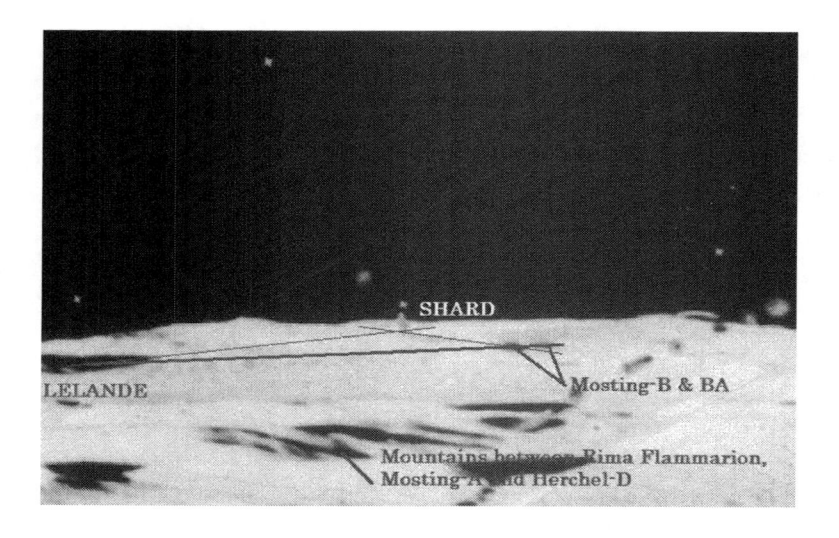

![Shard photo with labels: SHARD, LELANDE, Mosting-B & BA, Mountains between Rima Flammarion, Mosting A and Herchel-D]

Where in the lunar world is the Lunar Orbiter III-84M "Shard" located? This was one of many questions in my lunar surface studies that I found to be lacking in the Hoagland Moon-Mars Connection conference. And it was one of the most difficult questions to answer. The results revealed 3-4 possible locations.

Looking at the USGS Lunar maps of Sinus Medii area (Maps I-566 and I-548), I found myself at a loss when I first approached the question. Where was I to begin? What I needed were some location "beacons" for locating the Shard. I needed shadows, craters, boulders, mountains and other prominent geographical features for triangulating the location the Shard. The answer was found in the same "Shard" photo LO-III-84M.

Stellar Sterility

To locate the Shard, which was centered in the upper left corner of 84M along the horizon, I needed to identify as many of the craters and surface features as possible. These would be used to sight the Shard's longitude and latitude. The photo "footprint" map in the NASA publication SP-200 "The Moon as Viewed by Lunar Orbiter" helped me identify the photos and the areas covered by 84M. Along with the footprint map, the USGS maps helped to locate and identify the 84M craters and features. The footprint map helped to identify the following 84M area covered: III.84M Parameters: Blagg 1.1N. X 1.3E - 2.5N. X 0 degrees. Flammarion 2.4S. X 1.4W. (Random point) - 2.5N. X 7W. (Random point)

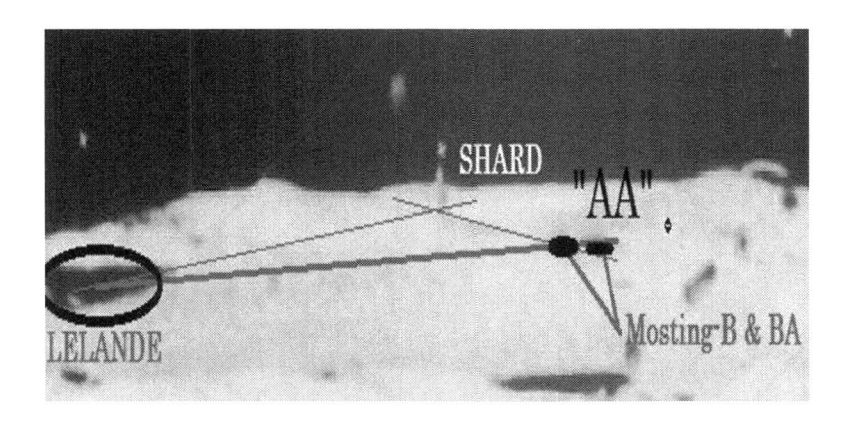

The geometry runs as follows: The 84M photo horizon appears to extend to Crater Lalande in the area South-west of Sinus Medii. Checking the 84M photo reaffirmed the extent of the horizon and the parameter in which the Shard rests. To begin locating the Shard, while drawing lines on the 84M photo and the USGS Maps, the next step was to locate and identify Crater Bruce as a beginning point to run lines

Stellar Sterility

out toward the horizon to pin-point the Shard on the USGS Map. This was easy. I drew a line (D) from Bruce to Crater Blagg which extended the first sight line to the horizon past Crater Mosting.

Another line (A) was drawn from Bruce to Crater Oppolzer-A (0.3S. X 0.2W.). From (A) was drawn a line outward towards and through a little mound (Z) north and north east of the Flammarion area (0.5 degrees S. X 2.0 degrees W.). My "E" line was extended further to a small crater (Flammarion-C) on 84M and the Map in the North of Flammarion (2.0 degrees S. X 3.5 degrees W.). The crater's location actually begins what is named the Rima Flammarion, a fault line which runs towards the Shard!

One can see on 84M the Rima Flammarion passing to the right of the Shard, where the Shard rests to the left on a small plains area or on top of a hill--Sighting the Shard somewhere between Lalande-R, Lalande Crater and the two small crater lets (which I name "AA", as circled on the photo blow-up of 84M (Mosting-A & BA., 2.4S. X 7.1W.). I thus sighted the location of the Shard somewhat beyond "Location-D" (3.1 degrees S. X 5.4 1/2 degrees W.), possibly at map Location-C (3.3 degrees S. X 6.0 degrees W.), or somewhere along my line from Location-C to B (4.1 degrees S. X 7.1 degrees W).

Stellar Sterility

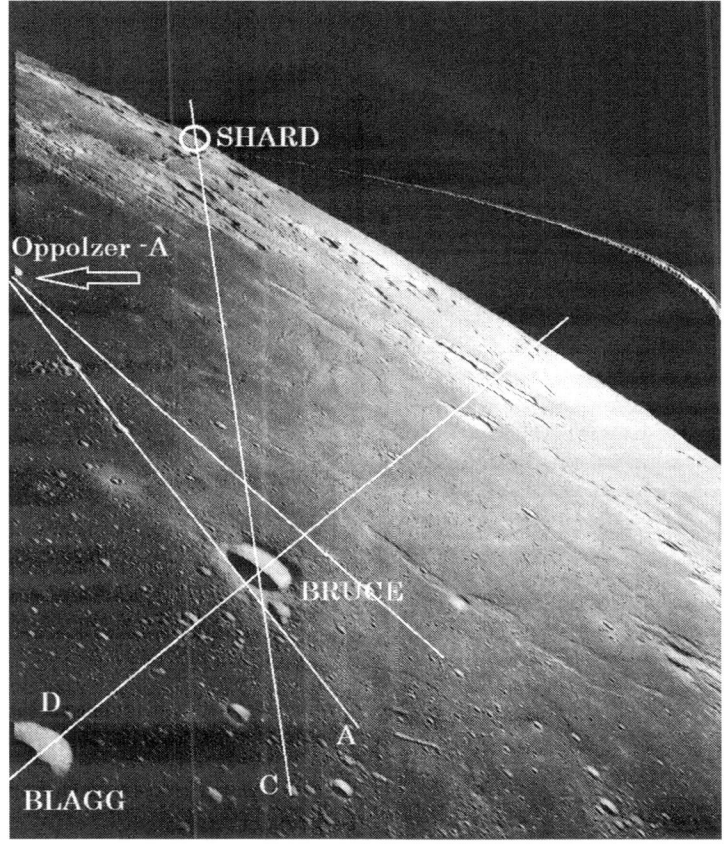

Triangulation to locate Shard

The most probable location for the Shard was found to be one of two prominent little mounds sitting to the east of Crater Lalande at Locations-A and E (4.3 degrees S. X 7.4 degrees W., and 3.5 degrees S. X 7.3 degrees W.). The Location Line-C, running from Bruce S.E. through Flammarion and past Mosting-A Crater, therefore runs through Rima Flammarion to Crater Lalande. The Shard, according to the 84M crater location sight lines, is thus located somewhere between Mosting-A and Lalande in a small area North of

Stellar Sterility

Lalande-R, somewhat north from Line-C. If one observes the little 84M photo (blow-up) at the end of this section, it will be seen that the Shard sits away from and beyond the double Crater "AA" or Mosting-B and BA, and Crater Lalande-C.

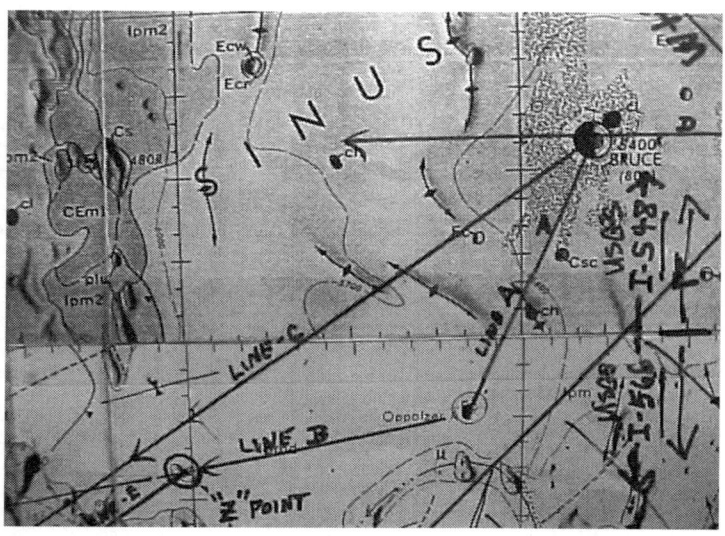

MAP-A

Behind it, according to the Map, sits a small scarp or hill. Past this is the big Crater Lalande. To the left there is a small horizon hill which is Location-A. The Shard thus sits at Location-E at 3.5S. X 7.3W. From Bruce to the Shard along Line-C, according to the USGS Map, is about 170 miles.

Stellar Sterility

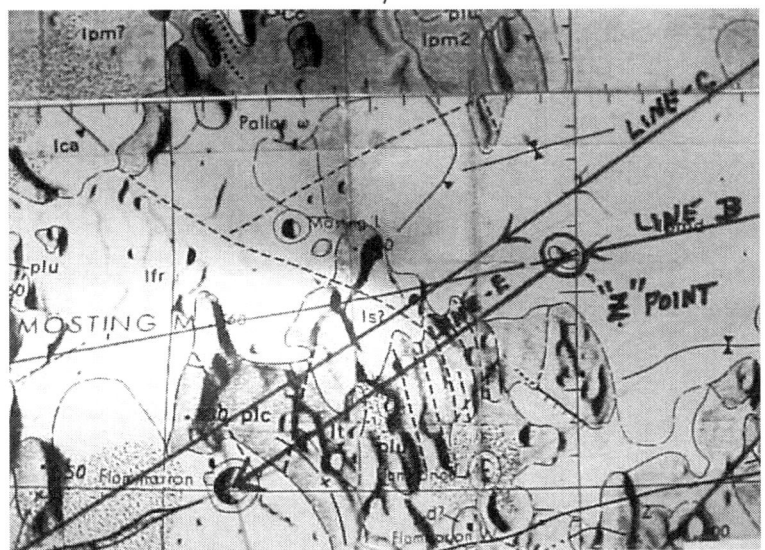

MAP-B

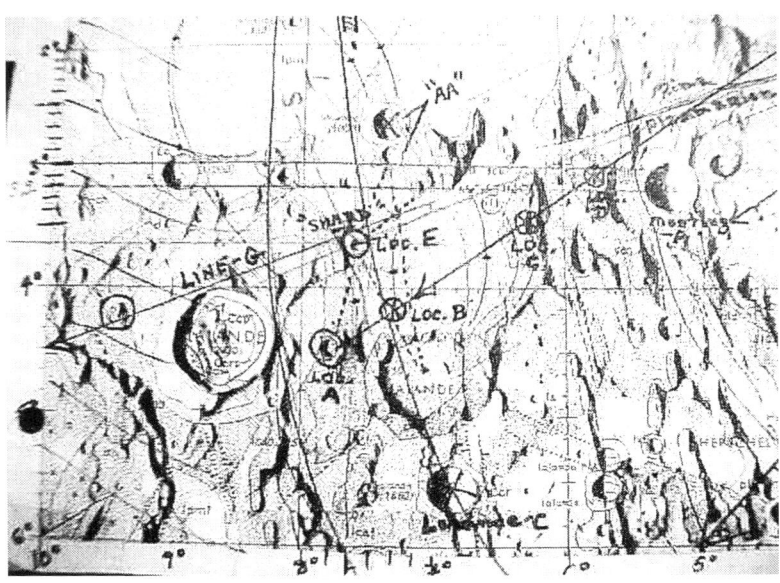

MAP-C

MAPS AND PHOTOS:

A. USGS Maps (I566 and I548) of Sinus Medii area with positions of locations and Shard.

B. Negative Print of LO-III-84M "Shard"

C. Negative Blow-up of LO-III-84M with location lines and identified craters with Shard in background. [Dotted line in sky is emulsion error.]

Crater Lalande-C F

Location-A F

E. Mosting-A & BA. My "AA"

G. Mountains between Rima Flammarion and Mosting-A & BA., Mosting-U on Map.

The Cube-Tower, which lies beyond the Shard, is still to be located. The problem in locating the Tower, according to 84M is that it rests beyond the 84M horizon, somewhere along Line-G. [Check out the mound at 4.1S. X 9.3W.] The lack of surface features for sighting presents a problem. Perhaps there will be other photos taken that will help locate the Tower?

The Missions numbers that I have compiled that cover the Lalande "Shard" area are, L.O. IV 113H, 114H, 108H and 109H. The maps covering the Sinus Medii area and the location of the Shard are, USGS Lunar Maps: Macrobius Quadrant I-799 (NW), Cleomedes I-707 (NE), Taruntius I-722 (SW) and Mare Undarum I-837 (SE).

As mentioned before, there are four possible locations for the Shard: Location A, B, C and E, discounting D. Locations A and B are located respectively 4.5S x 7.8W and 4,2S x 7.4W. Location C is at 3.5S x 6.3W. "A" is situated directly in-between Craters Lalande and Lalande-R. Shard position B is

Stellar Sterility

along line "C" at about 11 o'clock to Lalande-R. Location C is north-west of these at 3.5S x 6.3W, west of Mosting-A sitting on a large hill close to Mosting-U. This is south-west of Richard's specific location 3.2S x 354.8W, which I cannot find.

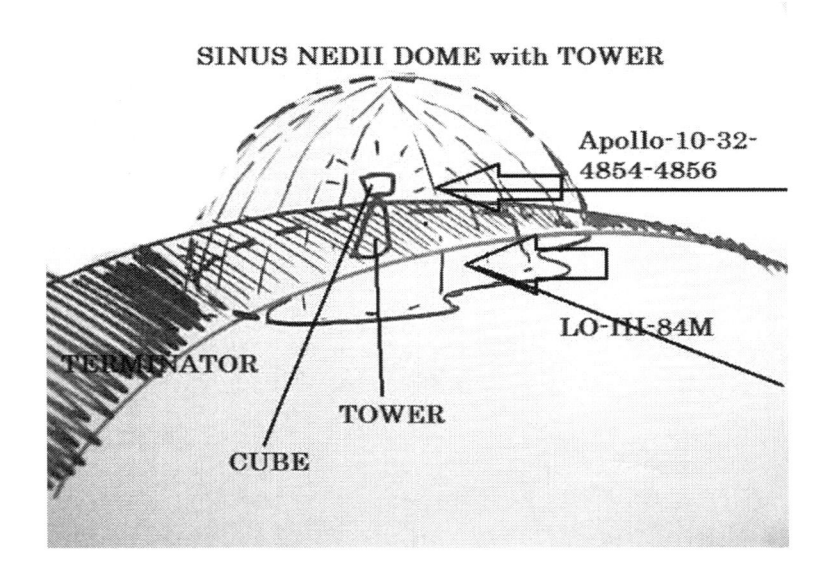

SINUS NEDII DOME with TOWER

Apollo-10-32-4854-4856

LO-III-84M

TERMINATOR

TOWER

CUBE

Possible location E is a very interestingly accurate site for the Shard. If one takes LO-IV-113-H3 and compares it with 84M (see chart) a very small crater will be found between Lalande and Mosting B/BA. Location line G runs through this point. Notice that the shard is a 1 and ½ mile high artifice and not a crater. In reviewing all the data and photographs, not one revealed any shard, tower or cube.

No alien shard, skyscraper or cubical shaped tower hotel ever jumped out from the frames except in 84M. Nothing. No clues, no peaks or parts or suggestions ever appeared. 84M frame is a lone ranger, the one and only photograph with the Shard. With the shadow it is casting, it appears so real and natural to the lunar surface, yet only in this one picture. Therefore, it is suspect to artificiality and photo processing development error and not being natural to the moon. It is not conclusive that it is real since it is not in any other photograph and the one photograph that it does appear in is choked with emulsion blobs and splatters.

According to the enlargement of the shard photo 84M, the shard appears situated behind and approximately in-between Lalande-R and Mosting-B east of the invisible Lalande Crater at about 3.7S x 7.7W at point D. Obviously, the first estimated locations A, B and C lineation's and cross references outlined on the map failed to truly locate the shard. Richard's location for the shard at 3.0S x 354.8W also fails as well to fit the surface details around the shard in 84M. His location places the shard too far into the foreground.

Accepting Hoagland's location "D" would place the shard Mosting-A, Flammarion-D and Mosting-B on the map. Yet, location E befits the LO-III-84M surface features and places the shard at 3.7S x 7.7W.

As mentioned before, none of the other possible photos of the area revealed any Shard. Looking for locations A and B in such frames as LO-IV-113-H3 is fruitless. Comparing craters Lalande and Lalande-R with the 84M revealed no shard. Also, none of the Apollo photos showed any super high structure in this area either. [20] Location C was blank as well. LO-IV-108-H3, 109-H1 photos of the Mosting-A crater area also revealed nothing of the shard as found in 84M.

What of the Flammarion area around Mosting-A? Would some of the other photos taken from different views reveal the shard? Search until your eyes cross.

There is no shard to be found. I need not display any photographic proofs when there is no evidence to be found. Anyone can check these locations on the Internet for themselves.

So, what is the conclusion regarding the existence of the shard? Is it truly there as 84M seems to show? This one and only photo does show an object and it seems to cast a shadow suggesting that it is really present on the Moon. As for any other evidence, no other photos seems to support the object.

There are many reasons why this is the case. The angle is wrong, the details are vague and the lighting is either too dark or too bright, and the object is too far away to be seen. 84M seems to be the only photo of the shard. Maybe the LROC (Lunar Reconnaissance Orbiter camera) will reveal something? [21] So far the LROC has not mapped this sight yet.

13
LUNAR RECONNAISSANCE ORBITER CAMERA PHOTOS

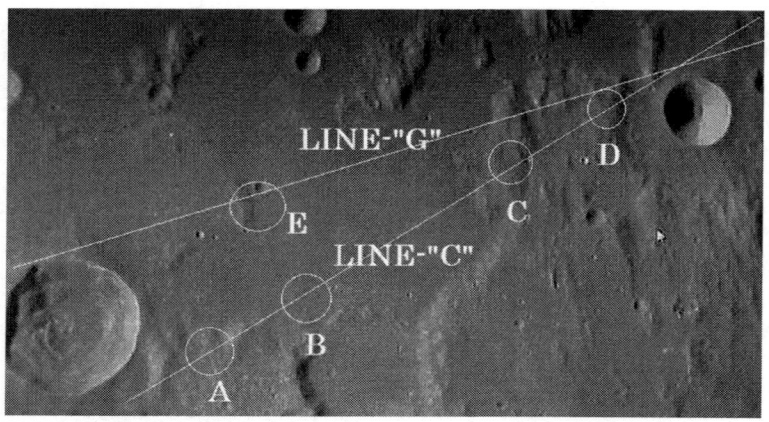

LUNAR RECONNAISSANCE ORBITER "Shard" Locations
[http://target.lroc.asu.edu/q3/http://target.lroc.asu.edu/q3/#]

The shard should be somewhere between Craters Mosting-A and Lalande, and around Lalande-R. Remember, this wild goose is supposed to be a 1 and ½ mile high tower like shard. Crater Mosting-A is about 11-12 kilometers or approximately 10 miles wide. Therefore, the 1.5 mile high shard should stick out like a sore eye somewhere between Craters Mosting-A and Lalande and above Lalande-R in the above enlarged area photograph. The Lunar Reconnaissance Orbiter photographs seem to be void of any alien manufactured super structure.

Stellar Sterility

Location - A Shard Missing

Location - B Shard Missing.

Location - C Shard Missing

Stellar Sterility

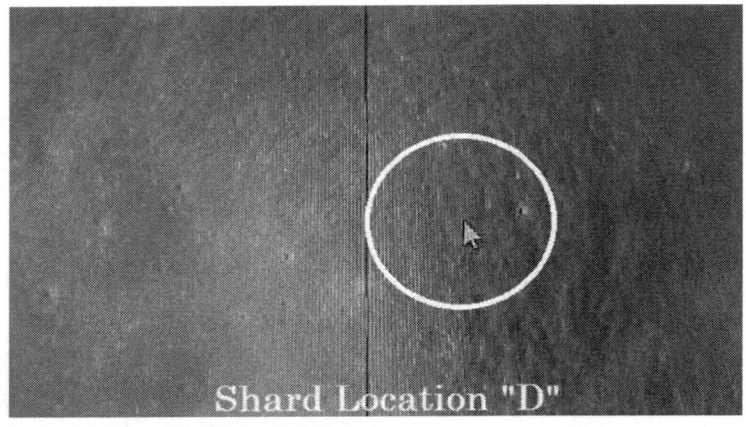

Location - D Shard Missing

Location - E Shard Missing

There is no tower, no shard, no playground park or swings, no swimming pools, no nothing! Nevertheless, maybe there is some other unallocated spot for the shard? - Have fun looking! And when you get tired, return to LO-III-84M for the only relief available. Could the shard also be just an oddly placed emulsion blob or splatter mark? Your guess is as good as mine or Richard Hoagland's. As far as this writers is concerned this mysterious alien artifact is just another Richard Hoagland charade.

Stellar Sterility

POSSIBLE LOCATIONS:

Bruce..1.1N X 0.2 1/2E
Oppolzer-A..0.3S X 0.2W
Z-Point..0.5S. X 2W
Flammarion-C (Line-E)............................2S X 3.5W
Location-A (Lalande Mound)...................4.3S. X 7.4W
Location-B (Possible Shard Location).....4.1S. X 7.1W
Location-C (Hill Top)..............................3.3S. X 6W
Location-D (Hill Top)..........................3.1S. X 5.4 1/2W.
Location-E (The Shard?).........................3.5S. X 7.3W.
Mosting-A & BA......................................2.4S. X 7.1W.
Shard's Most Likely Location. A little mound east of small scarp-hill ....3.5S. X 7.3W.

14.
REBAR BEAMS OF CRATER MANILIUS AS10-32-4822

Shot (enlargement) of Crater Manilius to the far right of Ukert. LO-III-84M

In looking for more evidence of crystal domes and alien structures and habitations, Mr Hoagland super enlarges a crater area in the top right corner of Apollo-10 photo 32-4822. The little crater is Manilius and is located about 7.5 N x 1.5 E on the USGS Maps. Our two-dimensional photographic forensic fanatic points out what he believes are *"extraordinary structures"* such as bridges, suspended crystals and rebar beams hanging around the atmosphere above the crater.

In observing closely around and above the crater, one will notice what looks like black streaks or thin multi-stranded features with some having three

Stellar Sterility

repeating parallel bands blatantly caught looping across the upper right hand top corner below the crater.

These extraordinary *"extraterrestrial suspension bridges"* arch several miles above the lunar surface. According to Mr Hoagland this is a *"prominent, 3-D sagging, three-stranded 'lunar-bridge' appearing in silhouette."* He continues to comment that on close inspection the photo shows what appears to be a series of right-angle *"brackets,"* parallel *"couplings"* (A) and a *"box-like framework"* suspended from the span (B). [22]

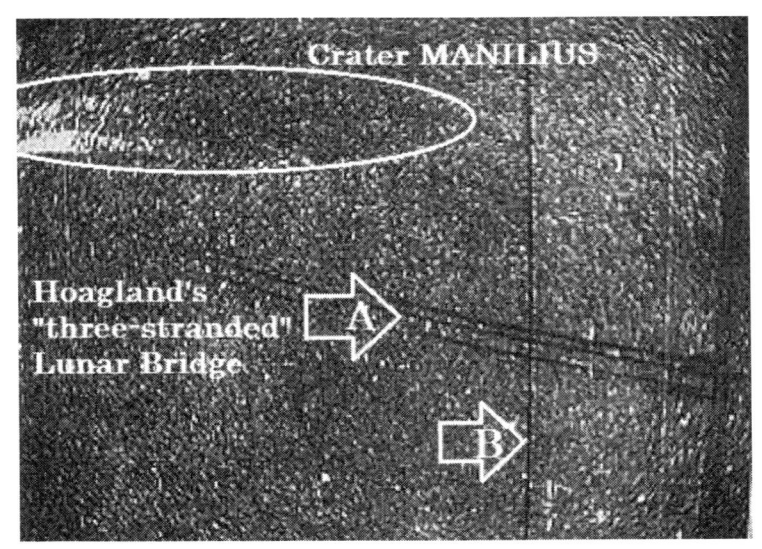

Crater Manilius and the Sagging Bridge

Our bridge builder goes on to propound that these *"artificial-looking structures, these remarkable 3-D details are increasingly difficult to explain as any type of 'natural lunar feature'... [Or as] any 'trivial explanations – such as 'scratches on the original NASA negatives or prints."* So, he asks, what are these *"mile-long, stranded, lunar structures," "sagging cable*

Stellar Sterility

cars" or *"hanging lunar bridges"* doing still clinging to equally incredible *"shimmering glass-like lunar towers"* - hanging miles above the Moon? He said, they are *"definitely not recognized, even in NASA, as natural geologic features."* Of course not, they are emulsion scratches!

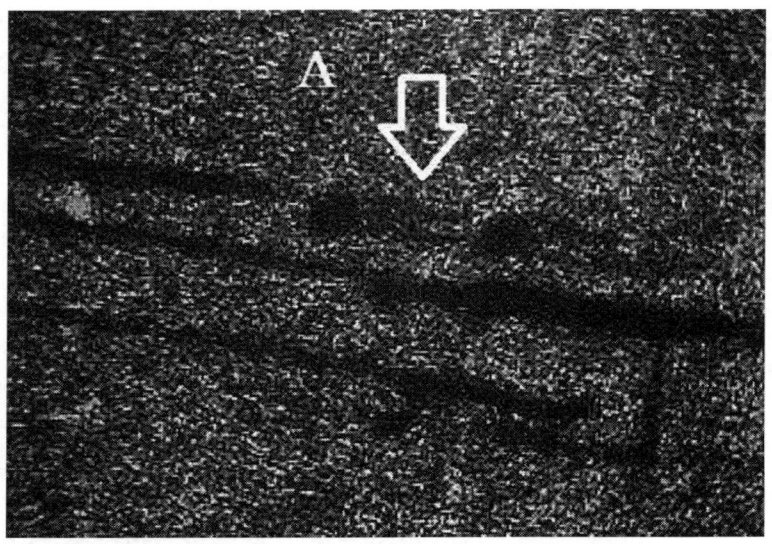

Close-up of three-stranded Bridge (A)

Our bridge builder goes on to propound that these *"artificial-looking structures, these remarkable 3-D details are increasingly difficult to explain as any type of 'natural lunar feature'... [Or as] any 'trivial explanations – such as 'scratches on the original NASA negatives or prints."* So, he asks, what are these *"mile-long, stranded, lunar structures," "sagging cable cars"* or *"hanging lunar bridges"* doing still clinging to equally incredible *"shimmering glass-like lunar towers"* - hanging miles above the Moon? He said, they are *"definitely not recognized, even in NASA, as natural geologic features."* Of course not, they are emulsion scratches!

On even closer inspection, one wonders where Richard gets such amazingly imaginative ideas. Are these black gouges in the photo really suspension bridge cables or actually scratches doctored over with black ink? There are only three choices to choose from. First, they are photographic scratches. Secondly, they are real suspension bridges. Thirdly, they are natural geologic features.

Which of these is the true explanation for these anomalous features? The third has been universally agreed upon as not being the case. This leave the other two choices. The first is obviously the conventional explanation and is discounted by Richard. The second is discounted by NASA but propounded as the truth by Richard.

Since Richard forgets to point out or purposely overlooks the fact that the larger suspension bridge cable (running counter and vertical to the three-stranded cables) runs the full length from way above Crater Manilius to way down below this area on a perfectly flat vertical plain – like someone took a knife and dragged it from the top to the bottom of the frame across the surface of the photo, we conclude that these cables are really emulsion scratches. Further inspection supports this as there are other "black cables" in other areas of the photo, in such odd places as to suggest that not even aliens would be so stupid as to put them there.

Another scratch appears 2/3's the way down, horizontally parallel to and to the right of Ukert Crater and to the far right in the photo. After some minor studies of these scratches, it is found that over 98% of the photo has these emulsion errors running vertically from top to bottom. The "black bridges" or suspension cables are scratches retouched and corrected with black ink by NASA technicians to salvage the photo for publication purposes. It is too beautiful a photograph to leave hideously scared. In fact, it is the only one of its kind unlike any other photo taken of the area. The filth, dirt, dust and scratches are common to most all NASA photos, with this one being effected more than others.

There are many more of these black lines. When the crater area is enlarged the inside of the crater presents a pattern. It also presents even more microscopic debris, hairs and scratches. Processing lines are visible within the yellow brackets. When the area around the crater is enlarged there can be seen many more black cables running oblique to the horizontal and vertical plains. Others (processing lines?) run vertically from top to bottom. Yet, in all the other Apollo 15 and 17 mission photographs of the crater the area shows no such "extraordinary" structures dangling in the upper atmosphere.

Processing lines, emulsion scratches and debris, hair and dust particles

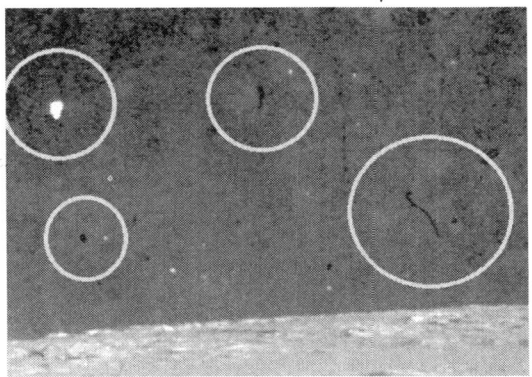

Stellar Sterility

More debris, hair and dust particles

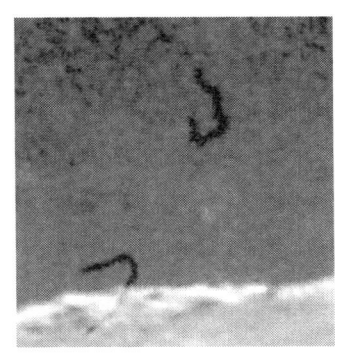

Example of "hair" from dust particles

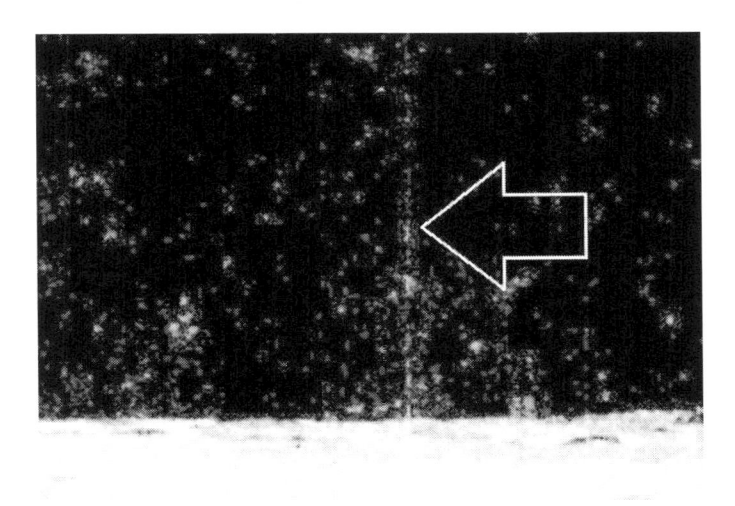

Stellar Sterility

Processing lines

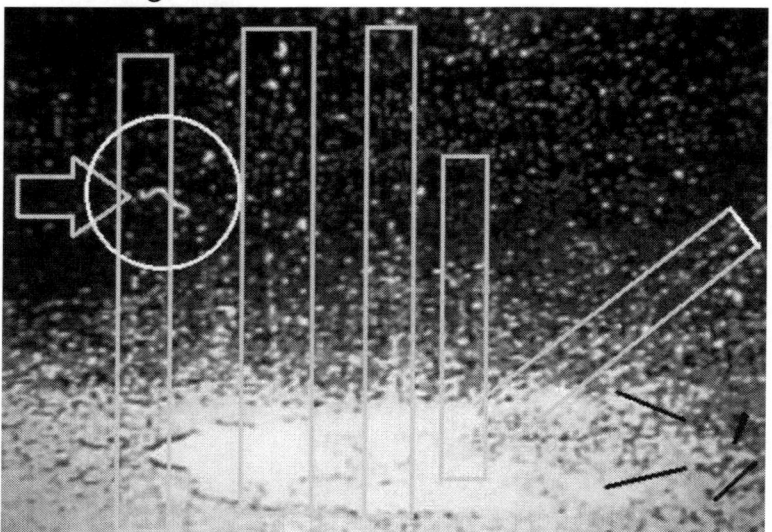

Processing Lines and blotches (enhanced)

Stellar Sterility

Processing lines and glitches

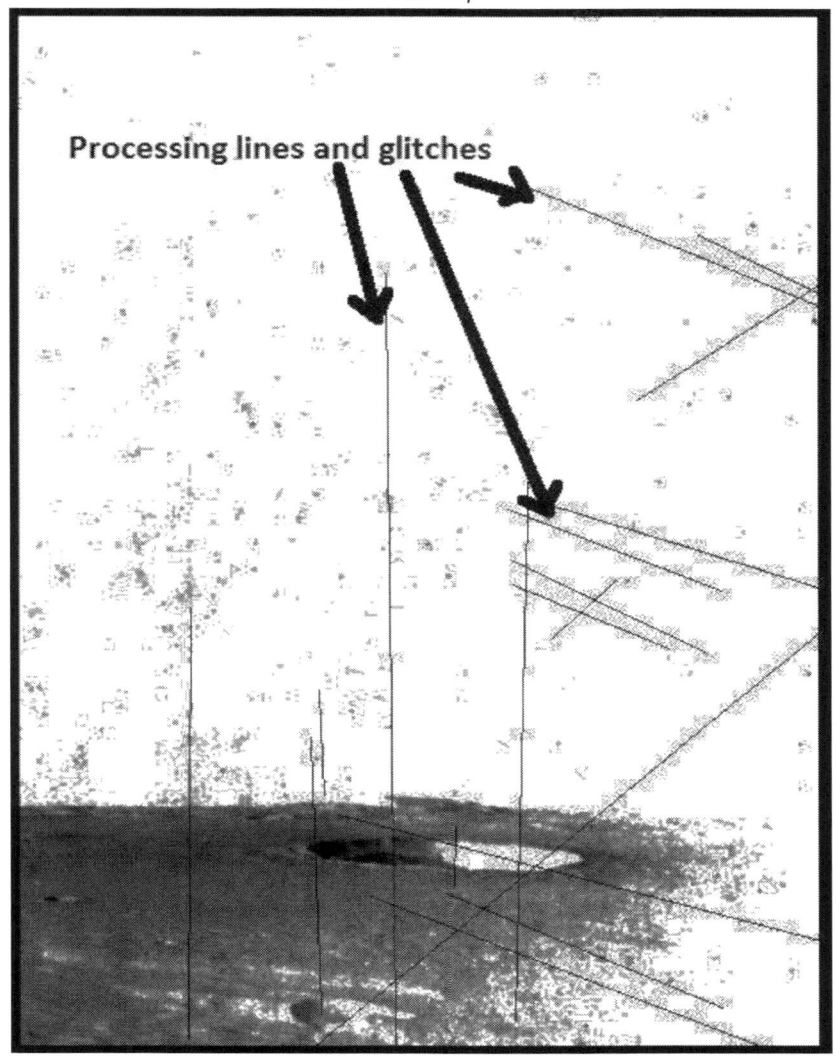

More scan and processing lines. Along these lines
can be seen dangling debris, hairs and other blotches.
(Negative Image)

15
THE "CASTLE" OF UKERT
(5.0 N x 5.0 E)

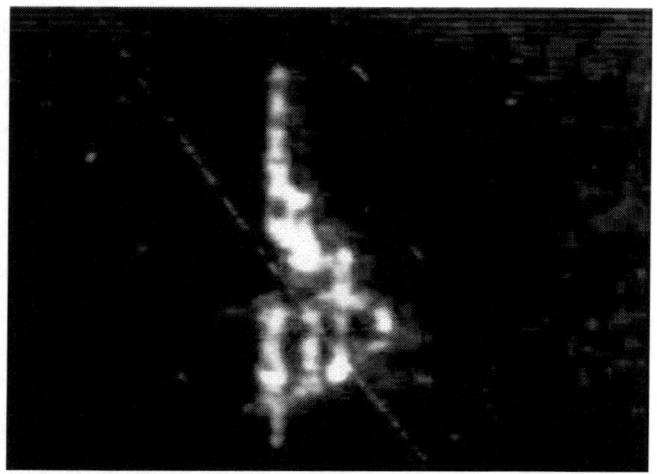

Hoagland's Famous dangling "Crystal Palace."

According to Mr Hoagland, there is much more evidence of an alien dome in the Sinus Medii area. Such so-called facts seem to be in total support of the concept that the Tower, the Cube, the Shard, and other atmospheric debris are but remaining fragments of a once far-larger, clearly artificial structure, once comprised of glass wired together and supported by some kind of darker structural framework. [23]

The Apollo-10 Mission shot a series of photos (Mag. S32) covering the Crater Ukert area North of Sinus Medii (A). The Photo AS10-32-4822 is one in the series covering this area. The series runs from frame 4809 to 4824. Photo 4822 is of special concern to Richard Hoagland. The remaining few "glass fragments," viewed in these photos are what caused Mr Hoagland

Stellar Sterility

to propose the one-time *"existence of an artificial, very ancient 'Sinus Medii Dome,' formerly completely enclosing this central region of the Moon."* The dangling glass crystal "Castle" is the last stop in Sinus Medii towards the final destination of the Mare Crisium dome complex.

In the video, "THE MOON-MARS CONNECTION" Richard reveals what he believes to be a great grid system or architectural structure of sorts, possibly indicating some kind of ancient lunar *"crystalline dome."* The dome lies somewhere around the Crater Ukert as he points out in the Ukert crater photo frame 4822. In this photo, scratches and all, he shows us "THE CASTLE," a small triangular shaped crystal structure dangling about in the air space, or vacuum space about 30-35 miles above the lunar surface! He locates it somewhere north and east of the crater Ukert.

The castle image of Mr Hoagland's MOON MARS Conference is located on a 20 x 24 inch b/w print, from the bottom right corner, at about 5 and 1/2 inches north by 3 and 3/4 inches to the left (west). Other photos that cover this area are AS10-32-4813, 4819, 4734 and L.O.III-M85.

Richard said, the "Castle" is the most important evidence (element) so far found, second to the Shard in significance of lunar geodetic dome construction. The crystal castle is one of the largest so-called *"bright slivers"* and *"glittering geometric fragments"* hanging in a supposed spider-web-like fragmentary framework of black invisible geodetic rebar support beams, high over the eastern end of Sinus Medii, estimated at 10-miles above the Lunar surface . We will soon see that 1000's of others seem to be hanging 110+ miles above the surface and along the surface of the photo.

523

Stellar Sterility

Above, beyond and further towards the horizon lies Crater Manilius with its acclaimed "suspension-wires" and "trolley-cars." The most mysterious observable anomaly about his photo (AS10-32-4822) is the lighting. The Sun is at a mid-morning position, thus lighting all the left side brightness of the surface features. Nevertheless, the photo shows a fading or veiling of the lunar surface from left to right, where the right side becomes darker and more obscure.

The little Castle or crystal structure hangs brilliantly over the Mare area within this blurry obscure area, along with other glittery aerial objects, black rebar-like-beams and emulsion scratches. Mr Hoagland gives us his extraterrestrial explanation of these objects in light of his lunar dome theory. He said, "*With no definable 'lunar atmosphere' – rain, fog or clouds or any form of familiar terrestrial 'optical absorption mechanism' – the only logical explanation for this 'obscuration and fuzziness'* (after the possibility of simple photographic defects to the original 4822 negatives has been eliminated – which it has) *is that the Apollo astronauts actually photographed the remains of some kind of remarkable, constructed 'optical anomaly"* - stretching over Sinus Medii: A Semi-transparent glass-like, mechanical medium, with remarkably focused optical properties."

Hoagland points out that this must be a fragmentary piece of glass-like material hanging in mid-air because it is being supported by an almost imperceptible wire-like slender cable, drooping and sagging under the weight of the object, very similar to the black suspension trolley-car wires about the Crater Manilius. Interestingly though, they are white rather than black.

Stellar Sterility

It might be, supposes Richard that one type of wire reflects light, while the other does not. Nevertheless, both types seem to support the suspension of these crystal-like objects as testimony of a once fully functional geodetic dome.

After reading all this super high technological babble and thumbing through reams of Lunar atlas photos, scanned, enlarged and enhanced images, it became necessary to order a NASA print of 4822 and check the fine details personally. The first contact at the NSSDC was the operator, whom kindly connected me with the librarian of the lunar image data collection.

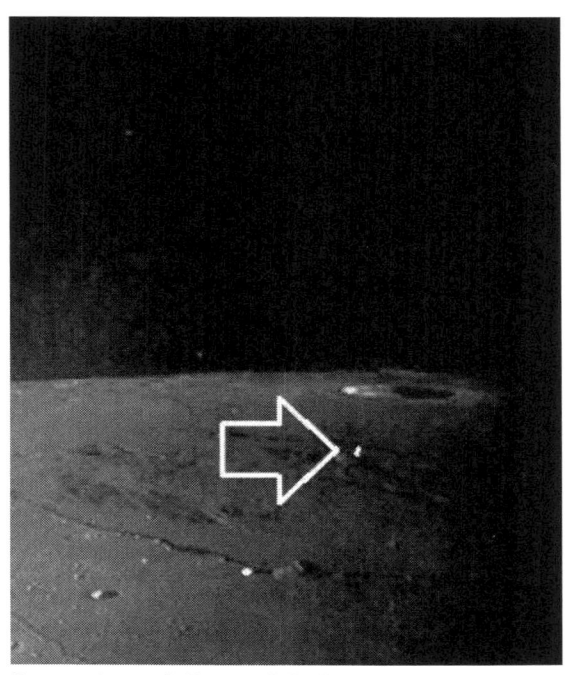

Lo cation of Crystal Palace on   Frame 4822

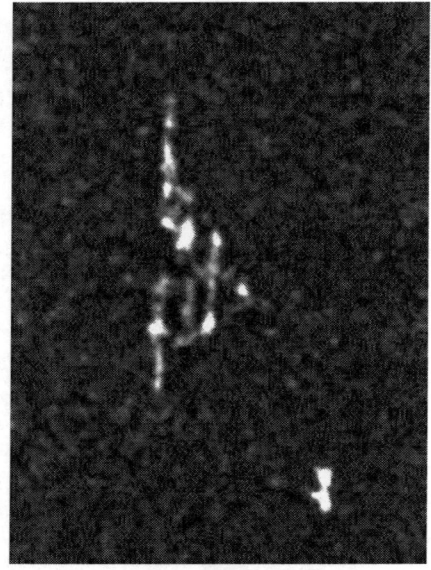

Closeup of processing glitch "Palace."

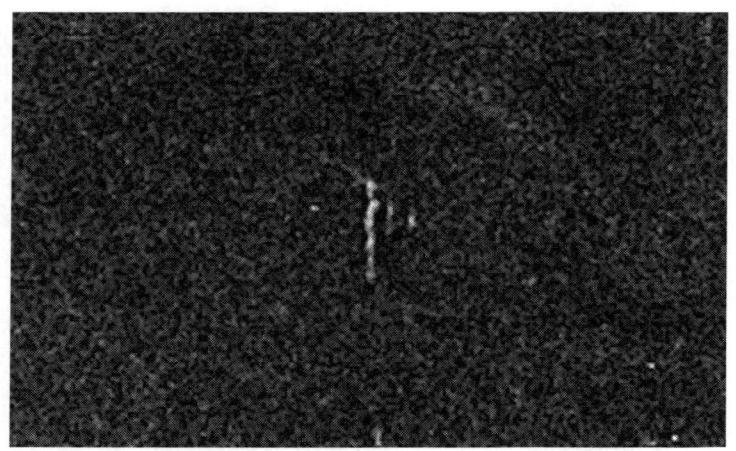

Hoagland's Famous dangling "Crystal Palace."

After some flapping about the frame 4822, Mr Bob Tice replied that he was very familiar with that photograph, since Mr Hoagland and every other artifact hunter had been ordering copies for the same purpose. In fact, Mr Tice more than happily filled me in on a fiasco that took place in the film department between Richard and his associates concerning so-called alien artifacts. The Enterprise team had descended upon the image library collection like locusts, ransacked the files, culled out what they believed to be the best photo proofs for their theory and then split. They were out the door before the check settled on the secretary's desk!

The event climaxed with an argument between Richard and his assistants as to whether certain evidences were really factual or not. One of the associates was overheard saying to Mr Hoagland, *"Richard! That's absurd and ridiculous! You can't pawn that off with that interpretation on your reading audience."* i.e. even your most devout believers will not swallow that. I cannot be too sure what exactly it was that they were arguing over, and neither can Mr Tice, but it must have been relative to these crystal "things" or some other such similar nonsense.

Furthermore, the scene also exposed the fact that they were more than aware of the processing errors, development debris and scratches, and were concerned with just how far Mr Hoagland might go in fabricating alien artifacts out of glitches. Needless to say, after a few more conversations and laughs, the frame 4822 and other photo negatives were ordered, received and eventually studied.

After a few days of analysis, it became apparent that what was NOT mentioned in the MOON-MARS Conference tapes were the many hundreds of other "Castles" and crystal structures scattered all over the frames, especially frame 4822. One crystal of equal interest is located at about 8 inches from the right side of the photo and 5 inches northward.

Stellar Sterility

This dangling derivative also displays the SAME brilliance and reflectivity as the Hoagland Crystal. Another interesting feature is both have similar highlights and shadows cast in the same direction. In the photo (contrary to the darker right half of the photo) the sun is to the right and is casting shadows from the surface features to the left.

The dangling object displays the same shadow and highlight phenomenon. By imaging and enhancing these and other surrounding areas, lines can be seen crisscrossing north and south, east and west like a grid system with some running diagonally and obliquely in different directions. Other 'scratches' can be found in many locations in the photo, from 75-90 miles high to 110 miles high if one calculates according to Hoagland's geometry.

Are these pieces of a huge crystal dome structure or emulsion scratches and debris in the development of the negatives? Every scratch and emulsion pattern flaw displays this same castle crystal-like structure to a greater or lesser degree. The following photos demonstrate the wide variety and placement of such errors as well as prove that the crystal-like dangling castle is not unique.

The dangling crystals and hanging shards of glass are found everywhere in and on the photo 4822. They are found in the middle, on the sides, scattered throughout space or what appears to be the surface of the photo. It seems that if the photo could be enlarged and expanded beyond the lunar image area the little crystals would also continue as well.

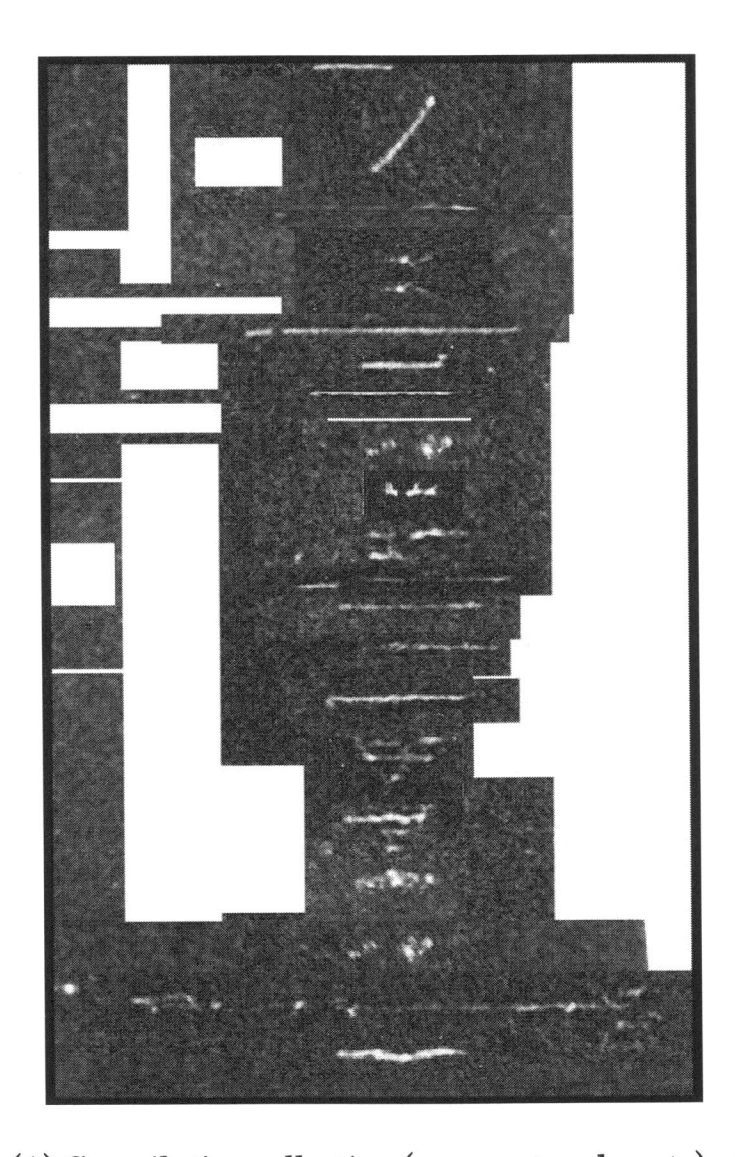

(A) Compilation collection (copy, cut and paste) of crystal-like structures taken from multiple areas of frame 4822

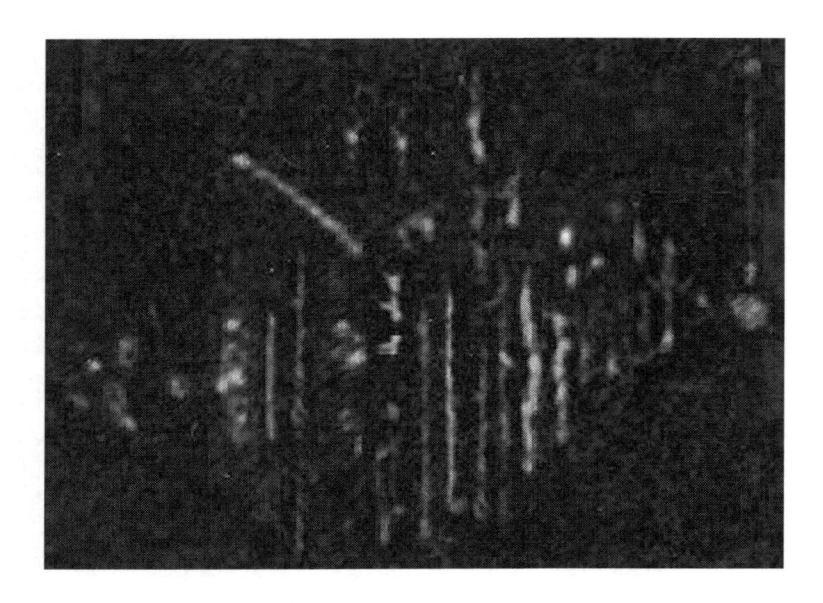

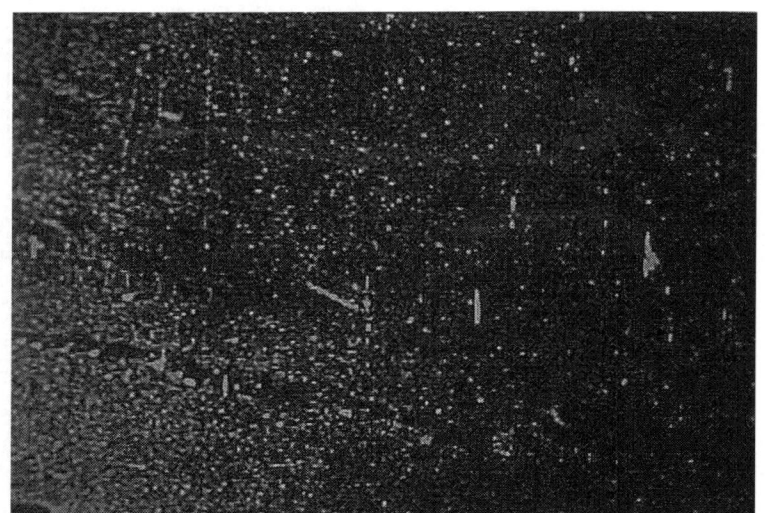

(B) **Compilation of other crystal-like objects.**

The following demonstration is the best explanation for the dangling crystal phenomenon. If we split the photo into two layers [See "A" and "B" above],

Stellar Sterility

the bottom layer "A" [demonstrated with negative image] would be the clean image of the lunar surface area around Ukert. The top plain "B" would be the surface area of the printed photograph with the development artifact inclusions. Notice that the scan lines and the so-called alien crystal-like structures run perfectly level and flat with the surface of the photo and not within the photo as if hanging "above" the lunar surface.

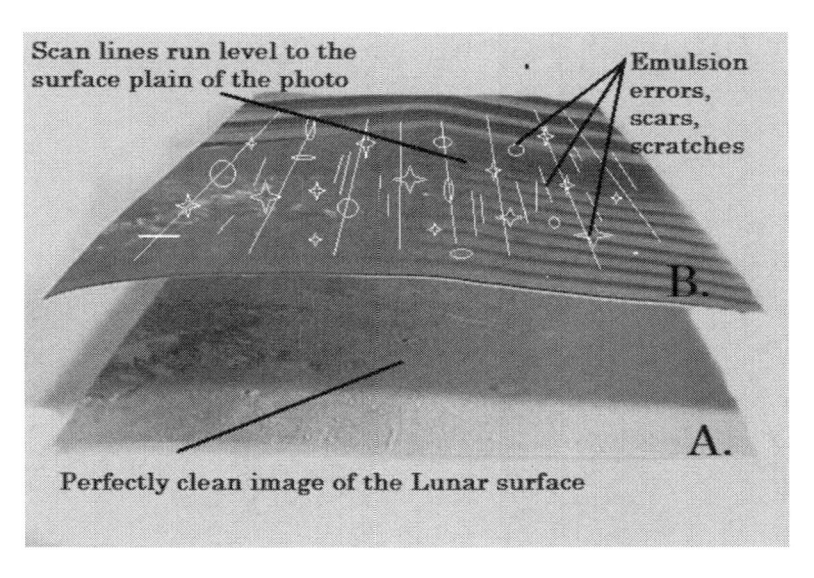

Scan lines run level to the surface plain of the photo

Emulsion errors, scars, scratches

B.

A.

Perfectly clean image of the Lunar surface

The above is the best explanation for the dangling crystal phenomenon. If we split the photo into two layers [See "A" and "B" above], the bottom layer "A" [demonstrated with negative image] would be the clean image of the lunar surface area around Ukert. The top plain "B" would be the surface area of the printed photograph with the development artifact inclusions. Notice that the scan lines and the so-called alien crystal-like structures run perfectly level and flat with the surface of the photo and not within the photo as if hanging "above" the lunar surface.

Therefore, it is now proved that the dangling crystal-like structures are not alien artifacts hanging above the lunar surface within ancient decayed structural support beams but are developmental artifacts embedded within the surface emulsion of the photograph. Furthermore, once these artifacts are stripped away and the observer is able to see clearly that there are no towers, support beams, fragmentary glass crystals, abandoned trolley cars, streets, highways, bus stops or any other alien manufactured device. The Lunar surface is finally exposed as sterile of all such alien nonsense.

16
THE MARE CRISIUM DOME AND CRYSTAL SPIRE (AS16-121-19438)

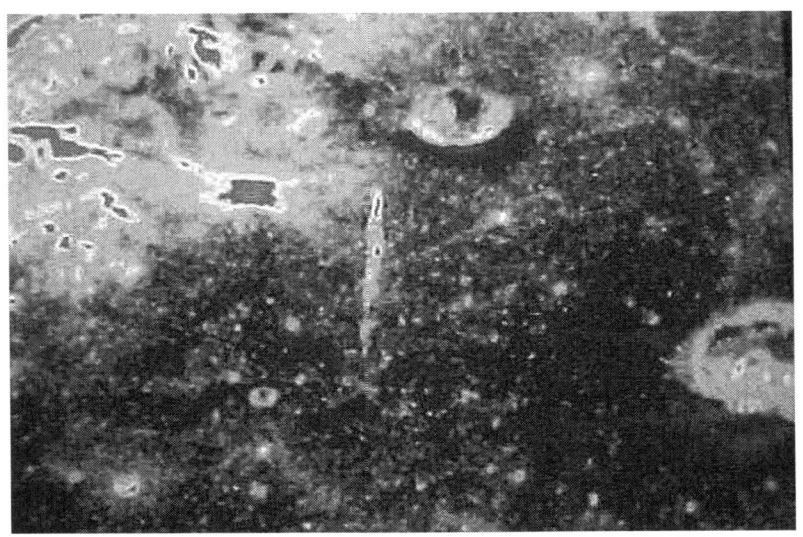

NSSDC PRINT of AS16-121-19438

The Sea or Crisis [Mare Crisium] is another location Hoagland leads us to in searching for massive alien constructions. Hoagland shows us an especially unique area, in fact the only photo example [AS16-121-19438] of a set of large circular light patterns within the Mare. If one bends their visual senses under extreme eye strain, as well as their rational senses, they will see what appears to be an enormous "spire" or tower rising from the lunar surface into the sky within one of these circular light patterns.

Stellar Sterility

When enlarged to an outrageous proportion this "spire" tower demonstrates the same layered "pixel-like" crystalline and cubic patterns as seen before in the Tower and Cube. The concentric circles of light over the surface are considered by the artifact hunters to be light reflection and refraction glistening through the remaining portions of the glass dome. The spire they say [25] is a 20 mile high version of the same crystal-like glass structure "shard" located directly west of Crater Picard.

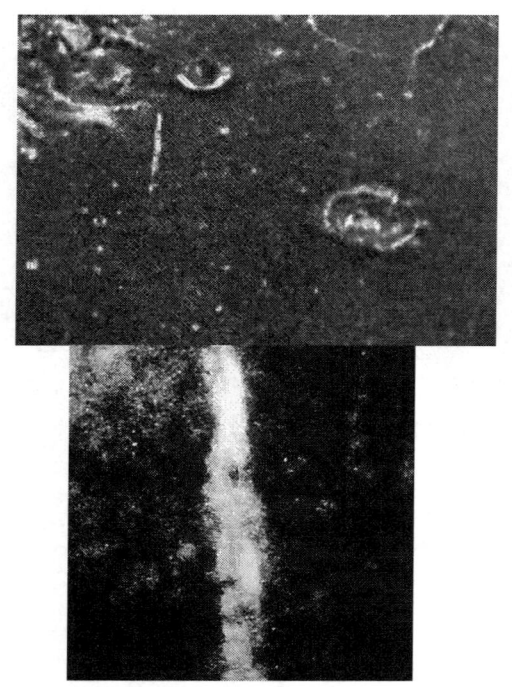

Close up of Spire showing the refractive properties of the so-called crystalline glass structures of what is left of a tower according to Richard Hoagland

Stellar Sterility

Location of the "Spire"   LO-IV-191-H3

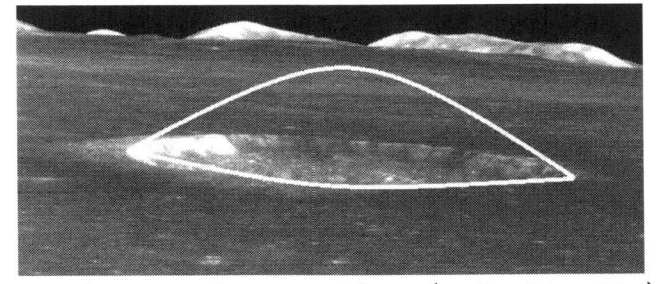

Picard crater and supposed dome (outlined in white) from Apollo frame AS10-30-4421

Telescope photograph

Stellar Sterility

Interestingly the spire cannot be found on any other NASA Lunar photograph no matter what angle or time of day the photos were taken. Comparison of the NASA print (AS16-121-19438) with earth based telescopic photographs and copies of NSSDC versions revealed several remarkable discrepancies. One extraordinary feature on the original NASA print turned up completely missing: a brilliantly illuminated spire, at least 20 miles in height, located directly west of crater Picard. [26]

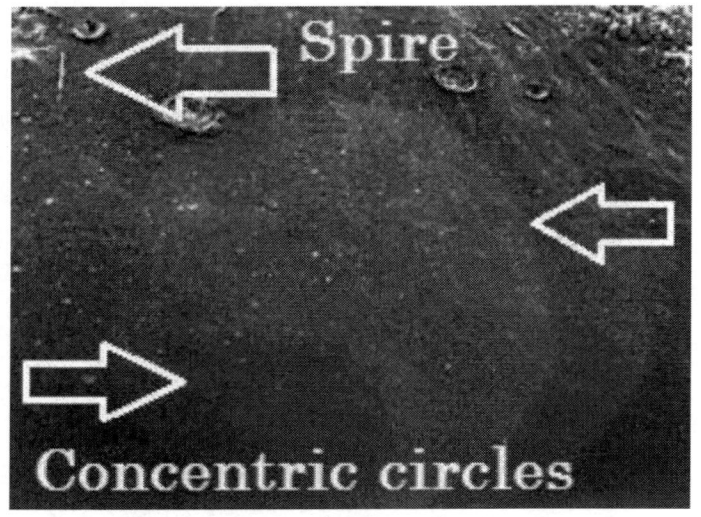

Possible larger "dome" over Mare Crisium

Stellar Sterility

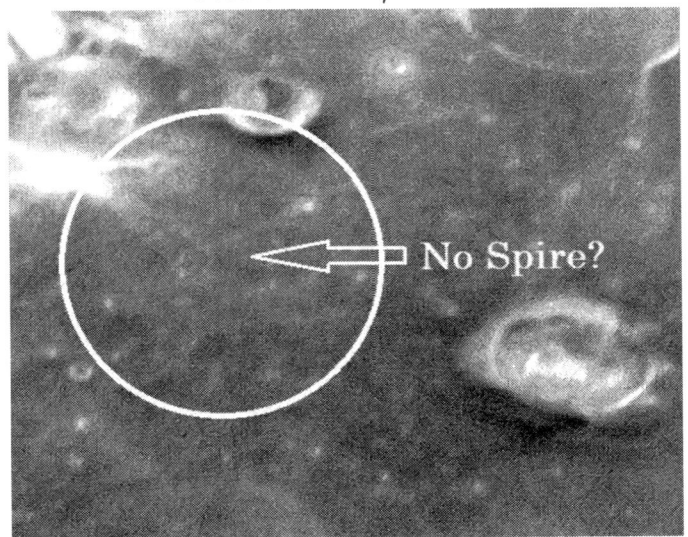

SPIRE is missing on the official NASA version of AS16-121-19438

Hoagland blurts out it is missing on the official NASA print (the original) in his Martian Horizons and exclaims, "*I kid you not!*" [See his figure 39 on page 51]. Of course he is not kidding us. It is a fact that only on the NASA copy at the NSSDC can the "spire" be seen. His excuse is that (choke!) the light has to be just right to refract and this has to happen just at the right time the astronaut is snapping the picture. Otherwise, the high allusiveness of the crystalline material makes it almost impossible to see.

Notice though, the same location shot by Apollo-11 of Crater Picard (AS11-42-6223). In this shot the Sun is to the right side causing shadows on the right insides of the local craters, while brightly lighting the left insides crater walls. Crater Picard is in the middle of this shot with no indication of any "spire" or 20 mile high tower. The spire should be seen in this shot sitting 20 miles southward in the frame, since the camera is

Stellar Sterility

much further away than in the other shot taken by Apollo-16.

None of the Lunar Orbiter Missions photographed the spire (LO-IV-191-H3). Apollo-11 did not capture the spire (AS11-42-6220-6223 and 6214).

Apollo-10 did not snap any picture of this spire either (AS10-30-4417, 4419 and 4421). Neither did Apollo-15 (AS15-M-0954) or the latest photography taken by the Lunar Orbiter Reconnaissance Camera (SPIRE and Picard Crater LROC QuickMap). The real crisis in the Mare is that there are no glass domes or spires and this is very dis-spiring to those who want aliens to be found on the Moon.

AS11-42-6214

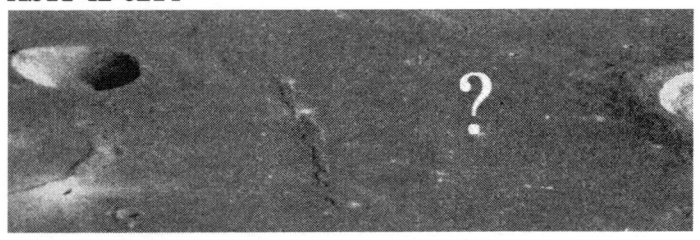

AS11-42-6223

AS11-42-6221

Stellar Sterility

AS11-42-6220

AS10-30-4417

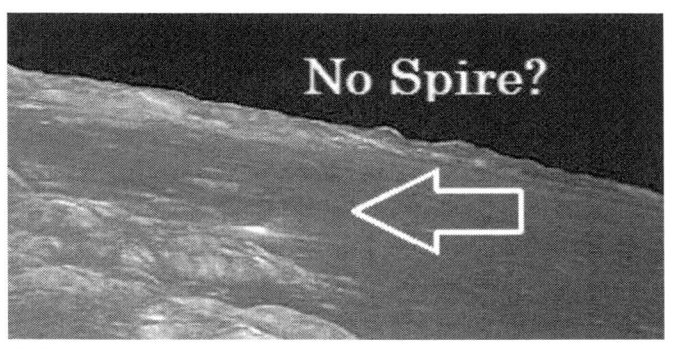

AS10-30-4419

Stellar Sterility

AS10-30-4421

AS15-M-0954

LROC QuickMap

Stellar Sterility

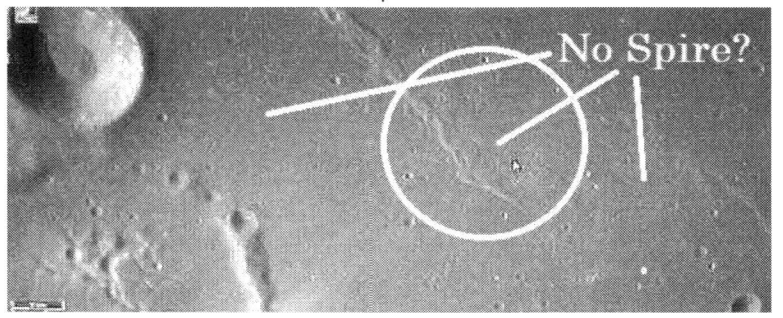

SPIRE missing near Picard Crater (LROC QuickMap)
Lat. 13.0 x Long. 54.0

Nevertheless, no matter what arguments are raised about the validity of the dome and the crystal spire, Richard continues to propound that on a closer examination of this spire through photographic enlargement and enhancement, it becomes readily apparent that it strikingly resembles a highly damaged cylindrical shaped tower frosted with glass and complete with "fins" as well! [26]

Well, after reading this cock-and-bull statement, which seemed fishy to me, I studied the photo myself and found, after "photographic enlargement and enhancement," that the "spire" in question rather strikingly resembled a huge, ugly and irritating development scratch.

While referencing the original NASA print with the NSSDC copy along with all the other photographs of the same area, it became "readily apparent" why the original print of 19438 and all the other photos of the area were missing the "spire," while the Greenbelt Maryland copy contained it.

This most "remarkable discrepancy" as Richard called it was no more remarkable than any other photo development artifact or film defect. The reason why the

Stellar Sterility

original lacked it while the NSSDC copy had it was that the copy got scratched somehow by mishandling in the process of usage. The other interesting thing is that this particular photo is one of the best, if not the best, shot of the area and was chosen for publicity purposes. Naturally, the scratch had to be inked over and corrected. Hence the reason why in many P.R. Photos the scratch is seen retouched and blotted out for visual purposes. Why the NSSDC did not obtain another clean copy is unknown. Maybe it was easier to just retouch what they had. I am sure Richard could make some conspiracy out of this if he has not already done so.

17
MORE UKERT CRYSTALS

Another area of alien interest for dumb dome seekers is the area northeast of Crater Ukert. It is a part of the hill behind (north of) the linear dome-shaped hill, which runs diagonally across the photo. It ends to the right at a small fault line or canal. In the foreground of this hill at Ukert crater, is a small square or diamond shaped field. Maybe an Alien baseball field for small alien children to play in? Or maybe foundations to walls that once existed? These two areas display unusual features suggestive of unnatural structures and surface anomalies.

The large area north of the diagonal hill contains regularly aligned rows of structures according to Dr. Bruce Cornet. There are more than a dozen small craters peppering the area, modifying the lunarscape or cityscape. At a distance, the rows appear like benches, platforms and building complexes, said Dr. Cornet.

Dr. Cornet suggests that, "*on earth, such a feature would be interpreted as the pattern produced by the eroded edges of layered rocks that dip below the surface*". Yet he argues, that "*on the Moon, there have been no physical processes that can account for such a regular geological structure (for) rilles and wrinkles on the surface of a cooling magma outflow do not form such a regular pattern as is evidenced in so many Mare areas on the Moon.*" His most interesting geological pattern point being, that this anomalous pattern has definite boundaries by and which it is absent. [24]

Stellar Sterility

Upon computer enhancement and magnification, the pattern takes on other characteristics. Rectangular features show up along the benches or rows, with some having gaps between them. Thin spires project up from the surface in several places along the rows. Some of the rectangular structures take on a form like buildings and high rise apartment buildings.

The whole area according to Mr Cornet, resembles what one might expect for a city the size of Los Angeles, that has been abandoned by its occupants and left behind, maybe after an atomic bomb drop. It looks like a decayed city complex! Too geometrically laid out to be natural and too artificial to be natural lunar surface features!

Surface processes such as meteoric bombardment have provided the abrasive forces that have destroyed and obscured much of the details. Yet, the ghostly image of an ancient alien, lunar-domed cityscape appears to stand out among the natural surface features and craters that surround Crater Ukert.

But, are these the ravings of an alien artifact hunter and the pseudo-scientific dribble of a retired, bored, burned out geologist? Does Richard Hoagland and persons like Dr. Cornet see things that are truly evident to alien occupation in some far distant past? Only the further imaging of these and other photos will tell us the truth, the whole truth and nothing but the truth. But, don't waste thousands of dollars on NASA negatives to find this truth out. Just look at these examples and save your hard earned money. I wish I had!

Let us look at some shots of the City in finer detail in these imaged photo enhancements. You will notice that there are many crystals, more than expected, ALL running along a photo development path, all causing the same emulsion pattern. Look at them ALL together and compare them. The whole photo from every square inch has these glitches and scratches. All run along a line contrary to some rounded dome shaped pattern, as suggested by Richard Hoagland. As to the City Complex and Dr. Cornet's theory of some abandoned dwelling place, there are many lunar surface areas with the same potential for alien city complexes. The whole surface of the moon may be said to have been occupied at one time.

Only a LUNATIC with a Mickey Mouse imagination would propose such a thing as these guys have. According to previous studies, Richard Hoagland and his followers do the same hoaxing as George Leonard and Fred Steckling! It has been said that Richard has been called out many times by his own technicians, that he was seeing things that should not be passed off on to the public: *"Richard, you should be ashamed of yourself, to think that others will see what you see! You cannot pass these interpretations off on to the public with a clear conscience!"* [Quote: A NASA employee who overheard them talking at the NSSDC photo library].

But, needless to say, Richard's conscience is as clear as his crystal glass domes he talks about, when it comes to pawning off artificial-looking cities, domed craters and dangling crystal-like glass trolley-car cable systems. I expect, as age creeps up on our fellow mythologist that we should expect seeing from him a blurry picture of a McDonald's in some crater (27).

18
NASA COVERUP OR TOUCHUP?
A Study of Lunar Shadow, Highlight, Light Reflection, Refraction, and Processing Anomalies:
(AS10-32-4734)

Supposed NASA cover up (inking-out) of alien crater habitations

One photo that mysteriously shows up supposedly censored by NASA (according to the Hoaglander's) is AS10-32-4734 (Apollo 10 Mission, Mag. 32., Photo Number/Frame 4734). The photo can be viewed in the NASA publication SP-246, "Lunar Photographs From Apollo 8, 10 and 11" [28]

The photo is a westward shot looking past Crater Triesnecker over and beyond Crater Chladni, further over the Murchison and Pallas-E areas, just south of Crater Pallas. The area location under discussion is 2.5 degrees North to 6.2 degrees North, by 2 degrees West to 2 degrees east.

The area includes part of Crater Pallas, Crater Chladni and can be seen in part in Lunar Orbiter

Stellar Sterility

Mission III, Medium Resolution Photo Number 85. The area is on the southern tip of the highlands that extend from the North into Sinus Medii. [29]

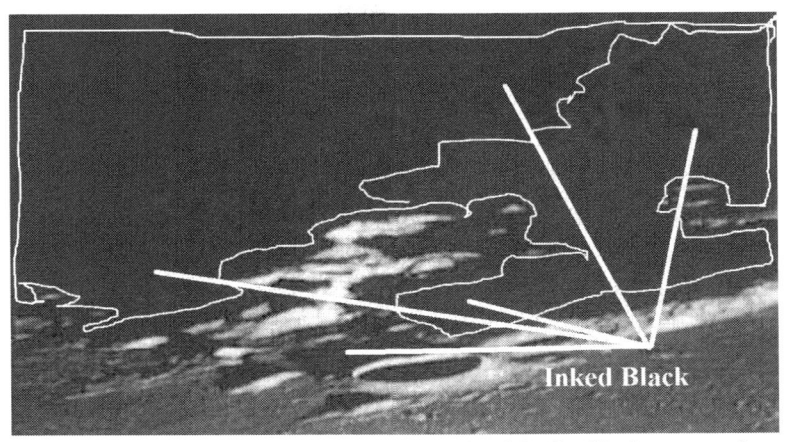

Logetronically processed retouched "inked" photograph

The obvious problem with this oblique (orbital) shot, in this photo, is that it has a lighter "gray" tone quality. Yet, mysteriously it has darker shadows on much of the lunar background landscape, unlike some of the other lighter gray shadows that lay around the lunar mare area. Remembering that the Lunar surface does not lend itself well to light refraction [almost none at all], because of the lack of atmosphere, it stands apparent, that the differences between the shadow's tones is an anomaly or an artist's touchup. It may be some kind of development process. All shadows should have (according to other scientists) the same-pitch black look with no light refraction.

In this photo, the two different shadow tones betray censorship, the conspirators say. No shadow should be dark black while others are light gray.

Stellar Sterility

Observe photo 4734 shot of Crater Triesnecker. The crater's shadow to the left is lighter gray. The inside crater shadow that lies along the crater floor is extremely dark black, like India Ink! Also, the shadow on the crater's floor conforms to the parameter of the floor rather than to the edge of the rim casting the shadow! (Somebody goofed?)

Next, take a look at the supposedly natural looking density of the shadow areas beyond Crater Chladni and compare them to the shadows about the Mare floor among the shadow areas of the various small ridges. This photo has obviously been "doctored" with jet opaque, black, retouch ink and some kind of retouch process has been performed on it. The ink job covers all the area from the horizon (which is in the Moon's terminator?), to the edges of the highland area where the Sinus Medii Mare area begins. One of the areas left out is the West Side shadow of Triesnecker. Has some NASA mole fixed up this particular photo? Has something been hidden from our view?

Now, the following questions arise. Why would NASA cover up half this photo with re-touch black ink? What would they be covering up? Are they covering up something? If they are, why publish the photo at all? If they are not, why retouch the photo, when the image could stand on its on? My sources tell me that the original negative does not have this retouching job on it. So, what is the reason for retouching or fixing up photo 4734, when all it would be showing is common lunar highland geography? For, the same area can be seen in LO-III-85 medium shot. [30]

Photo M85 is a Lunar Orbiter-III shot of the same Crater Murchison with Crater Ukert in the distant background. The area covered here, which is plainly visible, is the area "inked-out" in AS10-32-4734. Why

Stellar Sterility

black out one and not both? Maybe because 4734 is an oblique view looking across at a lower angle, while M85 is from a much higher angle, and 4734 would reveal allot more than M85? My god! how far should one take conspiracies?

My sources also tell me, since I do not have a copy of 4734 at the moment, that the untouched 4734 photographic negative depicts a good quality landscape shot of the surrounding Murchison area, which extends into the lunar terminator (dark-side). Could there be something more visible in the 4734 photo that is not seen in M85? But some will justly suggest that 4734 was touched-up for some purpose unique to the publication of SP-246. Yet, no good reasons can be given for inking-out surface features in a photo for this publication, short of hiding something! The photos and their shadows themselves should more than suffice in standing out on their own.

LO-III-M85

The SP-246 publication has other photos of less quality than 4734, and they seem to not be tampered with? Why not ink them or logetronically process them too? Why only 4734? Why retouch any photo? Is this not destroying research potential? Now! Have these NASA moles and conspirators really censored anything?

The Hoagland hoax theorists say, chances are, some others have been inked in, where they deal with Crater Ukert and Triesnecker, and the areas surrounding Northern Sinus Medii. Is something in the Murchison Crater area being purposely obscured by NASA? Are they hiding something?

Further studies and calling turned up more on this subject. I spoke to another friend of Bob Tice; a NASA photo-technician named Jay Freidlander, about this suspected fraud or re-touching. According to a conspiratorial mind, the arguments would sound somewhat suspiciously like a cover-up or something, but hearing the guy out led me differently.

His response actually answered all my questions. He said, that if you look at some of the other photos in the publication, SP-246, such as with AS8-13-2347 [on page 19], and AS8-13-2314 [on page-17] you could see the same similar effects. This effect has nothing to do with any cover-ups, as suggested by Hoagland, William Brian, Fred Steckling or George Leonard. The hiding of lunar alien bases, ancient cities, glass domed structures, crystal palaces or some other imaginative things just cannot be proved!

A large portion of the Apollo photos, they say, were processed two different ways: Photometrically and Logetronically. Photometrically, the photos retain their original look, as taken by the Hasselblad Camera.

Stellar Sterility

Logetronically developed, photos undergo image gain adjustments, where both the highlights and shadows are enhanced by high and low gain adjusting. High Gamma gain brings out the details in the highlight areas, whereas Low Gamma gain reproduction, the normal procedure, retains the original details, that is, if the photo is bad, the details are lost or are not very strong in details. Logetronic processed photos amplify where the result is the highlights are brighter and the shadows are darker: *"Different types of reproductions were made of the films. For Apollo-15 film, a logetronic process was used in which the reproductions were made with quality control on the density using exposure control and dodging techniques. A photometric procedure using a Niagara printer that allows photometric and photogrammetric (albedo) measures to be made processed Apollo-16 film and a second set of Apollo-16 film. The process used for Apollo-17 film involved gain adjustments, and both high gain and low gain reproductions were made. High gamma (gain) brings out details in the highlight areas. Low gamma reproduction is the normal procedure, but details in highlight and shadowed areas are lost.* [31]

This process was used in the early periods of photography during the Apollo Programs, for enhancing, before Computer enhancement began. The complete photos can be obtained either in Photometric or Logetronic processed formats. Early Logotronic enhancing is similar to modern computer line-art imaging. The Logetronic process was discontinued by NASA for obvious new technological reasons, as well as scientific and public complaints.

My contact also said the Apollo AS10-32-4734 photo in the SP-246 publication is a Logetronic processed photo and can also be purchased as a

Stellar Sterility

photometric, which will contain all the details and lack of details in its original form, which in some cases the logetronic version obscures. He said, there are absolutely NO conspiracies going on inside the NSSDC in hiding or altering photos, either to add or obscure any features. Just purchase the photometric versions and you will be studying photographs for their lunar geology and surface features.

I crossed referenced this location, which covers the area of Chladni, Murchison and Ukert, Craters located at the southern tip of the highlands extending into Sinus Medii, with other photos. The top portion extends into the dark area of Murchison. But is NASA supposedly hiding something here behind the logetronic darkening process? The darker area is covered in detail in the Lunar Orbiter photos LO.W-H-102 and 109. These cover the same area as seen in LO.III-M85, which is a top shot and is as clear as a bell in the Lunar Orbiter Photo album SP-200 NASA publication, page 56. [32] Therefore, no conspiracy to hid any weird or strange alien archaeology is evidenced here.

Later, as I studied the SP-246 photo album, I ran across a color photo with the same apparent 'logetronic' look, photo AS11A0-5921, a Hasselblad photo of the under-side of the Lunar Module, which only shows the exhaust bell and the blown away surface regolith under it. What is the problem? The problem is, only B&W photos were logetronically processed and this one is a color one!

One will notice the same crater shadow density different here in the rock in the lower left-hand corner. The rock's shadow is lighter than the shadows cast by the LEM's legs and exhaust bell. Also, looking up in the top left, at the equipment box (?) next to the other

Stellar Sterility

leg, the shadow of the object is gray, while the shadow is pitch black? Could the color photos also have been processed logetronically? Maybe the later technicians have forgotten? Or is this some kind of 'touch-up' work done with opaque black ink to enhance the photo for public consumer publication consumption in the SP-246 NASA publication?

Mr Freidlander said, no to both. But, the shadows are different and are of different densities. The LM's leg shadows are darker while tile rock's shadow and the equipment package are gray. One suggestion Freidlander said is that the process of printing the photo in the SP-246, in 1971 is, of course, a dot matrix printing. Tiny dots make up the color or B&W picture for mass printing. It is not a formal photographically developed picture. Order the photographs and compare them. [33]

Shadow and highlight play on the lunar surface is an oddity to the untrained and has its unique features and effects. Selenographers say there is a difference between refracted light and reflected light. Light does not refract very much at all in the lunar atmosphere, which is very thin. Light reflection is different. Light travels no matter what, and if it has a surface it hits against, it will light that surface up.

An example is the Lunar Module's gold colored reflection material for decreasing and reflecting the heat off the spacecraft. This is highly reflective material, whereas the rocks are flat with no shinny reflective surfaces. Hence, the rocks do not reflect much light back into their shadows, let alone refract any light, while the Lunar Module gets its whole dark side lit up from reflection.

Other higher reflective material surfaces are the manmade space suits, equipment, and others electrical

Stellar Sterility

devices used on the Moon's surface. All the organic materials, such as the rocks and minerals and surface dust are flat grays and are not very good reflectors of light.

I checked through my Apollo kinescope footage of Apollo-17. In one TV Camera shot as the astronauts walked up to the big rock and when they entered the area where the shadow was, the shaded side of the rock began to appear. There was some reflected light bouncing off the dark shadowed side, where the astronauts stood. When they backed off and as the Camera changed distance, the shadow darkened to pitch black again. Yes, there is even some minor reflected light going into the shadows of the lunar surface features. [34]

Yes, there may be some few touch-up jobs here and there, as any lab and distribution company would do. And yes, there may be some simulations or film editing for some time and cost factors, but almost every photo shop does touch-up work on people's faces, as they have done the craters. NASA may have touched up a few photos for publicity sake.

Every organization does this to impress the people. But, when one looks through all the 1000's of Apollo shots, they will see that a small percentage have that perfect studio look. Most are bleached out because of bright sunlight, or too dark for a lack of it. The astronauts were not your top notch photographers, and even if they had been, they were under such a heavy time restriction that, even a professional would have made mistakes.

As far as the touch-up work, the technicians in the lab have asked me, "*Would you like us to clean up some of the dirty photos, as the masters are getting old and*

Stellar Sterility

scratched up?" I said no, of course, for I preferred all the glitters, scratches and potentially "hidden" things.

Yet, overall, the NSSDC does not touch up, nor do they restrict you from purchasing any particular photo, as I have dealt with them personally and have ordered many lunar photos, films and video tapes. Actually, anything I request that I want, they will sell a duplicate of it.

No, I don't believe that there are glass domed ruined cities on the moon nor any mile high dirt digging machines, sucking machines or blowing machines lying around. As far as the Apollo missions, there are plenty of research materials available and I have seen plenty.

I cannot find any solid arguments of hoaxing or censoring or hiding going on. The NSSDC answered every question I had. Remember, NASA and the NSSDC may be "GOVERNMENT" facilities, but it ALL belongs to and is accessible to U.S. Citizens and professional scientists alike, as long as the small fees are paid for the processing of the data.

- FIN -

Stellar Sterility

FOOTNOTES

SECTION-I
INTRODUCTION

(1a.) See, "NASA's Alien Anomalies caught on film - A compilation of stunning UFO footage"
(http://www.youtube.com/watch?v=WlLN_Jcg1pc).

[1]. Letter to the Ecclesia in Rome, Chapter 11, verse 36 and Col. 1:16-17).

[2]. The Philosophical Impossibility of Darwinian Naturalistic Evolution by
Dennis Bonnette, Ph.D. www.godandscience.org/evolution/philosophy_darwinian_evolution.html]
.

[3]. "The Exoplanet Next Door: Nature News & Comments". Nature.com. doi:10.1038/nature11572

[4]. Wikipedia: "Rocket Equation" and "Russians to ride a nuclear-powered spacecraft to Mars: See, www.csmonitor.com/World/Global-News/2009/1029/russians-to-ride-a-nuclear-powered-spacecraft-to-mars

[5]. Wikipedia: "Nuclear thermal rocket"

[6]. D.F. Spencer and L.D. Jaffe. "Feasibility of Interstellar Travel." Astronautica Acta. Vol. IX, 1963, pp. 49–58.

[7]. General Dynamics Corp. (Jan. 1964). "Nuclear Pulse Vehicle Study Condensed Summary Report."

[8]. Ross, F.W. – Propulsive System Specific Impulse. General Atomics GAMD-1293 8 Feb. 1960

And, Eugen Leitl: Advanced nanotechnology - 3 new articles. Postbiota.org, December 1, 2007, retrieved July 18, 2011

(8a) Wikipedia, under "Project Orion."

[9]. "Destructive Physical Analysis of Hollow Cathodes from the Deep Space 1 Flight Spare Ion Engine 30,000 Hr. Life Test". Retrieved 2007-11-21. And, "The Prius of Space", September 13, 2007, NASA Jet Propulsion Laboratory.

[10]. "Dyon-fermion dynamics" Curtis G. Callan, Jr. 1982, and "Searches for Proton Decay and Superheavy Magnetic Monopoles" B. V. Sreekantan, 1984.

[11]. Wikipedia: "Bussard ramjet"

[12]. Wikipedia: "Beam Propulsion" and "Robert L. Forward"

[13]. See, "God and ETs", http://www.answersingenesis.org).

[14]. Wikipedia: "Interstellar Travel"

[15]. O'Neill, Ian (Aug. 19, 2008). "Interstellar travel may remain in science fiction". Universe Today.

[16]. "RADIATION HAZARD OF RELATIVISTIC INTERSTELLAR FLIGHT" By Oleg G. Semyonov. p. 2

[17]. "Starship pilots: speed kills, especially warp speed." Concepts by Valerie Jamieson, February 2010. www.newscientist.com

[18]. Wikipedia: "Telomeres"

[19]. Hazards of Spaceflight. Hubpages.com

[20]. RADIATION CANCER RISKS AND UNCERTAINTIES FOR DIFFERENT MISSION TIME PERIODS. Myung-Hee Y. Kim and Francis A. Cucinotta. COSPAR, 2012

[21]. Semyonov, p. 4

[22]. Semyonov, p. 5

[23]. Semyonov, p. 13-14

[24]. Wikipedia, under "Moon rock"

[25]. James Papike, Grahm Ryder, and Charles Shearer (1998). "Lunar Samples". Reviews in Mineralogy and Geochemistry 36: 5.1–5.234

[26]. Wikipedia, under "Lunar soil"

[27]. Ibid. paragraph 8

[28]. Taylor, Gerald. p. 1939. "Microbial assay of lunar samples" Proceedings of the Second Lunar Conference, Volume 2, pp. 1939-1948.; The M.I.T. Press, 1971

[29]. "The Sagan Criteria for Life Revisited." http://spacescience.spaceref.com/newhome/headlines/ast21may99_1.htm

[30]. Heiken et al. Lunar Sourcebook: A User's Guide to the Moon. Cambridge: University of Cambridge Press, 1991. See Section 8.8.

[31]. THE LUNAR SOURCEBOOK, p.634

[32]. Luna 16 and Luna 20 Drill, by David R. Williams, NASA Goddard Space Flight Center

[33]. Micro-organisms from the Moon discovered by the Russians in 1970

[34]. Many times, evolutionists have triumphantly announced 'proofs' of evolution against the Christian worldview, and the secular media uncritically broadcast them over the airwaves. However, later, further discovery discredits the evidence. We have only to remember Archaeoraptor, pushed as 'proof' of dinosaur-to-bird evolution by the influential National Geographic, but later exposed as a hoax (c). More closely related to the moon monuments claim is of course the alleged 'fossilized life' found in the meteorite ALH84001, supposedly from Mars. These claims are retracted later for a number of reasons, yet the media does not give retractions much publicity

Stellar Sterility

See, (d) "Life on Mars?" Separating fact from fiction and Mars claims weaken further.

[35]. Astrobiology: Biology Cabinet. September 26, 2006. 2011-01-17

(36). (A) Wikipedia, under "Life on Mars"

(37). Wikipedia, under "Mars soil" and "Phoenix lander"

(38). http://www.nasa.gov/mission_pages/msl/news/msl20120927.html

(39). http://www.cnn.com/2006/TECH/science/06/02/red.rain/index.html. The research papers of Dr. Godfrey Louis are available in the Los Alamos National Laboratory (LANL) archives.

(40). (A) Colored Rain: A Report on the Phenomenon. By S. Sampath, T.K. Abraham, V. Sasi Kumar and C.N. Mohanan, Centre for Earth Science Studies, Thiruvananthapuram 695031, India, Tropical Botanic Garden and Research Institute, Pacha, Palode, Thiruvananthapuram, India, November 2001.

(41). Courtesy: BBC News Online. http://news.bbc.co.uk/2/hi/science/nature/1466477.stm

(42). Precambrian Research 106 (2001) 15–34. "Life on Mars: evaluation of the evidence within Martian meteorites ALH84001, Nakhla, and Shergotty" by E.K. Gibson, Jr., and associates.

(43). Meteoritics and Planetary Science, 2007. Dr. Bell continues her meteorite research at NASA, where she has worked since 1994, curating meteorite samples collected by NASA teams in Antarctica. See, http://breakthrough.nsm.uh.edu/2008_11/bell.htm

(44). New Scientist 165(2228):21, March 4, 2000

(45). "Mars Meteorite Not Evidence of ETs" by Charles Q. Choi. See, http://www.livescience.com

(46). Once upon a time with Dr. Rhawn Joseph, in "Life and Death on the Red Planet". http://brainmind.com/Mars.html

(47). Dr. Rhawn Joseph, Journal of Cosmology. http://journalofcosmology.com/Life101.html#1

(48). SEDIMENTARY HEMATITE ON MARS AND ITS IMPLICATIONS FOR THE EARLY MARTIAN ENVIRONMENT. D. C. Catling and J. M. Moore.

Lunar and Planetary Science XXXII (2001)

(49). (A) A formation mechanism for hematite-rich spherules on Mars. By Chaojun Fana, Hongjie Xiea, Dirk Schulze-Makuchb and Steve Ackleya. Department of Geological Sciences, the University of Texas at San Antonio, TX.

(50). (A) Dr. Rhawn Joseph, in "Life and Death on the Red Planet" and from, http://brainmind.com/Mars.html

(51). "Abiotic synthesis of polycyclic aromatic hydrocarbons on Mars" by Mikhail ZolotovEverett Shock. Journal of Geophysical Research:

Stellar Sterility

Planets. (1991–2012). Volume 104, Issue E6, pages 14033–14049, January 1999. FULL QUOTE:

"Thermochemical calculations of metastable equilibria are used to evaluate the stability of condensed polycyclic aromatic hydrocarbons (PAHs) in cooling thermal gases and hydrothermal fluids on ancient Mars, which are roughly similar to their terrestrial counterparts. The effects of temperature, pressure, the extent of PAH alkylation, and the relative stability of PAHs and alkanes are considered. Inhibition of methane and graphite formation favors synthesis of metastable mixtures of hydrocarbons from aqueous or gaseous CO, CO2, and H2 below 200°–300°C. High-temperature quenching of H2 and CO in volcanic and impact gases and dynamic hydrothermal fluids also favor the synthesis of hydrocarbons. In addition, an excess of CO in cooling systems relative to equilibrium makes the synthesis from CO and H2 more favorable energetically than from CO2 and H2. Both the CO-H2 reactions through Fischer-Tropsch (FT) type processes and the CO2-H2 reactions could be catalyzed by magnetite. Volcanic gases and hydrothermal fluids related to mafic and ultramafic magmas and rocks are more favorable for FT type synthesis than those associated with oxidized Fe2O3-bearing rocks and regolith. We conclude that PAHs and aliphatic hydrocarbons on Mars and Earth could be formed without the contribution of biogenic carbon. Some PAHs could be formed because of pyrolysis of other hydrocarbons formed earlier by the FT type synthesis or other processes. If the PAHs found in the ALH 84001 Martian meteorite formed together with other hydrocarbons through FT type synthesis, it may be possible to bracket the temperature of the synthesis."

(52). "No, PAHs do not mean life on Mars – again and again." Science Magazine. 2012/05/23

(53). NASA Scientist Discovers Alien Life in Meteorites - Again! NOT! By Rich Deem. http://www.godandscience.org

(54). Gary Bates: http://creation.com/did-god-create-life-on-other-planets

(55). Allan H. Treiman, Lunar and Planetary Institute, Houston, Texas. FULL QUOTE: *"Purported biogenic features of the ALH84001 Martian meteorite (the carbonate globules, their*

submicron magnetite grains, and organic matter) have reasonable inorganic origins, and a comprehensive hypothesis is offered here. The carbonate globules were deposited from hydrothermal water, without biological mediation. Thereafter, ALH84001 was affected by an impact shock event, which raised its temperature nearly instantaneously to 500-700K, and induced iron-rich carbonate in the globules to decompose to magnetite and other minerals. The rapidity of the temperature increase caused magnetite grains to nucleate in abundance; hence individual crystals were very small. Nucleation and growth of magnetite crystals were fastest along edges and faces of the precursor carbonate grains, forcing the magnetite grains to be platy or elongated, including the "truncated hexa-octahedra" shape. ALH84001 had formed at some depth within Mars where the lithostatic pressure was significantly above that of Mars' surface. Also, because the rock was at depth, the impact heat dissipated slowly. During this interval, magnetite crystals approached chemical equilibria with surrounding minerals and gas. Their composition, nearly pure Fe3O4, reflects those of equilibria; elements that substitute into magnetite are either absent from iron-rich carbonate (e.g., Ti, Al, Cr), or partitioned into other minerals during magnetite formation (Mg, Mn). Many microstructural imperfections in the magnetite grains would have annealed out as the rock cooled. In this post-shock thermal regime, carbon-bearing gas from the decomposition of iron carbonates reacted with water in the rock (or from its surroundings) to produce organic matter via Fischer-Tropschlike reactions. Formation of such organic compounds like polycyclic aromatic hydrocarbons would have been catalyzed by the magnetite (formation of graphite, the thermo chemically stable phase, would be kinetically hindered)."*

(56). NASA: "The Conditions for the Emergence of Life were Present on Mars — Period, End of Story" (Today's Most Popular) February 17, 2012. See, http://www.dailygalaxy.com/my_weblog/2012/02/nasa-the-conditions-for-the-emergence-of-life-were-present-on-mars-period-end-of-story-todays-most-p.html

(57). "The Enigma of Methane on Mars." http://exploration.esa.int/science-e/www/object/index.cfm?fobjectid=46038f

(58). "Curiosity's detection of methane on Mars could suggest ET life." Jason McClellan, http://www.openminds.tv/curiositys-detection-of-methane-on-mars-could-suggest-ET-life-865

(59). "No Methane on Mars." Shawn Domagal-Goldman, http://www.astrobio.net/paleblueblog/?p=1689

(60). "Mars Methane? NASA's Curiosity Rover Finds None--Yet." Mike Wall: 11/02/2012 02:42 PM, SPACE.com

(61). Condensed from "Biology Cabinet Organization" by Biol. Nasif Nahle., http://www.biocab.org/Mars.html

(62). Ibid. http://www.biocab.org/LifeOnMars.html

(63). "Did Life Come from Outer space? http://creation.com/did-life-come-from-outer-space.

(64). Nature 404(6779):700, April 13, 2000. And, Boyd, R., "Sorry, but we are alone", The Courier-Mail, Brisbane, Australia, April 14, 2000, p. 10

(65). Hawking, Steven. "The Grand Design" and Dr. John Lennox, notes 11, 12, 13.

(66). Ref. http://www.changinglivesonline.org/evolution.html

(67). Newman, 1967, p. 662

Stellar Sterility

(68). http://www.christiananswers.net/q-crs/abiogenesis.html

(69). Five Arguments against the ET Origin of Unidentified Flying Objects JACQUES F. VALLEE. Journal of Scientific Exploration, Vol. 4, No. 1, pp. 105- 1 17, 1990

(70). Wikipedia: Under "Incubus"

(71). "On the likelihood of non-terrestrial artifacts in the Solar System"

By Jacob Hagg-Misra and Ravi K. Kopparapu, Nov. 2011

(72). Ref. "Scientists Suggest Moon Photos May Reveal ET Visitation."

http://www.pakalertpress.com/2012/01/04/scientists-suggest-moon-photos-may-reveal-ET-visitation

SECTION-2
GEORGE LEONARD
FOOTNOTES:

[1a] http://en.wikipedia.org/wiki/Olivineolivine,

http://en.wikipedia.org/wiki/Pyroxenepyroxene,
http://en.wikipedia.org/wiki/Plagioclaseplagioclase,

http://en.wikipedia.org/wiki/Ironferro-http://en.wikipedia.org/wiki/Titaniumtitanic

http://en.wikipedia.org/wiki/Oxideoxide

http://en.wikipedia.org/wiki/Ilmeniteilmenite.

(1b) Rukl, Antonin, Atlas of the Moon.

(1c) A New Photographic Atlas of the Moon, by Zdenek Kopal and Harold C. Urey (1971), Plate 211, page 290.

[1]. The Lunar Orbiter program photographed 99% of the surface of the Moon with resolution down to 1 meter. The Lunar Reconnaissance Orbiter Camera (LROC) consists of one Wide Angle Camera and two Narrow Angle Cameras to provide high-resolution images (0.5 to 2.0 m pixel scale) of key targets - a pixel scale close to 75 meters (246 feet). A single measure of elevation (one pixel) is about the size of two football fields placed side-by-side.

[2]. The origin of lunar crater rays: B. Ray Hawke. www2.ess.ucla.edu].

[3]. Location of Blair Cuspids in LROC. Center latitude 5.07, Center longitude 15.57. And, "Lunar Rocks," Ch. 6 - G. Jeffrey Taylor.

(3a). H.L. Hayward, Symbolic Masonry: An Interpretation of the Three Degrees, Washington, D.C., Masonic Service Association of the United States, 1923, p. 207; 'Two Pillars' Short Talk Bulletin, Sept., 1935, Vol. 13, No 9; Charles Clyde Hunt, Some Thoughts On Masonic Symbolism, Macoy Publishing and Masonic Supply Company, 1930, p. 101

Stellar Sterility

(3b). W. Wynn Wescott, Numbers: Their Occult Power and Mystic Virtues, Theosophical Publishing Society, 1902, p. 33

(3c). Haywood, quoted above, p. 206-7 and Rollin C. Blackmer, The Lodge and the Craft: A Practical Explanation of the Work of Freemasonry, St. Louis, The Standard Masonic Publishing Co., 1923, p. 94

[4]. Washington Post 11/23/66 and the Houston Sky, No.5, June/July 1995

[5]. Mysterious "Monuments" on the Moon. Argosy Magazine, August, 1970 ABAKA = a triangle. NOTE: Nowhere have I found the word "abaka" or "abaca" except when used for a Tropical plant native to the Philippines grown for its textile and papermaking fibre (Musa textilis) also called Manila hemp. Maybe Mr Blair has smoked too much and sees too much in these piles of rocks?

[6]. The Boeing News article, March 30, 1967

[7]. LO-III-214-M, a medium shot of Planitia Descendus the hopeful landing site of Luna 9.http://www.moonviews.com/archives/2009/06/lunar_orbiter_image_recovery_p_5.htmlwww.moonviews.com/archives/2009/06/lunar_orbiter_image_recovery_p_5.html

[8]. "New LO-IRP high res Lunar Orbiter image of western Oceanus Procellarum." Post Wed. June 10, 2009, Joel Raupe. Lunar networks.

[9]. "Boy, that sure looks like Luna 9!" Post Dec. 3, 2011, Joel Raupe. Lunar networks.

[10]. Moon Morphology, Shultz, p. 14

[11]. Conceptual Design of a Fleet of Autonomous Regolith Throwing Devices for Radiation Shielding of Lunar Habitats. Dennis Wells, 1992. NASA-CR-192078. P.10-11

[12]. Dome cities for extreme environments." Leonard, Raymond S.; Schwartz, Milton. Third SEI Technical Interchange: Proceedings.

[13]. a.) "The Sminthian Apollo and the Epidemic among the Acheans at Troy," by Frederick Bernheim and Ann Zener. b.) "The Cults of the Greek states. Vol.4. by Lewis Fernell. c.) "Pestilence and Mice," by James Moulton and A.T.C. Cree.

d.) "Apollo Smintheus, Rats, Mice, and Plague," by A Lang.

[14]. Apollo Moon Conversations. www.ufos-aliens.co.uk).

[15]. Wilkins, 1954

[16]. Evidence for Artificial Structures on the Moon. Lan Fleming, 1995

[17]. "GEOLOGIC MAPPING OF THE KING CRATER REGION WITH AN EMPHASIS ON MELT POND ANATOMY - EVIDENCE FOR SUBSURFACE DRAINAGE ON THE MOON." J. W. Ashley.

[18]. "LROC finds natural bridges on Moon." See,

http://lunarscience.nasa.gov/articles/lroc-finds-natural-bridges-on-moon/http://lunarscience.nasa.gov/articles/lroc-finds-natural-bridges-on-moon/ and,

http://lroc.sese.asu.edu/news/index.php?/archives/277-Natural-Bridge-on-the-Moon!.htmlhttp://lroc.sese.asu.edu/news/index.php?/archives/277-Natural-Bridge-on-the-Moon!.html#extended

[19]. "The Mare Tranquillitatis PIT"

http://lunarscience.nasa.gov/articles/lroc-captures-high-sun-view-of-mare-tranquillitatis-pit-crater/http://lunarscience.nasa.gov/articles/lroc-captures-high-sun-view-of-mare-tranquillitatis-pit-crater/

[20]. "The PIT" http://www.nasa.gov/mission_pages/LRO/news/first-year.htmlhttp://www.nasa.gov/mission_pages/LRO/news/first-year.html

[21]. NASA SP-168 "Exploring Space with a Camera." John McCauley of USGS, p.86

[22]. Wikipedia: "Ray system"

[23]. Hubble photograph of Moon pyramid.

http://alienanomalies.activeboard.com/f494751/lunar-and-iss-anomalies/

Also see

http://www.godlikeproductions.com/forum1/message2097592/pg1].

[24]. Apollo over the Moon: p.__

[25]. LROC. Frame M152390311LR

http://lroc.sese.asu.edu/news/index.php?/archives/633-Top-of-the-Landslide.html

[26]. NEXUS Magazine. Oct-Nov. 1995. p.47.

[27]. When Mr Saccheri mentioned that the librarian said that the lenses of the Lunar Orbiter cameras could zoom in on an anomaly when the on-board computer ran across a funny looking surface feature and snap high resolution frames, he was either outright lying or the (fictitious) librarian was ignorant. The on board computers had tiny RAM, and the lenses were unable to zoom in. This whole story is obviously a historical fiction promoted by an alien presence profiteer.

PHOTOS:

The following listed PHENOMENA descriptions were taken from the book "Somebody Else is on the Moon". NASA plate numbers and plate descriptions are freely available and listed on all NASA public moon plate photographs.

APPENDIX - 1
Leonard's NASA photo numbers, phenomena, and lunar locations - Cross-referenced to Johnson Space Center LO Numbers:

66-H-1293 lunar far side octagonal crater L. Orbiter I
= LO-I-1136-H3

66-H-1611 Western Mare Tranquillitatis

66-H-1612 South-eastern Mare Tranquillitatis
= LO-II-42-H1-3?

67-H-41 (plate-27) Mare southeast of crater Kepler
= LO-II-182-H1

67-H-187 (plate-28) Lunar Orbiter III
= LO-III-012-H1

67-H-201 Crater Kepler in Oceanus crater
= LO-III-162-M and "H1"

67-H-266 anomalous crater Surveyor I landing site

67-H-304 South of Maskelyne F

67-H-318 Oceanus Procellarum

67-H-307 West of Rima Maskelyne in Southern Mare Tranquillitatis

67-H-327 crater in Oceanus Procellarum

67-H-510 Crater Sabine D

67-H-758 Cratered upland basin taken Lunar Orbiter II
= LO-II-62-H3 and "M"

67-H-897 Northeast of Mare Imbrium near Alpine Valley

Stellar Sterility

= LO-IV-115-H3

67-H-935 Mare Orientale. Mare Veris and Rook Mountains
= LO-IV-187-H2

67-H-1135 Crater Vitello with Boulder Track
= LO-V-168-H2

67-H-1179 Tycho crater Tycho

67-H-1206 Tycho Crater

67-H-1400

67-H-1409

67-H-1651 Crater Tycho and northern highlands

67-H-8 lunar far side taken from Apollo 8
= AS8-13-2244

69-H-25 unnamed far side crater

69-H-28 Crater Humboldt and Southern Sea surrounding craters

69-H-737 Triesnecker crater

69-H-1206 Tycho Floor
= LO-IV-125-M

70-H-1629 Fra Mauro area

70-H-1630 Fra Mauro area

70-H-781 Taken by Apollo 14

70-H-1300

70-H-1765 Oceanus Procellarum and Herodotus

Stellar Sterility

70-H-834 North-western half of King Crater
= AS16-120-19263--19267

72-H-835 Mare Crisium, Mare Tranquillitatis and Crater Proculus

72-H-836 Far side King crater highlands
= AS16-120-19226

72-H-837 King Crater

72-H-839 Far side King Crater
=AS16-120-19228

72-H-1109 East of Mare Smythii

72-H-1113 Northwest of King Crater

72-H-1387 Lubinicky area
= LO-IV-125-H2

PHOTOS AND ILLUSTRATIONS:

Picture of George Leonard's Book: "Somebody Else Is on the Moon" Mare Tranquillitatis. Latin for Sea of Tranquillity: Named in 1651 by the astronomers Francesco Grimaldi and Giovanni Battista Riccioli in their lunar map 'Almagestum novum'

Lunar Orbiter II-042-H1 "The Rock Pile" Rukl's "Moon Viewed by Lunar Orbiter (p.49)9 NASA SP-206 Lunar Orbiter Photographic Atlas of the Moon. [www.lpi.usra.edu/resources/lunar_orbiter/book/lopam.pdf]

LO-II-042-H1 George's sketch of 66-H-1612 "Oval" shaped vehicles (p. 101)

(Kopal, Pl. 211, page 290)

NASA 67-H-304 A sharply defined, machine-tooled object is indicated in the area of Crater Maskelyne-F. (p. 177)

NASA PHOTOS:

PLATE 26, 70-H-1629, PLATE 26, Diggers, Dozers and Dirt-flingers.

LO-II-182-H1, 67-H-041 (Plate 27)

-H-1630 Plate 24, Control Wheels.

-H-1611 More crater wheels.

Surveyor 7. Crater Tycho Photomosaic. The central horizon hills are eight miles away from the spacecraft.

V-125-M Tycho.

Stellar Sterility

LO-III-194-H2 More craters with devices and debris. Possible "Play-boy Bunny Rabbit" face in bottom right picture.

Crater Tycho LO-V-125-M

Crater Tycho Photomosaic taken by Surveyor 7. The central horizon hills are eight miles

away from the spacecraft.

Tycho crater's central peak complex, shown here, is about 9.3 miles (15 km) wide, left to right (southeast to northwest in this view). (LROC); and TYCHO CRATER CENTRAL PEAK View of the summit area of Tycho crater's central peak. Boulder in the background is 400 feet (120 m) wide. The image itself is about 3/4ths of a mile wide. (Credit: NASA Goddard/Arizona State University)

69-H-1206 Leonard's Sketch, p.61 Some munched on pie chunks left by aliens.

LROC M190672344RC_pyr Lava flows and ground disturbances with collapsed crater rim.

Large picture of Crater Tycho and crater rays, medium shot.

Octagonial "oval" shaped covering with "PAF" glyph lettering. (p.120)

Leonard's sketches of "oval" spacecraft's. LO V-125-M and H2, and the LROC, Frame M129369888LC_pyr

LO V-125-H2 compared with LROC. Supposed "screw" in white box is not there.

Crater Tycho.

NASA 67-H-1651. An object in the Highlands, north of Tycho. (p. 124)

Tycho Floor A and B

LO-V-127-M. The only big "screw" observable in crater Tycho.

LO-V-126-M

LO-V-125-H2

LO-V-126-H2

Fingals_Cave_Staffa_Scotland. Organ-pipe type basalt lava.

Sleeping Pele is an example of "twisted lava" flow on Big Island, Hawaii.

Tycho Floor disruptions: Crater Tycho is littered with "alien" looking geometric shapes - all formed out of the geothermal disfiguring of the crater floor.

The BIG SCREW Photo and Leonard's sketch of Crater Tycho "screw device", p. 124

Ibid

Fingals_Cave_Staffa_Scotland. Organ-pipe type basalt lava.

Sleeping Pele is an example of "twisted" lava flow on Big Island, Hawaii; and More LAVA Shapes in Earth Nature.

Stellar Sterility

LO-III-162-M (NASA 67-H-201) An oblique view of crater Kepler.

THE CRATER KEPLER (LROC).

[nasa.gov/images/content/513108main_012511b.jpg]

Craters with crosses and regolith sprayings, p. 132. NASA 67-H-201

LO3-162-M Leonard's Latin shaped cross lying at angle. (p.62)

Crater Kepler area objects outlined.

BLAIRE CUSPIDS: Photo of area of RIMA ARIADAEUS and Crater Silberschlag.

Location topographic map for LO-II-62-M

LO-II-62-M

Enlargement of LO-II-62-M, NASA 86-H-758; LO-II-62-H3 (Negative image) Enlargements of medium shot.

LROC M159847595RC_pyr, Two small craters. BLAIR CUSPIDS (Closeup-1)

And, LO-II-062-H3 Cuspids in Aeriadaeus B, and large shadow on rim close-up.

-48 LUNA-9 photo of tower and shadow.

LO-III-67-H-187 and Leonard's sketch, p. 169.

LO-IRP Restored version of LO-III-214-M. Restored image of Planitia Descendus

LO-III-214-M

M132071202LC_pyr. Planitia Desscentus "Plain of Descent," borderland along the western edge of Oceanus Procellarum. Supposed Luna-9 Location.

M132071202LC_pyr

M132071202LC_pyr Enlargements

-H-187 (LO-III-012-H1) PLATE- 2, 1 L-BAR" is in square. Other "towers" in circles appear to be surrounding pyramidal structures to complement the "L". (Kopal, Pl.308, p.280)

-53 Tower sketches, perspective drawings with shadows.

LO-II-182-H1 (NASA 67-H-041) PLATE 27 Control Wheel in crater. (Drawing p. 179);

And 70-H -1630 Craters with wheels.

Leonard's sketches of Machine tools, T-bars and diamond shapes.

Rolling Stones: Moon As Viewed by Lunar Orbiter p. 97 and 119.

Crater Gruithuisen K. NASA 125-H1. Moo as viewed by LO p.69; and AS8-12-2052;

NASA 168-H3; and NASA 60-H2

LO-III-118-M, 118-M [2] and 118-H2.

67-H-935 (LO-IV-187-H2) (Leonard pl.29)

LO-IV.187-2 Medium shot of BLOB "dome" over crater rim; and close-up.

Stellar Sterility

LO-IV-187-2 Enlargement; and LO-IV-186-H3 and 187-H2 "Black crosses"

Extreme close-up of IV-186-H3 Pattern of "spheres"

PYRAMIDS of AS15-M-2087, 2088 and frame 2089

Leonard's drawing of the "domes" p.182

Domed habitat using crater for protection. ("Domed Cities." p. 4)

Regolith "lunar soil" covering dome for shielding purposes. ("Domed Cities..." p. 20)

ALPINE VALLEY construction site: 67-H-897 (LO-IV-115-H3) (See also LO-IV-H-116-H1)

LO-IV-115-H3 Dome, top left; 'sawhorse' dome top right.

Mare Crisium "bridge" on a clear night. (Compliments of Alex Norman, 'Anacortes Astronomy Club')

Mare Crisium Promontories at different times. (Compliments of http://www.the-moon.wikispaces.com/www.the-moon.wikispaces.com)

Mare Crisium "bridge." (Compliments of Alex E. Norman, 'Anacortes Astronomy Club')

Leonard's drawing, p.188

Sculptured "duck" and "skull" mounds.

AS16-121-19438 Apollo shot of northern rim of the Sea of Crisis -74 SEA OF CRISIS: Pictures of the "bridge" LO-IV-1191-H3-81 Mare Crisium: "Bridge" as see by Astronomy club

AS15-94-12753 Close-up of Mare Crisium, Apollo 15 photo; and AS15-94-12752 (Another section of Mare Crisium)

IV-061-H2 and

IV-191-H3

NAC M113168034R "The Bridge" Another amazing bit of lunar geology revealed by LROC!

M130863593L/R

M113168034R

M113168034R

M123791947L. See also, (M103725084L, M103732241L, M106088433L, M113168034R, M123785162L, and M123791947L).

M126710873R [NASA/GSFC/Arizona State University] "Bullet Hole" and the "Pit" in Mare Tranquillitatis

"THE PIT" Pits in Mare Tranquillitatis; Mare Ingenii and Marius Hills (LROC)

AS16-M-2493_med Bullialdus-Lubinicky Area

Ibid, AS16-124-19901

AS16-M-2496_med

M119062083MC. King Crater on the far side of the Moon – LROC

Stellar Sterility

-H-839 (AS16-120-19228). PLATE-12. Circles indicate several small craters in the process of being worked

-H-834 (AS16-120-19265) PLATE-10. X-drone w/no spraying

-H-836 (AS16-120-19226). PLATE-11. King Crater

Ibid, Gear teeth sticking out of rubble?

-94 Apollo-16 AS16-M-2496_med and Leonard's sketches of gears, teeth, 3-mile shaft and other unknowns.

Leonard's sketches of stitching, crater chains, LUBINICKY Area.

M119062083MC King Crater on the far side of the Moon – LROC

Apollo-16 photo 72-H-839 (AS16-120-19228) Leonard's PLATE-12.

72-H-834 (AS16-120-19265) PLATE-10. X-drone w/no spraying; and 72-H-836 (AS16-120-19226) PLATE-11. King Crater.

AS16-120-19228

72-H-834(a) (AS16-120-19226) PLATE-11. Top box shows the "sprayer" unit.

Leonard's attempt to draw an alien device in Crater King. Looks like a huge "cannon." (p.80)

LO-I-66-H-1293 "DINOSAUR CRATER" in Un-named crater on Far side of moon.

-106 Ibid. Images of mound icons "eagle", "dinosaur" and dot pattern; and sketch of X-drones.

CRATER KING AS16-M-0359_med Location of the X-Drones. NASA As16-M-0359, 0890-M,

LO-I-136-H3. Unnamed Crater on far side of Moon. Nickname: "Icon or Dinosaur Crater"

NASA Photo 66-H-1293

Enlargement of LO-I-136-H3

AS16-M-1322 Leonard's 72-H-1113 photograph on page 141 is of Lobachevski Crater; a crater NW of King Crater on the far side of the Moon at 10 degrees North by 112 degrees West.

Lobachevski Crater photographed by Clementine

-113 Ibid. AS16-P-5029_FULL_MED; and frames 5022 and 5029

71-H-781 (AS14-70-9686 and 9688) PLATE 6 Leonard's "super-rig"

LROC frame M1100495581L. Enlargement of Pasteur D crater rim and "Big-Rig" area.

-H-8 (AS8-13-2244)

AS17-151-23115. Crater Doppler area.

Apollo-8 photo 69-H-8 (AS8-13-2244)

Drawing of crater "rope-ladder" p. 175

LO-I-036-H3

LO-II-037-H1 Letters, numbers, clamps and cables.

LO-II-108-H1

Stellar Sterility

LO-II-162-H3 Just dirt and rocks! (LO-IRP enhancement: "MoonView.com")

-118 LO-II-162-H3 Just dirt and rocks! (LO-IRP enhancement: "MoonView.com"); and

AS10-33-4906; "Coffee Bean" Crater; and LO-III-213-H1-[2] Smoke Spiral Crater.

Snap_2011.07.13_11h25m39s_005. PYRAMID IN CRATER. (Supposedly shot by Hubble telescope, 2008)

AS15-M-1554_MED

AS16-121-19407 THE LOBACHOVSKY TOWER. Zoom-in on crater let impact point to what "looks" like a tower or standing object.

AS16-121-19407 and AS16-P-5024 Enlarged. Notice the rim is highlighted [1] and there is no mysterious object "tower" [2].

AS16-P-5024 Enlarged. Lobachevski impact craterlet enlarged.

LROC Photo M176899195LC. Crater Lobachevski from a direct 'top view'.

AS16-121-19407

AS16-P-4559_FULL_MED. Crater Kant P. Impact crater landslide, rim collapse slide... La Pérouse A Crater IMPACT LANDSLIDE

NAC mosaic (M152390311LR), image is ~4.5 km across [NASA/GSFC/Arizona State University].

Snap_2011.07.13_11h25m39s_005. PYRAMID IN CRATER. (Supposedly shot by Hubble telescope, 2008)

AS15-M-1554_MED

AS16-121-19407 THE LOBACHOVSKY TOWER. Zoom-in on crater let impact point to what "looks" like a tower or standing object.

AS16-121-19407 and AS16-P-5024 Enlarged. Notice the rim is highlighted [1] and there is no mysterious object "tower" [2].

AS16-P-5024 Enlarged. Lobachevski impact craterlet enlarged.

LROC Photo M176899195LC. Crater Lobachevski from a direct 'top view'.

AS16-121-19407

AS16-P-4559_FULL_MED. Crater Kant P. Impact crater landslide, rim collapse slide... La Pérouse A Crater IMPACT LANDSLIDE

NAC mosaic (M152390311LR), image is ~4.5 km across [NASA/GSFC/Arizona State University

Ibid. Craterlet LEFT: Enlargement as done by graphics expert. RIGHT: Enlargement of true angle – shows highlight at 45%

BIBLIOGRAPHY:

Bonnette, Dennis "Origin of the Human Species."

Burgess, F. and W. D. Whitley "Text Book of Hindu Astronomy."

Cortright, Edgar (1968) NASA SP-168 "Exploring Space with a Camera."

NASA, Washington D.C.

DiPietro, Vincent and Gregory Molenaar (1982) "Unusual Martian Surface features."

El-Baz, Farouk and L.J. Kosofshy (1970) "The Moon as Viewed by Lunar Oribiter."

NASA SP-200

Fielder, Gilbert (1965) "Lunar Geology." Lutterworth Press, London

Golden, Fred (April, 1985) "Discover"

Hoagland, Richard (1991 Fall Vol-1, No2) "Martian Horizons"

Hoagland, Richard (1991, Summer Vol-1) "Martian Horizons"

Hoagland, Richard (1992 Winter Vol-1, No.3) "Martian Horizons"

Hoagland, Richard (1994, Winter Vol-1, No.4) "Martian Horizons"

Hoagland, Richard (1995, Summer Vol-2, No.5) "Martian Horizons"

Kaysing, Bill "We Never Landed A Man on the Moon."

and "We Never Went to the Moon."

Kopal, Zdenek (1960) "The Moon." Chapman and Hall, London

Kopal, Zdenek (1971) "A New Photographic Atlas of the Moon." Taplinger Pub. Co.

Leonard, George (1976) "Somebody Else is on the Moon" (David McKay, Pub.N.Y.)

Loomis, Alden (1965 Oct. Vol-76, No.10) "Some geological Problems of Mars."

Geo. Soc. Of America Bulletin

Midnight (Feb 8, 1977)

Moore, Patrick (1953) "Guide to the Moon."

More, Patrick (1950) "The Planet Mars."

Mutch, Thomas (1972) "Geology of the Moon A Stratigraphic View."

Princeton University Press, N.J.

NASA (1968) "The Mars Book." SP-179, 1968 Edition.

NASA (1974) "Mars as Viewed by Mariner 9." NASA SP-329

NASA Washington D.C.

NASA SP-206 "Lunar Orbiter Photographic Atlas of the Moon."

David Bowker and Kenrick Hughes (1971)

NASA SP-241 (1971) "Atlas and Gazetteer of the Near Side of the Moon."

Guischewski, Kinsler and Whitaker

National Enquirer (Oct 25, 1977)

Nelson (1955) "There Is Life On Mars."

Schultz, Peter (1976) "Moon Morpohology" Univ. Tex. Press

Shneour, Elie and Eric Ottensen (1966) "Extraterrestrial Life: An Anthology and
   Bibliography" Nat. Acad. Of Sc.

Short, Nicholas (1975) "Planetary Geology." Prinston Hall, Inc.

Spurr, J. E. (1944) "Geology Applied Selenology." Science Press Printing Co.,
   Lancaster, Penn.

Spurr, J. E. (1949) "Geology Applied to Selenology." Volume - IV,
   Literary Licensing

Sreckling, Fred (1981) "We Descovered Alien Bases On The Moon."
   (G. A. F. International)

Surya Siddhanta

Technology and Youth (May 1968)

Thomas, Andrew (1971) "We Are Not The First" London, Sphere.

Wallace, Alfred Russell (1907) Is Mars Inhabitable?"

FOOTNOTES

SECTION-3
FRED STECKLING
FOOTNOTES

[1] See, 44,52,54,58,60,70,72,74,76,78, 80, 82, 84, 86, 88, 90, 92, 94, 98, 100, 102, 104, 106, 108, 112, 114, 120, 122, 124, 125, 126, 127, 128, 130, 136, 137, 138, 139, 140, 143, 144, 145; (151 is Fred's Pond Photo), and 154, 155, 156, 157-3, 158-3, 160-3, 169,183, 178-1.3, 178-1.4, 185-1.3, 185-1.4, 192-1.4

[2] See, 55-3, 82-1, 82-2, 86-3, 90-1, 91-1, 92-1, 92-3, 103-1, 121-1, 148-2, 149-1,2,3; 163-2,3; 164-1, 168-3, 182-1,2.

SECTION-6
RICHARD HOAGLAND
FOOTNOTES:

[1a]. Dr. David Morrison, Chief of the Space Science Division at NASA's Ames Research Centre, that Hoagland was largely "self-educated" in science. In an August 31, 1990, letter, Morrison told me

Stellar Sterility

that he knew of "no one in the scientific community, or who is associated with the NASA Mars Science Working Group, or who is working on Mars mission plans at such NASA centers as Ames, Johnson, or JPL, who ascribes even the smallest credibility to Hoagland or his weird ideas about Mars."

[1] "Cydonia - the face on Mars." ESA. September 21, 2006. http://en.wikipedia.org/wiki/Mesa

[2] "Planetary Names: Mars." Gazetteer of Planetary Nomenclature. USGS Astrogeology Research Program. And, "Planetary Names: Feature Types." Gazetteer of Planetary Nomenclature. USGS Astrogeology Research Program. http://en.wikipedia.org/wiki/Mountain

[3] Hoagland's Martian Horizons (M.H.) V2:5, p.3

[4] "We Found Alien Bases on the Moon" Fred Steckling.

[5] "Somebody Else Is on the Moon" George Leonard.

[6] M.H. V2:5, p.4

[7] http://www.lpi.usra.edu/resources/apollo/catalog/70mm/magazine/?32

[8] M.H. V2:5, p.29

[9] M.H. V2:5, p.15

[10] "Surveyor Observations of Lunar Horizon-Glow." By J.J. Rennilson and D.R. Criswell, The Moon, Volume 10, Issue 2, pp.121-142. 06/1974.

And, http://www.thelivingmoon.com/43ancients/02files/Surveyor_06.html

And, [http://en.wikipedia.org/wiki/Surveyor_6]

[11] http://www.thelivingmoon.com/43ancients/02files/Surveyor_07.html

[12] NASA Technical report 32-12162, Surveyor 6, Mission Report, part-III, JPL. Aug. 15, 1968

[13] Bruce Cornet. Misc. Correspondence, photo notes No. 16, 17 and 18. Interpretations, Correspondences to Richard Hoagland: April 24th, 28th, May 11, and July 4th 1994.

[14] The Moon has an atmosphere, but it is very tenuous. The Lunar Atmospheric Composition Experiment was deployed on Apollo 17. It was a mass spectrometer that measured the composition of the lunar atmosphere. On earlier missions, only the total abundance of the lunar atmosphere was measured by the Cold Cathode Gauge. The three primary gases in the lunar atmosphere are neon, helium, and hydrogen, in roughly equal amounts. Small amounts of methane, carbon dioxide, ammonia, and water were also detected. In addition, argon-40 was detected, and its abundance increased at times of high

seismic activity. Argon-40 is produced by the radioactive decay of potassium-40 in the lunar interior, and the seismic activity may have allowed escape of argon from the interior to the surface along newly created fractures.

http://www.lpi.usra.edu/lunar/missions/apollo/apollo_17/experiments/lace/

[15] Moon Dust: Every Apollo astronaut did it. They couldn't touch their noses to the lunar surface. But, after every moonwalk (or "EVA"), they would tramp the stuff back inside the lander. Moon-dust was incredibly clingy, sticking to boots, gloves and other exposed surfaces. No matter how hard they tried to brush their suits before re-entering the cabin, some dust (and sometimes a lot of dust) made its way inside. Once their helmets and gloves were off, the astronauts could feel, smell and even taste the moon. What is moon-dust made of? Almost half is silicon dioxide glass created by meteoroids hitting the moon. These impacts, which have been going on for billions of years, fuse topsoil into glass and shatter the same into tiny pieces. Moon-dust is also rich in iron, calcium and magnesium bound up in minerals such as olivine and pyroxene.

http://science1.nasa.gov/science-news/science-at-nasa/2006/30jan_smellofmoondust/

"First and foremost is just the fact that the dust just sticks to everything," said Jasper Halekas, a research physicist at University of California, Berkeley Space Sciences Laboratory in Berkeley, California. From gauge dials, helmet sun shades to spacesuits and tools, the "stick-to-itness" of dust during the Apollo missions proved to be a noteworthy problem, Halekas reported. Most amusingly, he added, even the vacuum cleaner that was designed to clean off the dust clogged down and jammed. Halekas recounted a technical debrief by Apollo 17's Gene Cernan after his 1972 Moon voyage. Cernan said that "one of the most aggravating, restricting facets of lunar surface exploration is the dust and its adherence to everything no matter what kind ... and its restrictive friction-like action to everything it gets on." The astronaut added: "You have to live with it but you're continually fighting the dust problem both outside and inside the spacecraft."

Although the lunar environment is often considered to be essentially static, Halekas and his fellow researchers reported at the workshop that, in fact, it is very electrically active. The surface of the Moon charges in response to currents incident on its surface, and is exposed to a variety of different charging environments during its orbit around the Earth. Those charging currents span several orders of magnitude, he said. Dust adhesion is likely increased by the angular barbed shapes of lunar dust, found to quickly and effectively coat all surfaces it comes into contact with. Additionally, that clinging is possibly due to electrostatic charging, Halekas explained. "I think it would behoove us to understand the lunar dust plasma environment as well as possible before we try to come up with detailed dust mitigation strategies," Halekas told SPACE.com. "This would mean characterizing

Stellar Sterility

the dust, electric fields and plasma around the Moon and understanding how they interact."

http://www.space.com/3080-lunar-explorers-face-moon-dust-dilemma.html

[16] [http://www.windows2universe.org/earth/moon/lunar_atm.html]

[17] [http://en.wikipedia.org/wiki/Rayleigh_scattering]

[18] LO-III-84M; LO-IV-97M, 101M, 101-H3, 102-H1, 102M, 108-H3, 109-H1 and 112-M

[19] AS16-121-19438, AS10-32-4854, 4855, 4856 and the previously discussed frame 4822.

[20] See Lalande photos: Virtual Moon Atlas, LOPAM_Lalande, LAC_LM, Lalande_LM77, Apollo Mapping Camera Lalande_A16 2. And, LO-IV-114-H1, 113-H3, 108-H3 and 109-H1.

[21] http://wms.lroc.asu.edu/lroc

[22] M. Hor. p.43 and 44

[23] M. Hor. p.22

[24] Cornet, Dr. Bruce: (Geologist) Handout notes published through NYWK SVCS Systems and MARS MISSION. 5/16/1994, p.6. [See PHOTO REFERENCES:

AS10-32-4734, AS10-32-4813, AS10-32-4819, 4809 and 4824. Also, L.O.IV.85M]

[25] Martian Horizons p. 51

[26] Ibid. p.51

[27] http://tomcornett.hubpages.com/hub/Apollo-Moon-landing--oh-really

[28] Lunar Photographs from Apollo. 8, 10 and 11. NASA SP-246. Printed by the

U.S. Govemment Printing Office. Washington, D.C. 1971. p.68

[29] U.S. Geological Survey Lunar Map #1511, Mare Vaporum Quadrangle, US Geological Survey, Denver, Co. 1968. (Northern Sinus Medii "Pallas" area).

[30] The Moon as Viewed by Lunar Orbiter. NASA SP-200. 1970 ed. by L.J.Fosfsky and Farouk El-Baz. p.56.

[31] NSSDC WDC-A-R&S Catalogue of Lunar Mission Data, July 1977, Page 85.

[32] Compare photos AS1042-4734, L.O.IV-H-102, 109 and L.O.III-M85.

[33] Phone conversation between and Bob Tice and Jay Freidlander of the NSSDC, August 17, 1995

[34] Apollo-17 TV Kinescope transmitted footage (Tape-2) of astronaut's suit, ALSEP Site Setup 347-03-[12] to (39), time code at 0.17.00m

and Tape video 3-4.
See, Photos and films studied and compared:
AS10-32-4734
L.O.IV-H-102 and H-109
L.O.III-M85
As8-13-2347
As8-13-2314
As11-40-5921
PHOTO REFERENCE OF MARE CRISIUM
N.W. Mare Crisium / USGS I-799
LO-IV-H-61, and H-66
Apollo-15 frames 9237-9230 and 9494-9490
S.W., S., S.E.M. Crisium, USGS I-837 and I-948
LO-IV 61-H, 54-H, 177-H, 191-H, 192-H, 184-H and 185-H, 191-192-H (Low Sun angle).
Apollo-=10-4500-4511
Apollo-11 6226-6220
Apollo-15 802-823 West, looking oblique
1371-1391 East looking oblique
1487-1504 North looking oblique
0358-0378 (Condorset)
0937-0957 (Auzont & Firemiscus)
1620-1640 (Apollonius)
Apollo-17 274-293 (North above 12'N to East above 14'N at West)
Apollo-15, 16, 17 Panoramas
S.W., Corner Mare Crisium, USGS I-722
LO-IV-61, 191, 192-H
Apollo-11 6230
N.W – N.E. Half M. Crisium, USGS I-707
LO-IV-H-61, 54, 177, 191, 192, 62, 55 and 67

BIBLIOGRAPHY:
Bonnette, Dennis "Origin of the Human Species."
Burgess, F. and W. D. Whitley "Text Book of Hindu Astronomy."
Cortright, Edgar (1968) NASA SP-168 "Exploring Space with a Camera."
NASA, Washington D.C.
DiPietro, Vincent and Gregory Molenaar (1982) "Unusual Martian Surface features."
El-Baz, Farouk and L.J. Kosofshy (1970) "The Moon as Viewed by Lunar Oribiter."

Stellar Sterility

NASA SP-200

Fielder, Gilbert (1965) "Lunar Geology." Lutterworth Press, London

Golden, Fred (April, 1985) "Discover"

Hoagland, Richard (1991 Fall Vol-1, No2) "Martian Horizons"

Hoagland, Richard (1991, Summer Vol-1) "Martian Horizons"

Hoagland, Richard (1992 Winter Vol-1, No.3) "Martian Horizons"

Hoagland, Richard (1994, Winter Vol-1, No.4) "Martian Horizons"

Hoagland, Richard (1995, Summer Vol-2, No.5) "Martian Horizons"

Kaysing, Bill "We Never Landed A Man on the Moon."

and "We Never Went to the Moon."

Kopal, Zdenek (1960) "The Moon." Chapman and Hall, London

Kopal, Zdenek (1971) "A New Photographic Atlas of the Moon." Taplinger Pub. Co.

Leonard, George (1976) "Somebody Else is on the Moon" (David McKay, Pub.N.Y.)

Loomis, Alden (1965 Oct. Vol-76, No.10) "Some geological Problems of Mars."

Geo. Soc. Of America Bulletin

Midnight (Feb 8, 1977)

Moore, Patrick (1953) "Guide to the Moon."

More, Patrick (1950) "The Planet Mars."

Mutch, Thomas (1972) "Geology of the Moon A Stratigraphic View."

Princeton University Press, N.J.

NASA (1968) "The Mars Book." SP-179, 1968 Edition.

NASA (1974) "Mars as Viewed by Mariner 9." NASA SP-329

NASA Washington D.C.

NASA SP-206 "Lunar Orbiter Photographic Atlas of the Moon."

David Bowker and Kenrick Hughes (1971)

NASA SP-241 (1971) "Atlas and Gazetteer of the Near Side of the Moon."

Guischewski, Kinsler and Whitaker

National Enquirer (Oct 25, 1977)

Nelson (1955) "There Is Life On Mars."

Schultz, Peter (1976) "Moon Morpohology" Univ. Tex. Press

Shneour, Elie and Eric Ottensen (1966) "Extraterrestrial Life: An Anthology and

Bibliography" Nat. Acad. Of Sc.

Short, Nicholas (1975) "Planetary Geology." Prinston Hall, Inc.

Spurr, J. E. (1944) "Geology Applied Selenology." Science Press Printing Co.,

Lancaster, Penn.

Spurr, J. E. (1949) "Geology Applied to Selenology." Volume - IV, Literary Licensing

Sreckling, Fred (1981) "We Descovered Alien Bases On The Moon." (G. A. F. International)

Surya Siddhanta

Technology and Youth (May 1968)

Thomas, Andrew (1971) "We Are Not The First" London, Sphere.

Wallace, Alfred Russell (1907) Is Mars Inhabitable?"

SOURCES AND LINKS:

1.) FREE PDF of book: "Somebody Else is on the Moon." By George Leonard. https://ia600404.us.archive.org/16/items/SomebodyElseIsOnTheMoon/SomebodyElseIsOnTheMoon.pdf

2.) https://www.metabunk.org/threads/debunked-alien-base-on-the-moon-triangle-of-dots-photo-artifact.2965/

3.) http://www.themortonreport.com/discoveries/paranormal/aliens-on-the-moon/

4.) http://ufodigest.com/article/who-else-could-be-moon

5.) http://www.godlikeproductions.com/forum1/message218714/pg1

6.) http://www.paranormalnews.com/article.aspx?id=1185

7.) http://gizadeathstar.com/2011/05/the-idea-that-will-not-go-away-bases-on-the-moon/

8.) http://www.thescienceforum.com/pseudoscience/381-3-different-moon-images-same-location-different-objects.html#post527548

OTHER SOURCES LINKS:

Books, Tapes, Videos, DVD's documentaries: at Weirdvideos.com

https://www.createspace.com/Preview/1143454

BOOK ORDER INFORMATION AND QUESTIONS:

c/o Ross S. Marshall P.O. Box 1191, Anacortes, Wa. 98221

ALIEN ARTIFACT SERIES:

Alien Artifacts, Volume-1

Sections 1-2 "Is Anyone Else on the Moon? The Search for Alien Artifacts According to George Leonard.

Stellar Sterility

PREVIEW: https://www.createspace.com/Preview/1144263
ORDER HERE! http://www.amazon.com/dp/1495987760

Alien Artifacts, Volume-2 Moon Mars Monuments Madness

Section – 3 FRED STECKLING Sticks It to Us! A Refutation of Fred Steckling's "We Discovered Alien Bases on the Moon"

Section – 4 JACK SWANEY'S "OBJECTS ON THE MOON" Intermediate Idiotic Incidentals of Intergalactic Garbage.

Section – 5 Mars Is Dead!

Section – 6 "Moon Mars Monuments Madness." The Continued Search for Alien Garbage on the Moon and Mars according to Richard Hoagland. A Polemic discussion and refutation of Mr. Hoagland's monuments supposedly found on the Moon and Mars.

SOURCES AND LINKS:

1.) FREE PDF of book: "Somebody Else is on the Moon." By George Leonard. https://ia600404.us.archive.org/16/items/SomebodyElseIsOnTheMoon/SomebodyElseIsOnTheMoon.pdf

2.) https://www.metabunk.org/threads/debunked-alien-base-on-the-moon-triangle-of-dots-photo-artifact.2965/

3.) http://www.themortonreport.com/discoveries/paranormal/aliens-on-the-moon/

4.) http://ufodigest.com/article/who-else-could-be-moon

5.) http://www.godlikeproductions.com/forum1/message218714/pg1

6.) http://www.paranormalnews.com/article.aspx?id=1185

7.) http://gizadeathstar.com/2011/05/the-idea-that-will-not-go-away-bases-on-the-moon/

8.) http://www.thescienceforum.com/pseudoscience/381-3-different-moon-images-same-location-different-objects.html#post527548

OTHER SOURCES LINKS:

Books, Tapes, Videos, DVD's documentaries: at
Weirdvideos.com
https://www.createspace.com/Preview/1143454

BOOK ORDER INFORMATION AND QUESTIONS:

c/o Ross S. Marshall
P.O. Box 1191
Anacortes, Wa. 98221

Made in the USA
Columbia, SC
24 September 2020